Diagnosing Wild Species Harvest

Resource Use and Conservation

Diagnosing Wild Species Harvest

Resource Use and Conservation

Matti Salo
Department of Biology, University of Turku, Turku, Finland

Anders Sirén
Department of Geography and Geology, University of Turku, Turku, Finland
Department of Geosciences and Geography, University of Helsinki, Helsinki, Finland

Risto Kalliola
Department of Geography and Geology, University of Turku, Turku, Finland

ELSEVIER

AMSTERDAM • BOSTON • HEIDELBERG • LONDON
NEW YORK • OXFORD • PARIS • SAN DIEGO
SAN FRANCISCO • SYDNEY • TOKYO

Academic Press is an imprint of Elsevier

Academic Press is an imprint of Elsevier
32 Jamestown Road, London NW1 7BY, UK
225 Wyman Street, Waltham, MA 02451, USA
525 B Street, Suite 1800, San Diego, CA 92101-4495, USA

British Library Cataloguing-in-Publication Data
A catalogue record for this book is available from the British Library

Library of Congress Cataloging-in-Publication Data
A catalog record for this book is available from the Library of Congress

ISBN: 978-0-12-397204-0

For information on all Academic Press publications
visit our website at elsevierdirect.com

Typeset by TNQ Books and Journals
www.tnq.co.in

Contents

Author Biographies

Matti Salo is a biologist and a PhD in Environmental Science. His fields of interest include governance, management and policy issues related to natural resources, biodiversity and conservation—with a particular emphasis in forest policies. Salo is a long-term Amazonia enthusiast and a member of the University of Turku Amazon Research team (UTU-ART). He has spent time in the region annually since the late 1990s, with a particular commitment to Peruvian Amazonia, but also working and traveling extensively in parts of Bolivia, Brazil, Colombia and Ecuador. In addition to academic work, Salo has published books and other writings about Amazonia and biodiversity issues directed to the general public.

Anders Sirén is a biologist and PhD in Rural Development Studies. He was a postdoc researcher at the University of Turku from 2009 to 2013, and is currently a lecturer in geography at the University of Helsinki. Sirén has spent over 10 years in Ecuadorian Amazonia, where he has conducted extensive field work for social and natural science research related to wild species harvest and land use change in indigenous communities, and also made shorter visits to Peruvian Amazonia. He loves fishing in the swift rivers of western Amazonia and dreams about saving Amazonian fisheries from the multiple threats of overfishing, habitat destruction and pollution.

Risto Kalliola is a professor of geography at the University of Turku. He has made a long career on biogeographical, ecological and resource management studies in Amazonia and northern Europe. He is interested in the role of scientific understanding in the use of renewable natural resources and in land use planning. Kalliola is one of the founder members of the multi-disciplinary University of Turku Amazon Research team (UTU-ART), which has over three-decades long research history in Amazonia.

Preface

Wild species are important resources supporting the livelihoods of millions of people around the globe. At the same time, the harvest of wild plants and animals is a major driver of worldwide ecosystem change. In some contexts, some people may consider such harvest to be something undesirable that should be stopped. Others may consider it to be something worth supporting for its socio-economic importance as well as for its supposed environmental benignity. For sure, wild species harvest is something that deserves being taken seriously. And that is why we have written this book.

Although many books deal with the harvest of wild species, most of them have focused on some limited set of resources or a particular type of harvest, such as fisheries, wildlife, timber, or the ill-defined concept of nontimber forest products (NTFPs). In this book, however, we wanted to take a comprehensive approach covering the harvest of all kinds of wild species, including plants as well as animals, and terrestrial as well as aquatic organisms. We believe that this is a useful approach because the challenges involved in maximising the benefits and minimising the harm from wild species harvest are often similar, although the harvested species themselves may be very different from each other.

Our intention with this book is to provide to the reader a complete reading experience carefully built to be explored from the beginning to the end – and back, if felt necessary. After an introductory part, covering the general importance of wild species harvest, and the meaning of some basic concepts, comes the second part, where each chapter presents a context-specific case of wild species harvest. These chapters are not independent scientific case studies but instead illustrating nonfiction stories that present a series of views on the world as it is, incomplete and sometimes even contradictory, but with countless pieces of information and nuances of realities from which lessons can be learned. These stories, we hope, will make readers acquainted with the realities of the lives of people who, in one way or another, make their living by wild species harvest, and will prepare readers for absorbing the theory presented in the third part.

We hope the theory will help the reader to scrutinise the cases from a set of complementing thematic perspectives and at the same time be, universal and applicable to any other case of wild species harvest one might encounter. Finally, in the fourth part of the book, we tie the theoretical perspectives together in a way that should help the reader to apply the message of the book to any case of wild species he or she may encounter, in the form of the Diagnosing Wild Species Harvest (DWiSH) Procedure. We believe that, when picking up this book and starting to read it, many readers may already have a case of wild species harvest in mind, being concerned about its impacts or hopeful about its potentials. We would like readers to see those cases of their own in parallel with the ones we present, and also reflect the case to all the theory and discussion found on the pages of the four parts of this book.

The main geographical focus of our examples is on Amazonia in South America. Resource overuse by overintensive extraction continues to be far too common in this area, yet we are convinced that there must also be solutions to this problem. Like elsewhere in the tropics, or almost anywhere on Earth, the harvested populations of many wild species of plants and animals easily become depleted because of their excessive or otherwise inappropriate harvest. Whether to prioritise biodiversity conservation or the production of goods for consumption is often a matter of intense debate locally, but these considerations can also have global significance. The complexity of this setting is particularly familiar to us in the Amazonian lowlands, where all three of us have conducted research during the last few decades. There, we have witnessed concrete situations of resource use and overuse, and a large part of our research has aimed at understanding the underlying factors affecting wild species harvest. Amazonia is, in fact, one of the parts of the world where the harvest of wild species seems to be almost omnipresent and indispensable for many people's livelihoods. An advantage of focusing on a clearly defined geographical area such as Amazonia is that we can be fairly sure that we have somehow covered most types of species resources whose harvest is significant there, in economic or ecological terms.

As our ambition is that the theory presented should be universally applicable, however, we also present outlooks to other parts of the world, in particular to the Nordic countries. We were all born in Finland, and as students, as professionals, and also as part of everyday life, we have become acquainted with many forms of wild species harvest practised here in the boreal landscapes, including berry and mushroom picking, fishing, hunting, logging, and reindeer herding. As the natural as well as the societal settings are so different in Amazonia and the Nordic countries, we believe that it can often be enlightening to compare the two, highlighting not only the differences but also the similarities.

We hope that scholars, students, and practitioners from a variety of disciplines or otherwise with different backgrounds will find value in the cases, in the theory, in the DWiSH Procedure and – more generally – in the ideas we lay out. We argue that understanding the complexity of wild species harvest from many different perspectives, instead of just a few limited viewing angles, helps to combat the vicious spiral of overharvest followed by ecological problems and socio-economic collapse. Concrete actions and decisions are needed by many different actors, from local stakeholder groups to big companies, policy makers, consumers, and nongovernmental organisations. Addressing such a multifaceted and diverse phenomenon as wild species harvest in one volume inevitably leads to a trade-off between breadth and depth of treatment. For those readers who wish to get deeper into some particular field of theory, an extensive list of bibliographic references is provided.

In Turku 25th of June 2013,
The Authors

Acknowledgements

We have received valuable inspiration, feedback, and inputs from a number of collaborators. Wild species harvesters in Peru, Ecuador, Brazil, Bolivia, Finland, and Sweden have shared their experiences and thoughts with us over the years; we are grateful for their friendly attitudes to our keen interest to learn what is harvested, how, and why. Our colleagues in the Amazon Research Team of the University of Turku as well as many other scholars at several universities and research institutions in Peru, Ecuador and many other countries have helped us to formulate our thoughts better. We also acknowledge three different groups of students at the University of Turku and one at the Pontificia Universidad Católica del Ecuador as they have evaluated and tested many of the approaches presented in this book, thus providing useful feedback.

Dr. Jukka Salo commented on an early version of this entire book in a way that is highly acknowledged, and he also wrote two thematic boxes into it. Several colleagues read and commented on draft versions of the whole manuscripts or parts of it. We are grateful to: José Álvarez, Juan-Camilo Cárdenas, Pedro Flores, Juha Hiedanpää, Miia Itänen, Mark Johnson, Sanna-Kaisa Juvonen, Juha Kotilainen, Martti Pärssinen, Kalle Parvinen, Aili Pyhälä, Henna Rouhiainen, Kalle Ruokolainen, Francesco Sabatini, Andrea Siqueira, Andrés Tapia, Hanna Tuomisto, David Wilkie, and the students of the aforementioned courses.

Many people also helped us during the recent fieldwork that was carried out, particularly in order to document some of the stories from the forest floor. We would like to thank in particular Natalia Aravena, Hugo Cabieses, Luciano Cárcamo, Alonso Córdova, Alfredo García, Ulla Helimo, Martín Huaypuna, Gareth Hughes, Sixto Luna, Petra Mikkolainen, Carlos Reyes, Juan Fernando Reyes, Róger Rumrrill, José Torres Vásquez, Nathan Vogt, and all the people mentioned in the stories from the forest floor. Some of the stories root on our fieldwork carried out over prolonged periods of time in the communities of Asocación Boberas, Teresa Mama, Chubakucha, Ishpingu, and, in particular, Sarayaku, and our deepest gratitude goes to all the numerous people of these communities who have supported, facilitated, and actively participated in these research activities. The same is true for many people in Iquitos, in particular at the Instituto de Investigaciones de la Amazonía Peruana (IIAP). We thank all the people who provided photographs for the book, and also Maiju Kähärä, who finished several of our figures and additionally drew the symbols that describe the seven thematic perspectives in Part 3. Finally, we would like to thank the Kone Foundation for funding the work necessary to carry out the project that led to this book (in the specially allocated grant call 'The Significance of Biodiversity, 2010) and the University of Turku for a stimulating working environment'.

Acronyms

ACTO Amazon Cooperation Treaty Organization
AFIMAD Asociación Forestal Indígena Madre de Dios
AMETRA Aplicación de Medicina Tradicional
CBD United Nations Convention on Biological Diversity
CIFOR Center for International Forestry Research
CITES International Convention on the Trade in Endangered Species of Wild Fauna and Flora
COINACAPA Cooperativa Integral Agro-extractivista de Campesinos de Pando
CPUE Catch Per Unit of Effort
DPSIR Drivers–Pressures–State–Impacts–Responses (framework)
DWiSH Diagnosing Wild Species Harvest
EBA Empresa Boliviana de Almendra y Derivados
EU European Union
FAO Food and Agriculture Organization
FSC Forest Stewardship Council
GIS Geographic Information System
GPS Global Positioning System
IIAP Instituto de Investigaciones de la Amazonía Peruana
IIRSA Initiative for the Integration of the Regional Infrastructure of South America
ILO International Labour Organization
INPA Instituto Nacional de Pesquisas da Amazônia
ITQ Individual Transferable Quota
ITTO International Tropical Timber Organization
IUCN International Union for Conservation of Nature
MCDM Multi-Criteria Decision-Making
MSY Maximum Sustained Yield
NGO Non-governmental Organization
NTFP Non-Timber Forest Product
OTCA Organização do Tratado de Cooperação Amazônica / Organización del Tratado de Cooperación Amazónica
POC Particulate Organic Matter
REDD Reduction of Emissions from Deforestation and Forest Degradation
RIL Reduced Impact Logging
SINCHI Instituto Amazónico de Investigaciones Científicas
SLAR Side-Looking Airborne Radar
TAC Total Allowable Catch
TURF Territorial Use Right
UNAP Universidad Nacional de la Amazonia Peruana
WCED World Commission on Environment and Development

WHO World Health Organization
WWF Worldwide Fund for Nature
ZEE Zonificación Ecológica Económica

Part I

Focus on Wild Species Harvest

In many places on Earth, renewable natural resources are often perceived to be poorly managed and their harvest deficiently coordinated, planned, regulated, and governed. This easily leads to overharvest and detrimental extraction, harming the targeted species' populations and causing collateral damage in the ecological systems of which they form a part. What are then left behind are exhausted resources and impoverished environments.

Ecosystem deterioration and problematic wild species harvest are common in many parts of the world, but the adverse effects of human interventions with nature are often particularly visible in tropical areas. This is not necessarily because of a more careless type of harvest in general, but rather because the effects of harvest are seen in ecosystems that are extremely diverse and at the same time still in better conditions than is common in many temperate regions. The challenges are not only ecological, however. Wild species harvest frequently also involves problems of social character, and people, in many tropical areas, appear to be trapped in vicious spirals of resource depletion and poverty – while seemingly living amidst a bounty of biodiversity-related economic opportunities. Fortunately, however, this is not the whole picture and there are also glimpses of hope that can be found by examining harvest situations in different contexts.

Although each case is unique, wild species harvest situations across the globe share many common characteristics. Thus, there is need for integrative approaches and further unification of concepts in order to focus on this phenomenon and give it a thorough diagnostic treatment.

Preview to the Chapters of Part 1

These chapters together provide the baseline understanding for the rest of the volume. Having the focus set to the diagnosis of the state of wild species harvest, it is important to introduce the overall context. First of all, by the word 'diagnosis' we refer to the task of identifying the status of wild species harvest as an activity, including both the state of the resources and the harvest systems in question. Furthermore it refers to the identification of the possible problems and their causes, as well as potential solutions in any particular harvest situation, anywhere on Earth.

But what do we mean by wild species harvest? And why is it important? Where does it occur? And how do we interpret and apply the key concepts and terms around this topic? By first making these issues clear, we hopefully not only inform the reader but also motivate you to further dive into the details of wild species harvest in Amazonia, in the tropics, and on Earth. This way, the reader will also be facilitated to explore the stories from the forest floor in Part 2, to digest the theoretical exposition in Part 3, and finally to be able to make use of the Diagnosing Wild Species Harvest Procedure (DWiSH Procedure), in Part 4 of the book.

Chapter 1 scrutinises the history and significance of wild species harvest as a worldwide phenomenon. Our human evolutionary history as hunters and gatherers implies that our ancestors lived from the harvest of wild plants and animals. A great number of valuable products have their origins in nature, and countless people rely on their harvest and use, including processing, transport, and trade. This implies that although the concrete action of wild species harvest is invisible to many consumers, it is still not only a part of history or restricted to those parts of the world considered 'poor' or 'underdeveloped' by many. Rather, wild species harvest and the products obtained therefrom are important or even essential to all contemporary lifestyles.

Chapter 2 provides the conceptual background to support the reading of this book and defines our interpretation of some important terms and concepts used, such as wild, species, harvest, and harvester, as well as biodiversity, ecosystem functions, ecosystem services, and sustainability. We also briefly explain how the harvester is in the juxtaposition of nature and society, something important that will be returned to in later parts of the book.

Chapter 1

All over the Earth, since the Dawn of Time

A VITAL RESOURCE BASE

Humans have always harvested wild species of plants and animals. In fact, until a few millennia ago, wild plants and animals provided humans with almost everything they needed, including food, medicine, and raw materials for making clothing, tools, and shelter. In some places on Earth, this still continues, at least partially. Although the relative importance of wild species harvest has declined in the economy of human societies ever since the advent of agriculture, and even more so since industrialisation took off, economic and socio-cultural practices based on extraction of wild species resources have anything but disappeared. On the contrary, wild species harvest continues to take place almost everywhere on Earth.

Products based on wild species form the basis of several important contributors to the global economy, particularly in the cases of fish and wood that globally move large sums of money and form important sources of direct and indirect employment from informal to formal and from harvest to processing and trade. In addition, these and many other types of wild species harvest continue to be fundamental elements of local livelihoods in many parts of the world, particularly for people inhabiting sparsely populated regions with vast extensions of natural landscapes such as the rainforests of Amazonia or Central Africa, the forests and tundra of the circumpolar North, and many ocean islands around the world. In some cases, selling of wild species products provides cash income; in others, they are used for subsistence purposes; and many times, they are used for both. In particular, harvesting wild species resources forms an important source of food for poor rural households, as hunting and fishing often provide a crucial contribution to the nutrition and physical survival of people. Moreover, wild species harvest is also an important part of local culture, including identity, tradition, language, and worldview. Similarly, although less important for daily subsistence, harvest of wild species is often a highly appreciated recreational activity and part of cultural habits also in industrialised and urbanised countries and regions.

The variety of species that are harvested by humans is enormous, as is the variety of their uses. Logging, fishing, and hunting are examples of wild species harvest likely to be well known to most readers. It is also commonly known

Diagnosing Wild Species Harvest. http://dx.doi.org/10.1016/B978-0-12-397204-0.00001-2

that people in many areas gather edible wild fruits, berries, seeds, and mushrooms. But how many are aware of the fact that wild brown algae (seaweed) are harvested in order to extract alginate, a polymer with numerous applications in the biomedical industry (e.g. Ugarte and Sharp, 2012)? Or how many know about the economic importance of *yartsa gunbu* (*Ophiocordyceps sinensis*), a parasitic fungus that infects certain moth caterpillars, finally killing them, after which the long-fruiting body (mushroom) grows out from their forehead? Well, the collection of these tiny mushrooms, which constitute a valued ingredient in traditional Chinese medicine, is actually a main source of income for rural households in the Himalayas, and represents no less than 8 percent of the Gross Domestic Product of Tibet (Winkler, 2009). Expanding the perspective to wild species harvest a little more, how many have dedicated thought to the grazing of natural rangelands as a type of indirect wild species harvest where humans manage domestic or semidomesticated animals such as cattle, reindeer, or yaks, which, in turn, harvest the vegetation and convert it to animal meat, hides, wool, milk, and other valuable products?

Usually, the harvest of wild species targets live animals or plants, but it may also concern dead ones. The wood of dead trees of some species, for example, may be very durable and is therefore highly appreciated. Similarly, the tusks of elephants, the source of the very valuable *ivory*, could, in principle, be harvested from dead animals, although this is seldom done in practice. There are also cases where the harvested matter consists of accumulated dead plant or animal material, such as the harvest of peat or coral. Peat is an accumulated organic material made up of dead plants on waterlogged land where the lack of oxygen prevents decomposition, and it is used in horticulture as well as burned as a source of energy, for instance in Northern Europe. Corals are marine animals that live in colonies and secrete calcium carbonate to form external skeletons that, with time, build up large coral reefs. These, in turn, are mined by people for use as construction material. Although peat and coral resources are no doubt formed by wild species, their regeneration and accumulation take a very long time. Whether or not it makes sense to consider the extraction of such resources as wild species harvest may be a matter of debate.

Wild species harvest is global and local; it can be motivated by commercial impulses, subsistence needs, or even pure pleasure; and it touches all of us directly or indirectly. Because it forms an intuitively clear-cut link between the integrity of natural environments and human needs, wild species harvest is commonly proposed to have potential to form part of a solution to the tropical deforestation crisis, while simultaneously being a means to alleviate poverty. It was the commercial value of many wild species products, rather than their use value for subsistence purposes, which gave rise to the market-based conservation strategies encapsulated by Timothy Swanson in 1992 with the catchy slogan 'Use it or lose it' (Swanson, 1992). The idea behind this slogan was that if the economic value of biodiversity can be appropriated and made useful by extraction and trade, natural ecosystems and the wild species populations they support will be seen as something worth maintaining in the face of pressure from competing land uses. Along these lines of thought, the proponents of market-based conservation argue that the potential to make economic profit by harvesting valuable resources from tropical forests provides an incentive against deforestation. Many have proposed that there is also the option to not remove any of the standing trees but to concentrate the extraction on other forest resources instead. According to these views, extraction of the so-called nontimber forest products (NTFPs) can be an environmentally more benign alternative to deforestation, making it possible to make money by using the forest without cutting it down.

Commodities based on wild species resources have also often had strategically important roles. Historically this has been seen, for instance, when timber was used for construction of vessels, and tar used in their sealing; when exotic spices stimulated the great European explorations; when wild rubber fed the Northern industrialisation; and so forth. In their search of wealth, hegemonic actors with interest in wild species resources have shaped entire societies and nations, exercising power over local populations and limiting their freedom of choice. Likewise, such historical events and processes as the great rubber boom of Amazonia (e.g. Barham and Coomes, 1996) have formed essential pieces in development processes within societies often far away from the resources themselves. Appreciation of the value and potential of wild species is also manifested by the fact that, particularly in the nineteenth century, many colonial powers sent naturalists to overseas destinations to seek new resources, and botanical gardens in the mother countries then helped to establish plantations of economically valuable species in new continents.

The emergence of global markets for products derived from wild species has often led to booms of extractive activity without much care for the regeneration of the species in question, rapidly leading to mining-like resource subtraction and consequent depletion of the resource base. Whaling, for example, brought several species of these marine mammals to the brink of global extinction by the mid-twentieth century. Although hunting of the most endangered species was later banned, the population growth is slow and whaling still sometimes occurs, resulting in the fact that many (if not most) whale populations have not recovered. Another giant, the big-leaf mahogany (*Swietenia macrophylla*), is one of the most valuable timber trees in the world. It has been logged for its precious reddish wood for centuries, for overseas trade first in Central America and then, progressively, throughout Amazonia at an incredible speed. The accelerating pace of logging has been made possible by improved technologies and infrastructure during the 1980s and 1990s, and up until the present time, leaving behind forests emptied of mahogany trees.

Transformations in socio-economic conditions, new harvest technologies, changing consumer demand, and improved access through infrastructure development, among other factors, continue to provide surprises, too. Problematic situations may arise when a species suddenly becomes valued and enters booming extractive economies before any systematic planning is in place. But this is not only in the sense that the species get subjected to excessive harvest. For instance, the unpredictably increasing demand for natural products such as new 'health foods', one of which is the fruit of the açaí palm (*Euterpe oleracea*), may help combat deforestation but also can lead to management practices with adverse environmental impacts, such as local loss of biodiversity, when the wanted goods are produced in monoculture-like plantations. Even resources that are abundant may become the centrepiece of conflict, as shown by the example of bilberries *(Vaccinium myrtillus)* and lingonberries *(Vaccinium vitis-idaea)* of northern Fennoscandian forests, where wild species harvest continues also in the context of industrialised societies (Box 1.1).

So, how should wild species harvest be seen, and what, if anything, needs to be done with it in different situations? Should it be seen primarily as a problem, emptying nature of some of its crucial components or, rather, as a solution supporting locally important economies while maintaining the natural land cover or aquatic habitats? Knowing how to approach these questions helps one choose whether the harvest should be promoted or suppressed in any particular case. Time, for instance, is a crucial factor that continues to be overlooked. It should, at least in theory, be possible to harvest wild species resources in a way that guarantees benefits over prolonged periods of time, efficiently and fairly distributing the socio-economic and cultural contributions of harvest in human systems but also conserving ecosystem functions and biodiversity. Yet, there are no universal models or solutions to follow, and wild species harvest must be examined case by case. The options and possibilities will vary from one case to another and, of course, depend on local values and norms. What is clear, however, is that the more one knows about the reasons for and consequences of wild species harvest, the better prepared one is to confront any particular situation, and, furthermore, the better prepared one is to adapt management strategies and practices to follow the inevitable change that characterises the environment and society.

Science, in a broad sense, therefore has an important role to play, and researchers in many fields directly related to wild species harvest have indeed made great progress in resolving some of the above questions. The intricate ecological consequences of wild species harvest are today much better understood than they were just a couple of decades ago, and, similarly, the social sciences have contributed much new understanding of the ways in which wild species harvest is embedded in, and hence directly affects, socio-economic and cultural contexts.

Still, however, scientific knowledge about wild species harvest remains quite fragmented as it is split between different disciplines. Timber harvest is the realm of foresters, hunting belongs to wildlife biologists, fishing to ichthyologists, NTFPs to ethnobotanists, areas to geographers, markets to economists, institutions to social or political scientists, and so on, and there is not as much exchange of ideas between these realms as there could be. What all of these different scientific ways of thinking often have in common is that

Box 1.1 Uses of Wild Species in the Nordic Countries

Logging timber in natural forests used to be an important component of the economy of the Nordic countries Finland and Sweden, but nowadays forestry operations mostly take place in heavily managed seminatural forests that sometimes resemble plantations more than natural forests. Commercial fishing, on the other hand, which once was a pillar of local economies in many rural coastal and archipelago regions, struggles with poor profitability, and the number of full-time fishers has dwindled to a minimal fraction of what it once was. While the above activities have become less common, other kinds of wild species harvest nevertheless continue.

Collecting wild berries and mushrooms is at least somewhat familiar to many – if not most – people in these countries. There is a strong cultural heritage for using these resources, which can be found in great quantities during the late summer and the autumn in most forests and bogs. There are about half a dozen species of berries that are widely used, and the number of frequently collected mushroom species exceeds 20. Their harvest is a very popular leisure activity, providing good exercise and an opportunity to be in nature. Some like doing it alone, while others may prefer to do the harvest together with family members or with friends as a social activity. Even many people who live in urban centres dedicate themselves to this work for a few days per year at least. A good motivating factor is that berries and mushrooms are highly appreciated at the household level. Bilberries (*Vaccinium myrtillus*) and lingonberries (*Vaccinium vitis-idaea*), among others, can be eaten raw on the spot, be frozen, or be used for the elaboration of pies, jams, juices, and the like.

The harvest of berries and mushrooms has also been an important subsistence activity in the countryside. It has provided welcomed additional income as their selling is tax-fee. A condition that also favours all these activities is 'everyone's right', or the public right of access to collect such wild species resources as berries and mushrooms (Figure 1.1(A) and (B)) from the wild no matter who owns the land and whether the motive of berry picking is household use or commercialisation.

Although an estimated 45,000 t of berries are harvested every year, the large majority (an estimated 95 percent) of the berries are left unharvested in the forest. Traditionally, the berries were picked by hand, but in recent decades the use of certain berry-picking devices that comb the underbrush for berries has become more common (Figure 1.1(A)). In the new globalised world, alongside the native harvesters there are also professional seasonal pickers from the Baltic countries, Thailand, and Vietnam, among others, as 'everyone's right' is valid also for non-resident people in Finland. They arrive in thousands each summer, often facilitated by professional organisers who not only help to arrange the trip to the site but also provide the basic logistics such as housing and transportation to previously identified good harvest places. The harvest work itself is extremely hard during the few months of harvest season, and the pickers cannot always take it for granted that they will make enough money even to cover their travel costs. Often, however, they can make decent earnings to bring home at the end of the berry season. This activity is generally welcomed because it guarantees that an increased proportion of the annual crop will get collected and sold in marketplaces or used in food

Continued

Box 1.1 Uses of Wild Species in the Nordic Countries—cont'd

industries. However, also problems have arisen when such nonlocal harvester groups have entered local people's traditional berry-picking grounds. Many say that the incomers should be guided to work in remote places where nobody else is, instead of coming to the neighbourhoods of the native dwellers.

In contrast to the picking of berries and mushrooms, fishing and hunting are subject to more regulation. The right to hunt is related, with some minor variations between countries, to land ownership. Land owners hold the hunting rights to their lands and can lease them to others. Hunting large animals such as elk (*Alces alces*) (Figure 1.1(C)) is, however, even more strictly regulated; annual quotas are conceded at the level of subregions, also directing hunting to specific types of individuals to regulate the size as well as structure of elk populations. Tens of thousands of elks are hunted every year. Similarly, fishing rights are also related to ownership of fishing waters, and there are detailed regulations and permit systems. However, some types of fishing are relatively free for anyone. For example, ice fishing using a simple line and hook (Figure 1.1(D)) can be carried out free of charge anywhere and by anyone in Finland, with some exceptions, such as rules restricting fishing in specific places in order to protect certain important and threatened fisheries.

they do not emphasise enough the various aspects of interlinked environmental, economic, and socio-cultural contexts in which wild species harvest occurs. Luckily, some progress has been made during the last couple of decades in integrating economics and other social sciences with ecology when studying wild species harvest, but that kind of science is rarely taught in universities other than in very specialised advanced-level courses. If ecologists often fall short of understanding economic and social phenomena, equally economists and other social scientists involved in the study of wild species harvest are often hampered by a lack of understanding of the ecological processes and interactions involved.

In sum, thus, there is a need for scientists and professionals who specialise in wild species harvest science, or the study of wild species harvest as a phenomenon taking place at the interface between human society and the natural environment. In effect, wild species harvest forms a bridge that tightly links social and ecological systems, transforming natural resources into social and economic assets and, similarly, reflecting social needs into ecological systems.

FUELLING ECONOMIES, FEEDING PEOPLE

The role of wild species harvest in the world economy is far from insignificant. For example, fishing has truly global repercussions: in spite of the rapid expansion of aquaculture during the last few decades, capture fisheries production is still almost 1.5 times that of aquaculture (FAO, 2010a, 2012). In 2011, global capture fisheries supplied over 90 million tonnes of wild fish, molluscs, and crustaceans, corresponding to a world average per capita consumption of

FIGURE 1.1 Wild species harvest is common as a recreational activity and for household use in the Nordic countries. (A) Harvesting lingonberries with a special berry-picking device in western Finland; (B) yellowfoot (*Cantharellus tubaeformis*) mushroom picking near the city of Turku, Finland *Photo: Hanna Tuomisto*; (C) elk (*Alces alces*) hunting in northern Sweden *Photo: Hans-Gunnar Norman*; and (D) ice-fishing in the archipelago of south-western Finland.

more than 10 kg of fish and seafood per year, and representing a value of almost US\$130,000 million. In fact, global capture fisheries increased almost fourfold from 1950 to 1990, although since then they seem to have reached a plateau (FAO, 2010a, p. 6) with some important fish species now showing signs of rapid decline and even threats of extinction. When concentrating merely on overall seafood landings volumes, however, the problems of specific species have been largely concealed by the fact that when some species have become increasingly scarce, other more common ones have replaced them in the catch.

Another globally important economic activity largely based on wild species is forestry. For instance, the global harvest of industrial roundwood from natural or seminatural forests is still almost three times that from tree plantations (Siry et al., 2005). This has important implications for biological diversity since many of these production forests, particularly in the boreal and temperate regions, are intensively managed and thus are becoming increasingly similar to plantations, losing a multitude of species in the transition. Industrial wood comes increasingly from planted or heavily managed seminatural forests, and globally over 20 million km^2 of forest are used for productive purposes (for either wood or other forest products) (FAO, 2010b). This is an area comparable to the size of

Russia and India combined, or more than twice the area of the United States, and it accounts for more than half of the total forest area of the world. Although the level and type of actual forest use and management vary considerably, it is noteworthy that up to 80 percent of this area is also covered by at least some kind of management plan (FAO, 2010b). The area and significance of plantation forests are growing, too, and globally these forests cover more than 2.6 million km^2 (FAO, 2010b), an area almost the size of Argentina.

Natural and planted forests often provide different types of wood products. For example, in many tropical areas, natural forests are the main source of sawn wood, whereas pulpwood in the tropics is predominantly produced in plantations, using non-native species. While the relative importance of extracted wild timber can probably be expected to diminish in the world market in the future, its role will remain central for years to come. Selective logging of tropical hardwoods has become a well-known example of resource overuse and detrimental harvesting methods, but the issue is fortunately getting increasing attention globally: export and trading restrictions as well as methods of certified production are gaining ground.

Timber species differ in their use potential and value. In natural forests, the presence of many different species often means both an opportunity and a challenge for harvesters. Traditionally, the most wanted and expensive species have been extracted for their exportation, while lower valued species have been used locally. Economic growth at the local level often means that an increasing share of valuable timber is going to domestic markets; this is currently the case in many tropical countries. Although Europe and North America certainly keep consuming considerable volumes of tropical timber, the traditional South–North flux of trade has been challenged, since the 1960s, by increasing South–South trade. All Asian countries and Brazil together are estimated to consume about 80 percent of tropical timber, while the 'old' industrialised countries in the North consume about 5 percent (Roda et al., 2005). Southeast Asia has rapidly lost many of its natural timber stocks, and whereas South America still has extensive stocks of many species, Africa has been the continent with the most prominent logging frontier yet since the 1990s.

According to FAO (2010b), the contribution of forest products to the world economy is on the order of US$120,000 million annually. This figure is based on country reports covering 85 percent of the world's forests for industrial wood and fuelwood, and 77 percent for NTFPs. Based on the above mentioned reports, the value of industrial wood represents the major part of the total economic contribution of forests globally (70% of the total value). This figure, however, can be an overestimate because the reported values of fuelwood and NTFPs (representing the remaining 30%) are most probably underestimated. Trade for both national and international markets supports a constant demand for timber, and forestry is typically among the most important sources of employment in forested regions. Many jobs in the forest sector in remote places like Amazonia are low-paid and informal, and because of widespread illegal logging and trade of timber as well as other wild species products, the formal revenues generated locally are often scarce. All this easily leads to a biased view of the importance of wild species harvest, particularly in those countries in which wild species resources are important in subsistence economies, remaining largely outside of formal statistics and thus under-reported or even unreported.

If we turn away from fishing and logging, and to other wild biological resources, global sums of tonnes or euros reported are of smaller magnitude. However, there are also other more shadowy ways in which wild species form part of the global economy; the illegal trade of wild plants and animals, or products based on wild species, is among the most important global black markets (Schneider, 2012). Even without taking into account the highly lucrative businesses of the illegal timber trade (Kishor and Lescuyer, 2012) and illegal and unreported fisheries (Flothmann et al., 2010), both with estimated annual values counted in thousands of millions globally, the illegal trade of wildlife products (including rhino horns; live parrots, falcons, and reptiles; whale meat; and tiger parts, just to name a few) operates on similar scales and often intertwines with such activities as illegal drugs and weapons dealing, as well as human trafficking. Because all of these businesses are illegal, informal, and unreported, their exact volumes, values, and forms can be only indirectly explored and estimated. It is clear, however, that illegal trade of wildlife is a global problem, and because the rarity of species commonly elevates the prices paid, exactly those species that are rare and endangered are those that are most affected.

On the other hand, harvest of wild species, even when informal and unreported, has also its positive roles at regional and local levels, being particularly vital for local subsistence (Bharucha and Pretty, 2010). Wild species serve as food, fodder, fibre, and raw material for construction and handicraft, as well as flavours, dyes, cosmetics, medicines, and poisons. Their harvest supports households in remote and sparsely populated regions where extractive harvest often associates with direct consumption. Moreover, wild species often have specific cultural value. Highly biodiverse ecosystems commonly coincide with high cultural diversity, and wild species use is intricately linked to local cultural adaptations to the particular useful species that can be found in the natural ecosystems surrounding local communities. Indigenous or other local know-how on harvestable biodiversity resources and their use is very context dependent and can differ considerably even among neighbouring villages or between the genders (Howard, 2003).

The selling of locally valued wild species–based products helps local people gain some income where few other income-generating jobs are available, hence providing an important economic safety net. Even in areas where the economy is largely subsistence-based, rural households need occasional income for reasons varying from healthcare to school fees, and from cooking utensils to clothes. In addition to the mere harvesting of wild species resources, there is also a potential for generating added value by processing the extracted materials into sellable products. Many wild species goods are consumed locally by the harvesters themselves as well as traded, often to middlemen in the market chain (Vormisto, 2002). Some products are of such high value that the local users choose not to consume them, but rather get the income through selling them, thus adding to household wealth (Clarke et al., 1996). However, sometimes it is also the case that people cannot capitalise the wild species goods extracted, and so they remain in subsistence use.

The economic significance of selling extracted products varies according to the conditions of their production, processing, transportation, and markets. Much depends on the local circumstances and the people in focus, and while for example some NTFPs or other similar wild species resources may present notable potential with a significant contribution to regional economies, truly successful cases are few, and even these do not have certain futures. Examples include ornamental fish, rattan, wild rubber, açaí fruit *(Euterpe oleracea)*, and Brazil nuts *(Bertholletia excelsa)*, of which the latter two are addressed in more detail in this book. Also, some plant extracts (e.g. chemicals for drugs) provide instances of economically successful cases, although the products originally collected from the wild in many cases have been later produced in plantations or even synthesised artificially in laboratories. Even so, there are constantly new discoveries being made, and there will probably for a long time to come be countless plant and animal species with potentially valuable chemical compounds waiting to become discovered.

This potential has encouraged active efforts to search for new economically interesting products from nature, called bioprospecting, particularly from species-rich areas such as tropical rainforests. Costa Rica is often referred to as a prime example of a country that has established appropriate infrastructure to motivate international companies to seek valuable products based on its biodiversity resources. A partnership agreement was made between Costa Rica's National Biodiversity Institute (INBio) and Merck, a pharmaceutical company, in 1991, and there were high expectations about its potential outcomes (Zebich-Knos, 1997). There is not so much evidence of major success in these activities, however. On the other hand, the role of extractivism in local consumption and in regional and national economies is often overlooked as rather marginal, and only a few products have gained larger interest in monetary terms, mainly because of their international importance.

In tropical regions, hardwoods still remain the prime extractive product from terrestrial ecosystems with economic importance from local to regional, national, and global levels; local markets are also important for products like fuelwood and charcoal, as well as timber for construction. This means that wild species are very often used as a component of mixed livelihoods, but much of this activity and 'value' remain invisible in economics statistics. Amazonian *ribereño* and *caboclo* cultures are a good example of such mixed strategies based on both subsistence consumption and local trade. Increasing urbanisation is also a general trend in the tropics, including Amazonia, and products based on harvested biodiversity resources are essential constituents in local and regional economies through urban–rural trading, many times reinforced by the fact that many urban dwellers maintain strong links to rural areas.

At the local level, the use value of many wild species can also be seen as a kind of subsidy from nature. A lot of different activities in the deep wilderness areas, including oil exploration, logging, and collection of NTFPs, have been 'subsidised' by the presence of food (wild meat and fish) and construction materials directly available from nature at a relatively low cost. This reality gets easily overlooked since these activities tend to escape common statistics of economy and welfare, making them also easily disregarded in development planning. Furthermore, wild species harvest often forms part of cultural practices that contribute to social well-being through values related to leisure and pleasure, as well as spiritual activities, rituals, or other similar habits. For example, picking mushrooms or berries, fishing, and hunting are in many places far from necessities, but they still thrive as they are practiced for the sake of pleasure or other cultural values.

The first and most direct level of wild species use is subsistence, the consumption that helps people meet their daily needs. Although people rarely rely on wild species only for their subsistence, where they have access to these resources, it is common that the poorest of them are most dependent on their use (Arnold et al., 2011; Kaimowitz and Sheil, 2007), and wild species are particularly important as a subsistence asset base for the rural poor. Thus, wild species harvest can be seen as a safety net for the poorest of the poor in those areas where such resources are available. Therefore, judging the economic importance of wild species harvest based on its monetary value alone may lead to somewhat doubtful conclusions. Whereas a meal of food for a wealthy citizen of a city in a highly industrialised country may cost tens of euros, the monetary value of a meal for a subsistence fisherman in Amazonia or the Congo basin is, in comparison, minimal. On ethical grounds, however, one may argue that providing a meal to a human being is always equally valuable, or even that it is of more value to provide a meal to somebody living in conditions of food scarcity than to provide it to somebody who has access to an excess of food.

Many people rely heavily on local fish and wild game resources for their nutrition. For example, in subsistence economies in Amazonia, where agriculture predominantly involves growing crops such as manioc and plantains consisting almost entirely of carbohydrates and dietary fibres, wild game and fish provide crucially important protein, in fact often in quantities well in line with, or above, those recommended as sufficient by the World Health Organization (WHO). As dietary fat often is in scarce supply, and the intake of it well below levels recommended by the WHO, the contribution of wild game and fish to fat intake, and as a concentrated high-energy and low-fibre energy source, is also very important (Sirén and Machoa, 2008).

Wild species products also provide welcomed input to the diets of urban populations, including those living in some of the most industrialised countries. In a special mention of this, the Food and Agriculture Organization (FAO) of the United Nations has estimated that approximately 1000 million people around the globe use wild species as part of their daily diets (after Bharucha and Pretty, 2010). Much of this consumption is based on wild fish from the sea, but there are also plenty of other uses. In Brazilian Amazonia, the average consumption of fish, which is almost exclusively captured wild, has been estimated to be 110 g per day per capita among urban populations (Isaac and de Almeida, 2011).

Wild plants usually make up a small part of food intake but may make significant contributions to the intake of micronutrients (Ogle, 2001). Access to wild species as an additional source of diet therefore has several advantages compared to relying only on agriculture and industrially processed food. As for food security, many wild foods are of excellent nutritional quality, and when a wide array of species is available, problems of seasonal or catastrophic scarcity of staple products can be overcome (Arnold et al., 2011). Moreover, wild species of animals and plants also form a source of traditional medicines that are essential for the primary healthcare of many people both in rural areas and in urban centres (Elujoba et al., 2006; Sheng-Ji, 2001).

WHY CARE?

As seen in this chapter, wild species products continue to be used and appreciated by practically all cultures worldwide, by rich and by poor, and in spheres of society ranging from formal to informal and from legal to illegal. The wild species goods consumed include a wide variety of fish, game meat, fruits, seeds and palm hearts, timber, fuelwood and charcoal, vines, palm leaves, latexes and resins, feathers, hides, bones, bamboo, insect larvae, molluscs, and mushrooms. The availability and continued regeneration of the wild species resources depend on the integrity and functioning of the corresponding ecosystems. Humans get a multitude of benefits from nature in the form of diverse resources and processes, among them the provision of these useful wild species resources for harvest and consumption. This emphasises our strong interest in what happens to the biosphere within which we developed, of which we form a part, and on which we depend. Caring about ecosystems and their integrity is to care about ourselves, too.

As with any other use of nature, wild species harvest inevitably intervenes with the targeted ecosystems and causes some degree of change in them. Harvest removes whole individuals of targeted species, or their parts, and while it is possible that certain levels of harvest can even improve the state of the resource (e.g. moderate levels of extraction of leaves may stimulate the growth of a plant), more often the result is the opposite and the targeted individuals may die or their viability may get reduced. When harvest levels are consistently higher than the rate of recovery or regeneration, this ultimately impacts the individuals and, further, the populations of the targeted species, often finally reducing their abundance. Sometimes, through chain effects, harvest may induce changes in the entire ecosystem, reducing its integrity and affecting the ways in which it functions.

Despite the value that wild species products have today, unmanaged or poorly managed harvest is still far too common, often resulting in serious depletion of the resource stock and even outright extinctions, whether locally, regionally, or in extreme cases even globally. This would be nothing new. For instance, evidence supports the view that our early ancestors sometimes contributed to the extermination of mega-faunal species in the late Pleistocene, beginning as early as 200,000 years ago in Africa and culminating on the American continent between 12,000 and 10,000 years ago, although also climatic change may have played an important role. Notable is also how the development of seafaring technology in past centuries led to the extinction of many animal species that were endemic to ocean islands, some of them directly affected by harvest for human use. While this was sometimes a relatively slow process, the rather sudden emergence of new technologies and international markets during the last few hundred years has led to increasingly rapid processes of resource depletion or species extinctions in a variety of geographical locations. The rates of human-caused global extinctions cannot be assessed with much precision, but even with conservative estimates it is beyond doubt that species extinctions at local and regional levels affect the functioning of ecosystems (Stork, 2010).

Sometimes people not only harvest a species but also engage in purposeful activities in order to increase the stock of the species of interest. This may involve, for example, stocking, replanting, feeding, removal of predators or competitors, or even introduction of exotic species. In such cases, the net effect may be an increase, rather than a decrease, of the abundance of the target species. The effect on the ecosystem as a whole may, however, be considerable, sometimes even greater than the effect of harvesting only. If such purposeful management is very intensive, talking about wild species harvest becomes misleading, as it then rather becomes a question of cultivation of species in human-made ecosystems. This is often the case today, for example, in industrial-scale wood production in heavily managed production forests, or intensive large-scale fish farming. Just as there is no clear-cut dividing line between natural and human-modified ecosystems, there is no clear divide between what constitutes the harvest of wild species and what constitutes the harvest of farmed plants and animals. Rather, there is a continuum between these extremes.

Among other types of human intervention, extractive activities shape many characteristics of the biosphere, of which biological diversity is a vital one. For this reason, the danger of major biodiversity collapse is one of the main concerns of our times. In a historical perspective, perhaps with the exception of global climate change, the significance of this concern probably overrides any other human activity of the late twentieth and early twenty-first centuries. This point was made strikingly clear by the world-famous Neotropical botanist Alwyn Gentry (1945–1993) in a lecture he gave when visiting the University of Turku in Finland in early 1992, shortly after the collapse of the Soviet Union.

At that time, when the ending of the Cold War was the hottest of all possible news, this scientist advised the audience to understand the big picture instead of momentary details. Future generations will look at our times because so many species went extinct, hardly remembering that there had been a Soviet Union or a Cold War overall. The further away in history that referred moment remains, the wiser Gentry's notion begins to look. More recently, scientists have started to warn about the danger of a global-level ecosystem state shift that may occur as a result of excessive planetary-level human influence, with unanticipated effects on the biota, including us (Barnosky et al., 2012). Biological systems may then shift from the state where they exist at present to a different state in a way that is both rapid and irreversible.

Species-rich areas such as Amazonia are particularly critical in this respect. The loss of their forests would have serious consequences with cascade effects both within the tropics and at the global scale. Their high biological diversity means that there are innumerable interactions of myriad kinds among the different species and their individuals. This makes it difficult to predict the consequences if systems are increasingly changed through human intervention. Whether the excessive extraction of even a single species turns out to be a great problem or not is hard to anticipate in such complex systems, and the ecological chain effects caused by wild species harvest may be easier to comprehend in species-poor ecosystems. But in the complex multispecies tropical forests, we are only scratching the surface when it comes to understanding ecological functions and interactions. Naturally, targeting only one or a few species in a species-rich system is likely to induce less dramatic consequences than those activities that would cause complete habitat conversion. Therefore, as long as deforestation does not take place, the sustainable harvest of selected wild plants or animals hardly threatens the existence of most rainforest species in any direct way. Indirect consequences may, however, be surprising, particularly when so called keystone species are being depleted. For example, flying foxes (large fruit-eating bats) are the most important (and sometimes the only) seed dispersers for many plants with large seeds in tropical Pacific islands. Where over-hunting has decimated their populations, not to extinction but below a certain threshold determined by the social behaviour of the bats, their efficiency as dispersing vectors has been shown to dramatically drop (McConkey and Drake, 2006). Because many of the trees depending on bats are long-living, it may take a long time before the populations of tree species start to decline, but they will.

Over-exploitation of wild species resources, as well as habitat destruction, may also diminish the locally available development assets when benefits from harvest cannot be obtained on levels similar to those before. Due to interlinked natural and human systems, ecological feedback may easily become an economic issue, and vice versa. For example, after selective logging, a species-rich forest may still remain, just missing some of its previous species. But if the missing species were the most wanted ones, then the remaining forest is of lower value, be it economically, socio-culturally, or even ecologically. Consequently, arguments for its preservation in the face of deforestation pressures may be weakened. In a similar fashion, over-hunting has often led to forests without large-bodied animals. This not only has consequences for ecosystem functioning in the long run but also forces hunters to find other game animals, usually smaller sized.

Ecosystem-level changes also impact many important ecosystem functions. Despite the fact that many species benefit from human activities (e.g. the many successional species favoured by human disturbance), the biosphere would survive very well also without our presence – as it has done before our times – and will also survive if and when we are gone (an interesting thought-experiment about this is presented in the book *The World without Us* by Alan Weisman, which was published in 2007). But it is in the interest of our present and future generations to guarantee that no single species changes this planet too much. Therefore, we need to seek possible pathways to try to overcome the problems of short-sighted resource use by replacing it with more sustainable use. This can take place only on the basis of a better understanding of natural and social processes from local to global levels, and through actions committed to minimising the adverse ecological effects of our activities. Wild species harvest, despite its long history and worldwide occurrence, needs to be accompanied by ethics and a sense of moderation, and societies need to learn to better understand, use, manage, and preserve these valuable resources for future generations.

Chapter 2

A Conceptual Primer to Wild Species Harvest

HARVEST, WILD, AND SPECIES

Environments representing a wide continuum from wilderness to heavily human-modified landscapes form a warehouse of harvestable wild species resources. Individuals of wild species can be targeted by humans because of a diversity of motives, most commonly including one or more of the following: (1) the value of the biological matter found in these organisms, (2) the value of the cultural practices of harvest in itself, and/or (3) the harm that some specific organisms are perceived to cause to humans.

Often, it is a combination of more than just one motive that makes people extract wild species. For example, in recreational extraction of wild species such as recreational hunting and fishing, although the extracted matter does have a use value, it is not the main motivation of harvest but rather it is the value of the mere activity itself or of being in nature. Likewise, in nonconsumptive harvest such as catch-and-release fishing, the organisms are actually just temporarily extracted from the ecosystem and then returned back to their natural habitat. Although we dedicate little space to such cases in this book, much of the theory presented is valid also for cases of recreational and nonconsumptive harvest.

Wild species are also killed or extracted in spite of not providing products of much value in themselves. In those cases, extraction may be driven by the perceived adverse effects that these species have when alive and present in nature, from the viewpoint of some people. This is the case, for example, with some species of insects and snails that act as vectors for diseases, predators threatening domestic or game animals and exotic invasive species that harm the native biota (Figure 2.1). However, we consider removing or killing individuals of these species to be an act of combatting wild species, not harvesting them, and this is thus not particularly addressed in this book. Wild organisms often are also harvested accidentally when they are extracted collaterally, for example as side catch. Similarly to combatting wild species, although this kind of extraction does not fit our definition of wild species harvest, often it has important ecological implications.

The concept of **harvest**, which is the focus of this book, we define as the extraction of wild organisms or parts thereof from the ecosystem, with the main

Diagnosing Wild Species Harvest. http://dx.doi.org/10.1016/B978-0-12-397204-0.00002-4

FIGURE 2.1 The American mink (*Neovison vison*) is an exotic species in Finland, and it is hunted with traps and dogs in order to protect the native bird fauna from predation. *Photo: Tommy Arfman.*

motivation for this being the perceived value of the extracted matter, in addition to the cultural value of the practice itself. For instance, if a fruit has use as food, this edibility results in a value, be it for subsistence or for further trade of the fruit, and extraction therefore takes place in order to obtain a desired feedback (a need fulfilled) (Figure 2.2).

In our definition of wild species harvest, the term **wild** refers to organisms that carry out their life cycles predominantly through natural processes without major conscious human intervention. We acknowledge the ambiguousness of this definition, as there are many cases where animals or plants are only 'kind of wild' (meaning that they are managed by humans to some extent), but we do not see this as a problem. Much of what this book tells about wild species harvest certainly also applies to situations in which the 'wildness' of some species is not complete. For example, the countries of the Baltic Sea region have released millions of smolts to support natural salmonid populations. To what degree the populations experiencing such human interventions are wild or natural is a relevant question, but we trust that the reader will be able to judge the applicability of the theory presented in this book to such cases. Wild species are, according to this definition, not restricted to wild nature but may occur also in ecosystems heavily modified by humans. Ultimately, almost nothing on this planet is completely independent of human influence.

A common definition of **species** is that it is a group of organisms that are similar to each other and can interbreed to produce fertile offspring. This definition is, however, not applicable to organisms that have only asexual reproduction.

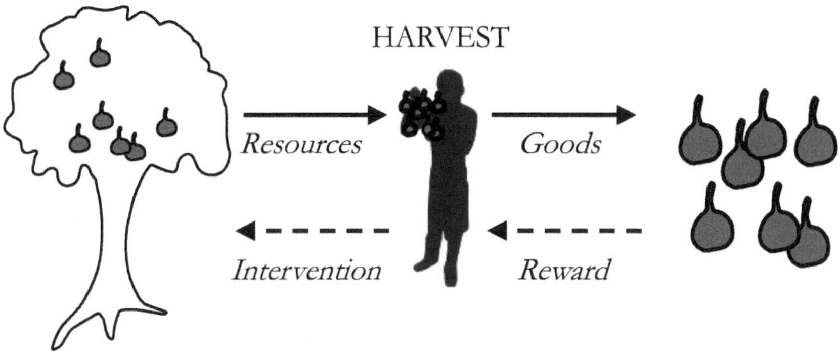

FIGURE 2.2 Essential components of wild species harvest. A fruit-bearing tree provides a resource for the harvester, and once the fruit are extracted, they can be further delivered as goods that have specific utility in society. The harvest intervenes with nature through the removal of matter, and the harvester, in turn, gets a reward in the form of, for example, a payment or just a full stomach.

It also becomes problematic e.g. for some plant groups, such as orchids, where hybridisation is common; the offspring are not necessarily infertile, but they may have somewhat reduced fertility – enough to maintain the two parent species separate. In most cases, however, it is relatively easy to distinguish different species from each other. In this book, we use the term 'wild species harvest' not only because it is often very relevant to study this phenomenon at the level of species, but also simply because this term is short and handy to use.

As a multifaceted phenomenon discussed within different academic disciplines as well as outside academia, wild species harvest cannot be addressed without employing a series of concepts and terms from across disciplines. Confusion can thus easily occur in cases where one and the same term has an established and very specific meaning within one school of science, but another meaning in other sciences, politics, or everyday speech. In other cases, there are terms and concepts that are so new that their exact meaning remains a matter of debate even within the discipline where they originated. Clarity of thought, however, requires clarity of concepts and terms, and therefore the rest of this chapter is dedicated to clarifying the meaning of some key concepts and terms as used in this book.

FROM BIOLOGICAL DIVERSITY TO BIODIVERSITY – FROM SCIENCE TO POLITICS

The value of biological diversity as a structural characteristic of nature has been widely recognised during the last few decades. Increasing knowledge about the different species, their characteristics, and their interactions also adds to the way we see the available provision of living natural resources. Diversity is an inherent characteristic of nature that is important in its own right, but it also has very concrete implications on the harvest of wild species. The more species-rich an area is, the more there are potentially useful species. In species-poor conditions, in turn, there are typically a smaller number of useful species, but one or a few of them may sometimes be very abundant. This example invites us to explore biological diversity more in detail.

One hectare of pristine rainforest in Amazonia typically hosts between 200 and 300 tree species (e.g. De Oliveira and Mori, 1999), an order of magnitude more than the number of tree species in the entire country of Finland. This amazing variety of life forms was vividly described by the nineteenth-century British naturalist Alfred Russel Wallace in his account from inside the Amazonian forest:

If the traveller notices a particular species and wishes to find more like it, he may turn his eyes in vain in any direction. Trees of varied forms, dimensions and colours are around him, but he rarely sees any one of them repeated. Time after time he goes towards a tree which looks like the one he seeks, but a closer examination proves it to be distinct. He may at length, perhaps, meet with a second specimen half a mile off, or may fail altogether, till on another occasion he stumbles on one by accident.

(Wallace, 1878, p. 65)

This reality has very significant implications for wild species harvest as an economic activity. On one hand, there is an enormous number of different species to harvest, out of which many may be useful for humans for a wide variety of purposes. On the other hand, most species occur at very low densities, making harvest complicated and costly. Also, managing ecosystems for harvesting a multitude of different species is much more challenging than is the management of ecosystems where only one species is of interest, and often also dominant.

A common consequence of human activities is that species are lost and nature becomes more uniform. This may be due to conscious choice, for example when the various tree species in a natural forest are replaced by a monoculture plantation of just one fast-growing tree species, or it may be the unintended outcome of resource extraction, such as when animal species are hunted to extinction. Both cases are, however, examples of what scientists during the twentieth century started to call the loss of biological diversity. Even up to the 1970s, the term **biological diversity** was unknown to the general public and only rarely used in scientific journals. Its use became somewhat more common in the 1980s, as the general public became increasingly aware of the problem of species extinctions, and as scientists needed a concept that would embrace the problem of species extinction at large, rather than just discussing the extinction of species one by one.

The abbreviated term **biodiversity** was then coined as late as 1986, when the United States-based National Academy of Sciences and the Smithsonian Institution arranged a National Forum on Biodiversity. One of the participants in the forum, Lester Brown, formulated the need to reach out to the general public about the importance of biodiversity, stating:

We've got to move the issue from the scientific journals into the magazines and the popular press, so that maybe someday Jane Pauley will say, 'And today we have a scientist who's going to discuss biodiversity. THAT'S RIGHT, biodiversity'.

(Brown, 1988)

And soon, indeed, biodiversity became part of the everyday vocabulary far outside the academic circles, and it even found its place on the agenda of international politics. Only six years later, at the Earth Summit in Rio de Janeiro in 1992, the United Nations adopted a Convention on Biological Diversity (CBD), where signing parties affirmed that conservation of biological diversity is a 'common concern of humankind' and pledged to elaborate national strategies and action plans in order to conserve biological diversity (Heywood, 1995). Although the Convention in its wordings carefully stuck to the term 'biological diversity' for a 'scientific flavour' (Kaennel, 1998), the word's much more catchy neologism 'biodiversity', with essentially the same meaning, was already taking over in many other contexts.

Today, little more than two and a half decades after the aforementioned biodiversity forum, biodiversity not only is discussed among the political leaders of the world, but also is a common term in mainstream mass media, forms part of primary school curricula, and is a standard ingredient in the discourse of indigenous leaders struggling to protect their homelands in Amazonia and around the globe. Hence, in hindsight, the organisers of the forum in 1986 can, for sure, be very satisfied with how the biodiversity concept has been popularised ever since. Likewise, as a consequence of the CBD, biodiversity concerns have been mainstreamed to almost all levels of political decision making. The Convention's three main objectives – conservation, sustainable use, and equitable sharing of benefits – are closely related to wild species harvest as it is explored in the present book.

Many topics related to wild species harvest are also being constantly debated in CBD-related processes and initiatives worldwide. Many of these are found under the umbrella label of 'cross-cutting issues under the Convention on Biological Diversity' addressing biodiversity from various angles, including concerns of how biological diversity contributes to development and poverty alleviation; how access and benefit sharing should be organised in relation to genetic resources; how awareness building, information sharing, and communication can contribute to biodiversity use and conservation; and how trade should be regulated, just to list a few (Box 2.1).

The Convention on Biological Diversity recognised three different levels of diversity: diversity within species, between species, and of ecosystems. Diversity within species is basically the same as **genetic diversity**. This is important in that it enables genetically different individuals of a species to cope slightly differently with changing external conditions such as disease outbreaks or environmental change. If all individuals in a population are genetically identical, the population may easily get wiped out by a disease or sudden environmental change, but if

Box 2.1 Convention on Biological Diversity: Hopes, Achievements, and Disappointments

Like the Convention on Climate Change, the Convention on Biological Diversity (CBD) is a binding international environmental framework agreement under the umbrella of the United Nations. Currently, the Convention has more than 190 parties (member countries and organisations), and it is the major international undertaking for directing policies and practices of conservation and sustainable use of biological resources globally. CBD entered into force in 1993 and has a permanent secretariat in Montreal, Canada.

The Convention recognised for the first time in international law that the conservation of biological diversity is 'a common concern of humankind' and is an integral part of the development process. The agreement covers all ecosystems, species, and genetic resources. It links traditional conservation efforts to the economic goal of using biological resources sustainably. It sets principles for fair and equitable sharing of the benefits arising from the use of genetic resources, notably those destined for commercial use. It also covers the rapidly expanding field of biotechnology through its Cartagena Protocol on Biosafety, addressing technology development and transfer, benefit-sharing, and biosafety issues.

The main outcome of CBD has probably been the catalytic effect of carrying out national analyses on the conservation status of biological diversity in the member countries. Subsequently, most of the member countries have also established national strategies on biodiversity and promulgated legislation and regulation on trade-related issues, most notably on genetic material.

The CBD Nagoya Protocol on Access and Benefit Sharing was adopted in 2010. The Protocol aims at fair and equitable sharing of benefits arising from the utilisation of genetic resources, thereby contributing to the conservation and sustainable use of biodiversity. The underlying logic of CBD and the Nagoya Protocol is to increase the economic value of genetic resources, species, and natural ecosystems. Low economic value of the diverse ecosystems is one of the key factors driving deforestation of tropical forests and their conversion into agricultural lands. When originally adopted, CBD was thought to be able to create an economic base for the sustainable use of genetic resources by the biotechnological and pharmaceutical industries and to provide effective tools to eradicate poverty in rural areas, especially in tropical countries.

What was not foreseen during the early years of the Convention was that most drug development is nowadays made by 'rational drug design' (i.e. by developing the molecules by modelling in a laboratory, instead of using the natural compounds present in plants and microorganisms). It has been estimated that only 1 out of 10,000 samples collected by bioprospecting of natural plants leads to further development of the substances found. Most of the promising pharmaceutical plants and the plant material needed in plant breeding are already in national or international collections, causing low interest in time-consuming and expensive bioprospecting in natural habitats.

In the early years of CBD, the common saying was 'Biodiversity – use it or lose it', indicating that only if the economic value of biological resources increases,

Box 2.1 Convention on Biological Diversity: Hopes, Achievements, and Disappointments—cont'd

there will be increased efforts to protect natural habitats, especially in conditions where legislation and regulation are not effectively protecting ecosystems of high biological diversity. The lesson learned during the 20 years of CBD has shown that the value of genetic resources may not be high enough to promote the protection of natural ecosystems.

By Jukka Salo

the population is genetically diverse, the chance is greater that at least some individuals are resistant to the disease or are able to adapt to the stress or change and survive. **Ecosystem diversity** is a term that has been used with different meanings, sometimes describing diversity *within* an ecosystem, in terms of species composition as well as ecological interactions (Jizhong et al., 1991), and other times to describe diversity *among* ecosystems, at any spatial scale from local to global. Attempts to quantify ecosystem diversity in this latter sense have actually often been based on the composition of species assemblages in the different ecosystems. Perhaps most commonly, however, the term 'biodiversity' refers to **species diversity**, which refers to the composition of an assemblage of species found in an ecosystem, including the relative abundance of different species.

The most intuitive way of quantifying species diversity is simply counting the number of species present in a given area, which is called **species richness**. Note, however, that this measure does not take into account how abundant each species is. Imagine two forest areas, each with 1000 trees comprising 100 tree species. In one area – let us call it Area A – there are 10 individuals of each species, whereas in the other, Area B, 901 trees belong to one and the same tree species, and the other 99 species are represented by only one individual each. Obviously, our intuitive feeling looking around in the two forests is that Area A is more diverse than Area B. In Area A, we would probably have a hard time finding two individuals of the same species reminding us of Alfred Russel Wallace's experience from the Amazonian forest, quoted above. In Area B, however, almost every tree we see belongs to the same species. Nevertheless, the species richness is the same: 100 species in both samples.

In addition to species richness, species diversity therefore consists of yet another component, species **evenness**. In the example of the two forest areas, although both have the same species richness, Area A has higher species evenness, all species being equally abundant, and therefore it also has the highest species diversity. This example was a relatively clear-cut one (because both areas had the same species richness and differed only in evenness), but what if one area has higher species richness but lower species evenness than the other? Which one then has the highest species diversity? To answer this question, one must first mathematically define how to weigh richness and evenness together in order to create one single measure of diversity.

It turns out, however, that there is no single 'best' way to go about this. In fact, ecologists have invented many different ways to calculate **indices** of species diversity, and great confusion has reigned. A unified concept of species diversity, has, however, been suggested and is referred to as the **effective number of species** (Hill, 1973; Jost, 2006; Tuomisto, 2010), which is calculated using a general formula that can be slightly varied in order to give more or less weight to species richness or evenness, respectively, depending on what the analyst chooses as adequate in each particular case (Box 2.2).

Box 2.2 Effective Number of Species (based on Hill, 1973; Jost, 2006; Tuomisto, 2010)

The effective number of species has been mathematically defined as the inverse of the weighted generalised mean of the proportional abundances of the species, these proportional abundances being used as the weights. The formula is

$$^{q}D = \frac{1}{\sqrt[q-1]{\sum_{i=1}^{S} p_i p_i^{q-1}}} = \left(\sum_{i=1}^{S} p_i^q \right)^{1/(1-q)}$$

where p_i is the proportional abundance of the ith out of S species present in the sampling unit, and q is the **order of diversity**. This is an arbitrarily chosen parameter set between zero and infinity that defines which mean is used in the calculation. The lower the value that is used for q, the more importance is given to species richness, compared to species evenness, in quantifying ^{q}D. Conversely, the higher the value of q used, the more importance is given to any species with high proportional abundance. ^{q}D is called the **effective number of species** because, *if all species are equally abundant*, the result of the calculation equals the actual number of species.

Several of the commonly used diversity indices turn out to be just different transformations of special cases of this equation – using some specific value of q – for the effective number of species. Setting $q=0$ (which corresponds to using the harmonic mean) yields the diversity value ^{0}D, which is equal to the number of species (i.e. species richness). Setting $q=1$ (which corresponds to using the geometric mean) yields the diversity value ^{1}D, and the logarithm of this is called the *Shannon index* of diversity.

Setting $q=2$ (which corresponds to using the arithmetic mean) yields the diversity value ^{2}D, which is equal to the inverse of the probability that two individuals drawn at random from a sample (with replacement of the first individual before drawing the second) belong to the same species. This is also known as the inverse Simpson index. For example, if there are 100 equally abundant species in an area, the probability that two randomly drawn individuals belong to the same species is 1/100, and the inverse Simpson index, or order 2 diversity, is then $^{2}D=1/(1/100)=100$. If there are only 50 equally abundant species in another area, then $^{2}D=50$. Regardless of how many species there are in total, if only one species is extremely dominant, making up almost 100% of the individuals, again, ^{2}D would be just slightly above unity.

Species diversity can be measured at any spatial scale of one's own choice, and species numbers are, in practice, always somewhat higher when large areas are covered than at the local scale, because within large areas there may be many different environments with different species assemblages, due to differences in, for example, soils, local climate, or ecological processes. In broad terms, local species diversity has often been called **alpha diversity** (α), whereas the overall species diversity within a large region has been called **gamma diversity** (γ). Finally, diversity of local habitats with differing compositions of species within a landscape has often been called **beta diversity** (β). There has, however, been considerable discrepancy about how to actually define these concepts, and only recently a set of stringent mathematical definitions has been developed that is gaining increasing acceptance (Box 2.3).

For the reader who is not a fan of theoretical ecology and mathematics, it is not necessary to get into the details of Boxes 2.2 and 2.3. The point we want to make is that alpha, beta, and gamma diversity are terms that have specific mathematical definitions which however, require that the spatial scales of interest are also provided. It is therefore recommendable to reserve the use of these terms to when referring to the mathematical properties of data sets based on field inventories at defined spatial scales. When referring to less specific observations and phenomena, it is preferable to use colloquial terms such as local-scale species diversity, habitat diversity, and regional-scale species diversity, which are somewhat analogous, but not synonymous, to alpha, beta, and gamma diversity, respectively.

The question of spatial scale is very important when discussing the impacts of wild species harvest and other human activities on biodiversity. An intermediate level of disturbance of ecosystems on a local scale actually often increases species diversity, but it would be erroneous to assume that species diversity therefore increases also on a regional scale if the spatial scale of such disturbance

Box 2.3 Alpha, Beta-, and Gamma Diversity

Gamma diversity (γ) is the total species diversity of any chosen taxonomic group (e.g. trees, ferns, or mammals) of any defined study area. **Alpha diversity** (α) is calculated in a similar way as gamma diversity, the difference being that species diversity first is calculated for each one of a number of subunits, or sampling plots, and then the average, weighted according to the number of individuals in each subunit, of these diversity values for each subunit is the alpha diversity of the whole study area. **Beta diversity** (β), finally, is the gamma diversity divided by the alpha diversity, and it is a measure of the diversity in terms of habitats within the whole area of interest. Thus, the total, or gamma, species diversity in an area is the product of the local, or alpha, species diversity and the diversity in terms of habitats (i.e. the beta diversity): ${}^qD_\gamma = {}^qD_\alpha \times {}^qD_\beta$ (Tuomisto, 2010).

This implies that whereas gamma diversity has meaning only when the limits of the study area are defined, alpha and beta diversity have meanings only when also the size of the subunits (sampling plots) is defined.

becomes regional. On the contrary, it is more likely that such disturbance causes a decrease of regional species diversity.

Whereas it can be useful to know the definitions of quantitative measures of species diversity, one should not discard the less precisely defined and more colloquial (sometimes also value-laden) uses of the term 'biodiversity'. It should be noted that whereas species diversity (as defined in this chapter) could be increased (at least temporarily) by introducing exotic species, this would hardly be considered a positive thing by anybody involved in 'protecting biodiversity', and it could easily in the long run lead to decreased species diversity as native species get outcompeted.

Focusing back on wild species harvest, biodiversity as a whole is an important topic for many different reasons. First, the biological matter to be harvested comes from some selected species only – which are from a pool of many co-occurring species. The more species-rich the system, the more there are likely to be plants and animals that are potentially useful or valuable for humans. In such conditions, it is crucial for the harvester to know species well. Second, it is important to note that the other (nonharvested) species that occur in the same habitat can play an important role for the targeted species through ecological interactions. Together, all the co-occurring species make up the total system with its complex interactions. This deserves to be carefully attended to in order to guarantee the long-term well-being of the system even under conditions of continued harvest pressure on only one species.

Third, the different spatial scales of diversity help us understand important concepts such as **species distribution** and **population density**. For example, the concept of habitat diversity (which can be quantified by studying beta diversity at appropriate spatial scales) calls attention to the mosaic of different habitats with their corresponding species assemblages. As an example, the distributions of different habitats within a region can be used as surrogates to map patterns of species distributions and ecosystem functions. This is particularly valuable in species-rich areas, where it can be extremely difficult to make full inventories of all species present, let alone provide reliable distribution mapping of all possible species. In such cases, mapping edaphic variability and site characteristics can be much more efficient.

ECOSYSTEMS AND THEIR SERVICES

Just as biodiversity is an inherent and essential feature of nature, a myriad of other structures and processes of ecosystems maintain the existence of life on Earth. They contribute to the maintenance of life through, for example, primary production, by offering habitats, and by regulating vital biogeochemical cycles (e.g. De Groot et al., 2002). These structures and processes, sometimes also called **ecosystem functions**, exist independently from humans (Escobedo et al., 2011). Plants use solar energy to convert water, carbon dioxide, and nutrients to a wide variety of substances that enter the food webs, providing the basis for the existence of many physical things that, once taken from nature, can be used to satisfy material or other human needs, such as food, raw materials, and energy.

On the other hand, natural systems also provide habitats where wild plants and animals can live and reproduce, and regulate essential processes such as local, regional, and global climates; water runoff and river discharge; the formation of soils through rock weathering and accumulation of organic matter; and soil retention due to vegetation cover and root matrices in the soil, to name a few. The habitats and the regulation functions are essential for the maintenance of each other and of the primary productivity of ecosystems. And wild species harvest can affect all of them.

Humans are part of the biosphere and the ecosystems therein, and while many characteristics of the natural world are both vital and irreplaceable for our existence, others that benefit us are nevertheless to varying degrees substitutable by something else – at least momentarily or in a specific location. Also, highly anthropogenic ecosystems, such as agricultural, pastoral, silvicultural, or aquatic ecosystems that are designed and managed by humans with the purpose of producing some specific goods, usually depend to a great extent on natural processes that sustain habitats and regulate the functioning of the ecosystem through such things as pollination, soil decomposition, and gas and nutrient cycles. Truly sustainable use of ecosystems implies the maintenance not only of their productivity but also of their functions related to habitats and regulation of ecological processes.

All structures, processes, and functions of ecosystems discussed here can result in benefits for humans. In one way, the provision of these benefits can be perceived as a service (i.e. something that is 'done' by someone for someone else). Of course, there is no person that realises these services, but rather it is nature itself. Therefore, when a specific value from the human perspective is attached to them, the above benefits are called **ecosystem services**. This is a strictly anthropocentric concept, meaning that only when humans are present to assign a value to something provided by nature can it be called an ecosystem service (Carpenter et al., 2009; Burkhard et al., 2010). The Millennium Ecosystem Assessment (MEA, 2005) classifies ecosystem services in four categories: supporting (general life-supporting processes such as primary production and nutrient cycling), regulating (regulation of such things as climate and water systems), provisioning (provision of natural resources and energy), and cultural services (how humans experience nature culturally).

While all these are very important for wild species harvest, first and foremost it depends directly on the provision of harvestable resources from nature, and thus on the **provisioning ecosystem services**. The physical result of this kind of services are tangible things, such as water, plants, and animals. Whether cultivated or wild, they include fish, timber, and other types of wild species resources that are treated in this book. These tangible results of provisioning ecosystem services are often referred to as ecosystem "goods". In this book, however, we do not use this term but prefer to refer to them as "resources". This is because we want to distinguish resources, which are things that are still in their natural state, from goods, which are things that have already been harvested by humans, as we detail further in the next section.

TRANSACTIONS CONNECTING NATURE AND SOCIETY

Wild species harvest is an action that directly connects the natural environment with human society. Attaching a value to components of **biological matter** in nature turns it into a **resource**, and the subtraction of that resource from nature and its incorporation, or 'addition', to human society is a process that converts it into an economic **good**. The harvest action thus consists of transactions with nature (between the harvester and the natural environment) as well as transactions with society (between the harvester and the larger society) (Figure 2.3).

The harvester's **transactions with nature** build on the **provision from nature** of harvestable matter that has become a resource (about how 'resources are not; they become', see Zimmerman, 1951). The **subtraction from nature** of the resource takes place through an **intervention with nature**. In short, the provision from nature results from a series of ecosystem functions without

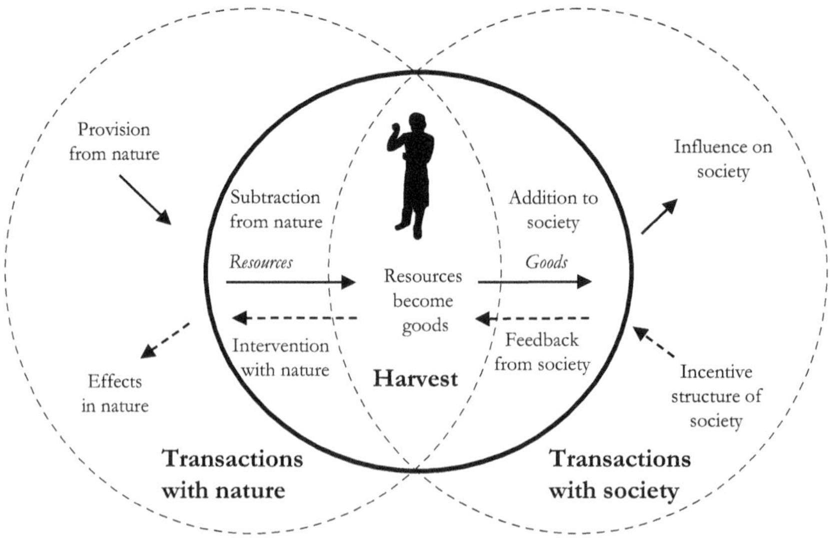

FIGURE 2.3 Wild species harvest is an activity that takes place at the interface of the natural and human systems. Nature provides harvestable resources that, in the course of the harvester's work, become goods. The harvested goods can have multiple influences on society, and also produce a feedback to the harvester. However, a harvest implies an intervention with the natural system and can have different kinds of direct and indirect effects. Matter and impacts are constantly exchanged between different components of the natural and human systems. Thus, the continuous transactions with nature and society are essential for the sustainability of harvest. Full arrows show flows of extracted matter from nature to society and dashed arrows indicate the direction of feedbacks from society towards nature.

which there would be nothing to harvest. The intervention with nature includes the very act of subtracting the desired matter from nature, but it may sometimes be considerably more extensive than that, for example when harvest involves such activities as the construction of logging roads or the use of explosives or poisons to catch fish. All this may lead to a wide range of direct and indirect **effects in nature**, which are changes to the ecosystem induced by the action of harvesting. Commonly, such ecological effects affect the very provision from nature of the desired matter, and may compromise the sustainability of harvest. In sum, the transactions with nature thus mean that there is always a 'price' that has to be paid for today's harvest in the form of the provision of harvestable matter available in the future, as well as in the form of other ecological effects. Even when measures are applied to compensate for the loss of the subtracted resource, intervention may induce changes in other components or functions of the ecosystem.

The harvester's **transactions with society**, on the other hand, include not only the **addition to society** of the goods based on the harvested matter but also the **feedback to the harvester**. In its simplest form, this feedback may be merely the pleasure felt by the harvester when for instance eating a tasty wild fruit. It may also be in the form of money received when selling the harvest. In the long run, the feedback is expected to be positive, and if it turns negative harvesting is often discontinued. The feedback may, however, oscillate between positive and negative, for example in the form of occasional fines that the harvester may have to pay whenever he or she is caught harvesting wild species illegally. The harvester's transactions with society also include the **influence on society** that the presence of the harvested matter implies. For example, the availability of wild fish or timber in a society can make quite a marked difference, such as through the development of trade and processing-related commodity chains and associated institutions. The transactions with society also consist of the **incentive structure** present in the society that affects harvesting behaviour in one way or another, including the demand from consumers, as well as policies, regulations, taxes, and subsidies, among others. Finally, even the term 'society' needs some clarification. It may be used to refer to a group of individuals that are related with one another in some way, ranging from the household level through to villages, cities, regions, countries, and all the way up to global society as a whole. In this book, the transactions with society mainly consider the immediate interface between society and harvester, understanding that any particular societal entity is simultaneously part of a larger whole.

Finally, the people who harvest wild species we call **harvesters**. Even this term is, however, more complex than it might first appear. In its simplest form, the harvester is the person who physically performs the act of harvesting. Often, he or she is free to make the choice of whether to harvest or not. Yet this is not always the case, as the person physically performing the act of harvesting the biological material from nature in some cases may be subordinated to some other person who actually makes the decision. In any case, the decision to harvest is critical for any transactions to take place, and in order to fully comprehend the problems and potentials related to wild species harvest, it is important to focus not only on the action of harvesting but also on the decision making related to this action – decisions that may take place far from the operation site itself. Therefore, we can consider the harvesters as one or more individuals who are involved in any particular harvesting activity, whether as workers in the field or as direct decision makers such as field supervisors or property rights holders, but not, for example, as those people who are indirectly affecting these decisions, such as policy makers, legislators, and consumers.

In sum, wild species harvest is an ensemble of actions and transactions that, on one hand, involves subtraction of biological matter from nature and, on the other hand, addition of that matter into society. Both subtraction and addition have further implications in the ecological and social systems involved; in other words, harvesting causes effects in both nature and society. Both of these processes occur in space and in time, the latter requiring some further conceptual clarification, which we present in the next section.

SUSTAINABILITY – THE CAPACITY TO ENDURE IN TIME

For wild species harvest to continue over prolonged periods of time, a number of conditions must be met. These conditions relate to the idea of what we often refer to as **sustainability**, a central concept of this book. The concept of sustainability was first explicitly formulated in the eighteenth century in the German forestry literature as the principle of *Nachhaltigkeit*, derived from *nachhaltig*, meaning 'lasting for a long time'. In 1804, a German forester wrote

You cannot expect or imagine a lasting forestry, if timber extraction from the forest is not based on sustainability. Every wise forest-administration has to start immediately an evaluation of the state owned forest and utilise them as intensively as possible, but in a way such that following generations will receive an equal amount of benefits as the one that the now existing generation appropriates.

(Hartig, 1804 [translated from original in German])

During the twentieth century, this sustainability concept became important also in wildlife and fisheries management, and in the 1970s, the concept was also applied to aspects relating to the functioning of world society as a whole (e.g. Basler, 1977, cited by Wiersum, 1995). The sustainability concept then came up on the political agenda and rapidly gained the status of a key concept in global affairs. This was particularly when, in 1987, the World Commission on Environment and Development (WCED), created by the United Nations, released the report *Our Common Future* (WCED, 1987), launching the concept of **sustainable development**, which it defined as 'development that meets the needs of the present without compromising the ability of future generations to meet their own needs'.

After the term 'sustainability' entered the arena of political discourse, its meaning changed and broadened and became a subject of intense debate. Somehow, the concept of sustainability – related to durability in time – often became confused with the concept of development, which is related to satisfying people's needs in a constantly improving way. There is nothing wrong here if it is about how to assess the potential to sustain the continuous search of how to get more of what we value; the problem is when the term 'sustainability' began to be used as a synonym for anything 'good', 'desirable', 'morally right', or 'politically correct' without notions to its original meaning, which is essentially related to **time**.

Even in academic circles, the term has not always been used with a meaning related to something enduring in time. Calling profitability of businesses 'economic sustainability' (Hamadeh et al., 2001) is not so far removed from the original meaning of the concept in natural resource management if it is about the long-term viability of firms. But when an article in a prestigious social science journal refers to 'social sustainability' as social support, social cohesion, and community participation among community residents (Ho and Cheung, 2011), one begins to wonder whether the term has lost its original meaning. The relation to the original meaning of the term is even less obvious, for example when the impacts of tall buildings on urban landscapes are discussed in terms of 'visual sustainability' (Tavernor, 2007), or when fuzzy logic is used to calculate 'gender sustainability' as a macroindicator of the performance of firms (Addabbo et al., 2009).

Sometimes, identical compound terms, including the word "sustainability, have been used with entirely different meanings, as, for example, when a professor of landscape architecture comes up with the term 'cultural sustainability' to mean 'long-term ecological health that is perpetuated by cultural values and behaviours' (Nassauer, 2004), while the same term is used by others in the sense of cultural continuity among ethnic minorities (Bekerman and Kopelowitz, 2008). A similar lack of clarity can be seen across disciplines and sectors, with examples such as 'political sustainability', 'legal sustainability', 'ethical sustainability', 'linguistic sustainability', 'hygienic sustainability', 'moral sustainability', 'corporate sustainability', 'chemical sustainability', 'military sustainability', 'intellectual sustainability', 'religious sustainability', or 'emotional sustainability'.

In order to understand and discuss issues of natural resources and environment, however, sustainability in its strict sense relates to durability in time. In this book, we use the term **sustainability of wild species harvest** to exclusively mean the potential of such harvest to endure in time. Sustainability, in this sense, is not the only criterion for improved wild species harvest situations; it is one of many issues that deserve to be taken into account. While achieving sustainable levels of extraction in the long term is a necessary condition for prolonged harvest, other aspects of nature and society are important too, such as biodiversity conservation, justice, and social equity. However, we prefer to refer to these other aspects calling them by their own names, reserving the word 'sustainability' for its original meaning as a time-related scientific concept.

It must be emphasised that sustainability is not a synonym of desirability. For example, there is no doubt that hunting whales can be done in a sustainable way, if the hunting levels are low enough to not threaten the long-term persistence of the whale populations. Whereas many people would find this to be something perfectly acceptable, others, however, consider whale hunting highly unethical in itself, just as they might also consider the hunting of chimpanzees *(Pan* spp.*)*, gorillas *(Gorilla* spp.*)* and other highly intelligent animals. Science can help bring clarity about what practices are sustainable or not. Determining what is right or wrong, however, is a matter of personal and collective values.

A good metaphor that helps one to understand the concept of sustainability is the 'sustain' button on an electric guitar amplifier. The further to the right one turns the button, the longer time the tone lasts when one picks the strings (Figure 2.4). Nothing in the world is eternal, and even the Sun will die one day. So something being 'sustainable' does not mean that it can last an eternity. It can be sustainable only within a certain time frame. Defining this time frame may not always be easy, but, when evaluating the sustainability of some activity, one should at least be clear on which processes one takes into account and which not. Compared to the potential of the harvester's transactions with society to endure in time, which can hardly be evaluated in quantitative terms, the

FIGURE 2.4 A musician sustaining a tone on his electric guitar. *Photo copyright: Thomas Lilley.*

sustainability of the harvester's transactions with nature may be more straightforward to assess empirically.

For example, harvesting leaves from a tree may be sustainable at the level of the individual tree, as long as the tree can continue producing leaves, year after year, at the same rate at which we continue harvesting the leaves, with no significant effects on the survival chances of the tree. However, one could also use a more restrictive definition of sustainability, involving longer time scales and taking into account the regeneration of the harvested tree species, requiring that the leaf harvest does not harm the overall viability of the tree to the point that seed production is decreased and therefore regeneration is compromised in the long run. Leaf harvest may possibly also constitute a selective pressure leading to evolutionary changes in the gene pool of the harvested population, which is another criterion for assessing sustainability. So, even when we stick to sustainability's original meaning related to durability over time, it is still a rather complex concept that can be approached from different angles and assessed using different criteria, depending on the question, and level, in focus.

We may consider the meaning of sustainability still more restrictively by expanding the assessment to include effects on the surrounding ecosystem. Removing leaves from the ecosystem year after year may potentially, in the long run, reduce the amount of nutrients in the soil, which in turn could affect the biological production of leaves. Alternatively, leaf harvesting may indirectly affect other species present in the ecosystem, which may, through ecological interactions, affect the species of our interest. Even if this would not significantly affect the abundance or productivity of the particular species of interest to harvesters in the short term, one may argue that harvest is not sustainable if it causes continuous deterioration of the environment. Finally, even harvest that has persisted over a long time and appears to be sustainable may actually be unsustainable because it decreases the ability of the ecosystem to maintain its integrity or to rapidly recover when subject to disturbance, a property referred to as **resilience** (Holling, 1973; Ludwig et al., 1997). This characteristic varies between different types of ecosystems.

As we can see, the sustainability of wild species harvest is in essence dependent on how the extraction affects the environment, including the harvested species and their populations. However, when discussing the potential of the harvested resources and the environment to bear continued subtraction and associated human intervention with its chain effects, it is important to remember that harvest practices largely depend on the harvester's transactions with society. Even if we could identify harvest methods and levels of extraction that would be sustainable, this would be of little help if we do not know and understand what factors in society influence or determine the harvest levels and methods used by the harvesters.

Ecologically sustainable harvest can be made to endure in time, whereas harvest that is unsustainable sooner or later comes to an end. Many ecologically unsustainable harvest situations nevertheless continue for a long time despite ecological evidence showing that this happens at the cost of future benefits.

Unsustainable harvest levels will lead to an initially weakened and eventually destroyed resource base, but harvest can continue as long as there is anything left to harvest. Many fisheries are an example of this kind of situation, raising the question of what it is in a society that can consciously maintain harvest at levels that are clearly destructive in the long term.

Sustainability is, thus, also fundamentally a societal question. If decisions about how to establish sustainable harvest levels and methods are based on only the ecological side of the equation, the picture is biased and greatly simplified because harvest is essentially an activity that links the natural and the human systems and does not exist without both. Natural and social sciences have only relatively recently started to build better integrated theories to address these kinds of coupled **social-ecological systems** (e.g. Ostrom, 2007). However, because both the natural environments and societies with their cultural conventions and institutions are constantly changing, no harvest situation is ever static. How harvest systems cope with changes in nature and society depends on their capacity to reorganise. This relates the sustainability of wild species harvest to the resilience of entire ecosystems and human societies combined (e.g. Folke, 2006). In an ideal world, we would live in resilient societies and extract wild species in a sustainable way, while conserving biodiversity and enjoying a full range of ecosystem services. How attainable that is is an issue we will explore in the rest of this book.

Part II

Stories from the Forest Floor

Having placed the harvester in the middle, connecting nature and society, the diagnosis of wild species harvest is greatly enriched by opening one's eyes for a look around with the aim to see examples of the conditions in which harvesters live and work. There is always someone, one person or more, within these everyday livelihoods, with dreams and necessities, options and commitments, working under variably rewarding conditions, maybe suffering from the lack of alternatives. His, her, or their realities are right in the interface where nature and society meet and harvest decisions are made.

These stories from the forest floor take the reader to a tour across Amazonia and seek to portray splinters of the lives of Amazonian people involved in a variety of wild species harvest situations. We meet men, women, and children who are engaged in the harvest, management, or processing of wild species and the products derived from them. Seeing things from the harvester's perspective may help readers to realise how people earn their living by harvesting wild plants and animals from the environments they live in.

Each story from the forest floor is tightly tied to a specific time and place and therefore shows how harvest takes place in this specific context. Thus fishing, hunting, logging, or any other form of harvest may be different in another moment or elsewhere. Stories are incomplete and do not explain everything – but that is how the world looks like for most of us. Our capacities to absorb and describe the world are limited. However, harvester perspectives as they are portrayed in this part of the book, and in general, tend to share some common features with other stories from other locations in the past, in the present, and most likely in the future too.

Preview to the Chapters of Part 2

Chapters 3 to 12 present 10 different scenes of wild species harvest in Amazonia. Rather than a collection of case studies in the traditional sense, this assemblage of stories is largely made up of bits and pieces of the lives of people engaged in wild species harvest in contemporary Amazonia, with some selected references to the scientific literature. **Chapter 3** is slightly different from the others in that it provides a review of the history of wild species harvest in Amazonia as a whole. **Chapter 4** is about hunting and the efforts of the people in an indigenous village to improve the sustainability of their hunting practices. **Chapter 5** deals with commercial and subsistence fishing, and the problems of managing a migratory resource. **Chapter 6** presents river turtles as an extraordinary example of recovery of a harvested resource. In **Chapter 7**, the harvest and management of palm leaves for roof thatch are discussed as an example of how locally designed management systems may be only partially effective. **Chapter 8** is about the Brazil nut, a wild species product that is well established and traded in considerable volumes on the international market. **Chapter 9** treats the changes in Peruvian forest laws and policies, and how these affect logging practices. **Chapter 10** focuses on the experiences of an attempt to make business out of Amazonian medicinal plants. **Chapter 11** presents the story of the fruit of the açai palm in the Amazon estuary, which used to be produced in wild stands only but today is produced in intensively managed stands, such that it provides an intermediate case between a wild and a farmed resource. **Chapter 12**, finally, addresses the roles of knowledge, science and planning in the use of land and resources in the context of Peruvian Amazonia.

These chapters can be read as an appetiser for Part 3, which deals with wild species harvest theory. To facilitate this, we use symbols in the margins of the stories as links to the seven theory chapters in Part 3 of the book. When encountering any of these symbols in the margin, it means that the text touches upon some issue that will be treated from a theoretical perspective in the corresponding theory chapter in Part 3.

Resource Dynamics. This symbol indicates aspects about the natural boundary conditions to harvest. It also draws attention to ways in which harvest interventions can affect the individuals and populations of the harvested species – as well as ecological communities, ecosystems, and even the whole biosphere.

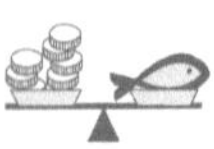

Costs and Benefits. This symbol refers to issues that the harvesters of wild species resources face when they make decisions about harvest, weighting the costs and benefits related to the activity against each other.

Management. This symbol refers to purposeful measures by people aiming at improved productivity of the resource and also at controlling any negative effects of harvest.

Governance. This symbol has to do with societal arrangements that regulate the behaviour of different actors that have a stake in wild species harvest. These arrangements include direct incentives, such as laws, and rules, but the symbol also refers to issues in the larger societal context that affect the harvest activity as a whole.

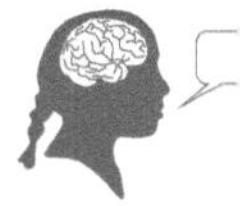

Knowledge. This symbol refers to the roles of different kinds of knowledge, from indigenous to scientific, in relation to harvest. It also deals with the ways in which information flow among different stakeholders keeps the harvest system in movement.

Spatiality. This symbol refers to the ways in which spatial settings processes and variability affect wild species harvest, including how harvesting practices are affected by such issues as environmental heterogeneity, distance and accessbility, and resource mobility.

Legacies. This symbol refers to how events and processes in the past have shaped the present options for wild species harvest, and how current events shape the options available in the future.

Chapter 3

Millennia of Wild Species Harvest in Amazonia

GENESIS OF THE AMAZON BASIN

Once upon a time, there was no Amazonia. Until some 140 million years ago, the land mass that we today know as South America still formed part of the supercontinent Gondwana which had only 'recently' been separated from the yet larger supercontinent Pangaea. Approximately where the Amazon River today flows from west to east, there was a completely different river system with a westward flow from somewhere in what today is Africa. Then, Gondwana started to break apart (Figure 3.1A), and South America slowly drifted westward from Africa, gradually giving rise to the southern part of the Atlantic Ocean. As the tectonic forces moved the continental plate westward, it forcefully crashed against the oceanic plates of the Pacific. Some 65 million years ago, the compressional forces of this collision started the formation of the modern Andes, with the main drainage in the Amazon Basin turning towards the northwest (e.g. Hoorn et al., 2010).

On the western fringes of the continent, water running off the eastern flanks of the Andean cordilleras increasingly got trapped between the mountains and the shield areas consisting of old remnants of the American plate in the east, forming an enormous sedimentary basin. The humid and warm climate of this equatorial zone allowed lush and diverse vegetation to develop, and since at least 55 million years ago, the northern parts of South America began to be dominated by rainforests (Maslin et al., 2005). As time passed, the basin experienced many distinctive phases of large-scale landscape formation, but always with tropical rainforests as a central component of the region's nature. In the lowlands to the east of the Andean mountains, in the course of time, even periods with widespread marine and estuarine conditions existed, with water connection extending from the forelands of the Central Andes to the Caribbean in the north from about 23 million up to 10 million years ago (Figure 3.1B). Finally, about some 10 million years ago, the present Amazon River system with its drainage to the Atlantic Ocean got established (Hoorn et al., 2010).

Vegetation and fauna kept developing and adapting to these slow but significant landscape-level changes. If the rising of the Andes was a drastic phenomenon in geological terms, the lowland areas near the Andes also got their share of the forceful tectonic changes. For example, some parts of the Andean forelands locally started

Diagnosing Wild Species Harvest. http://dx.doi.org/10.1016/B978-0-12-397204-0.00003-6

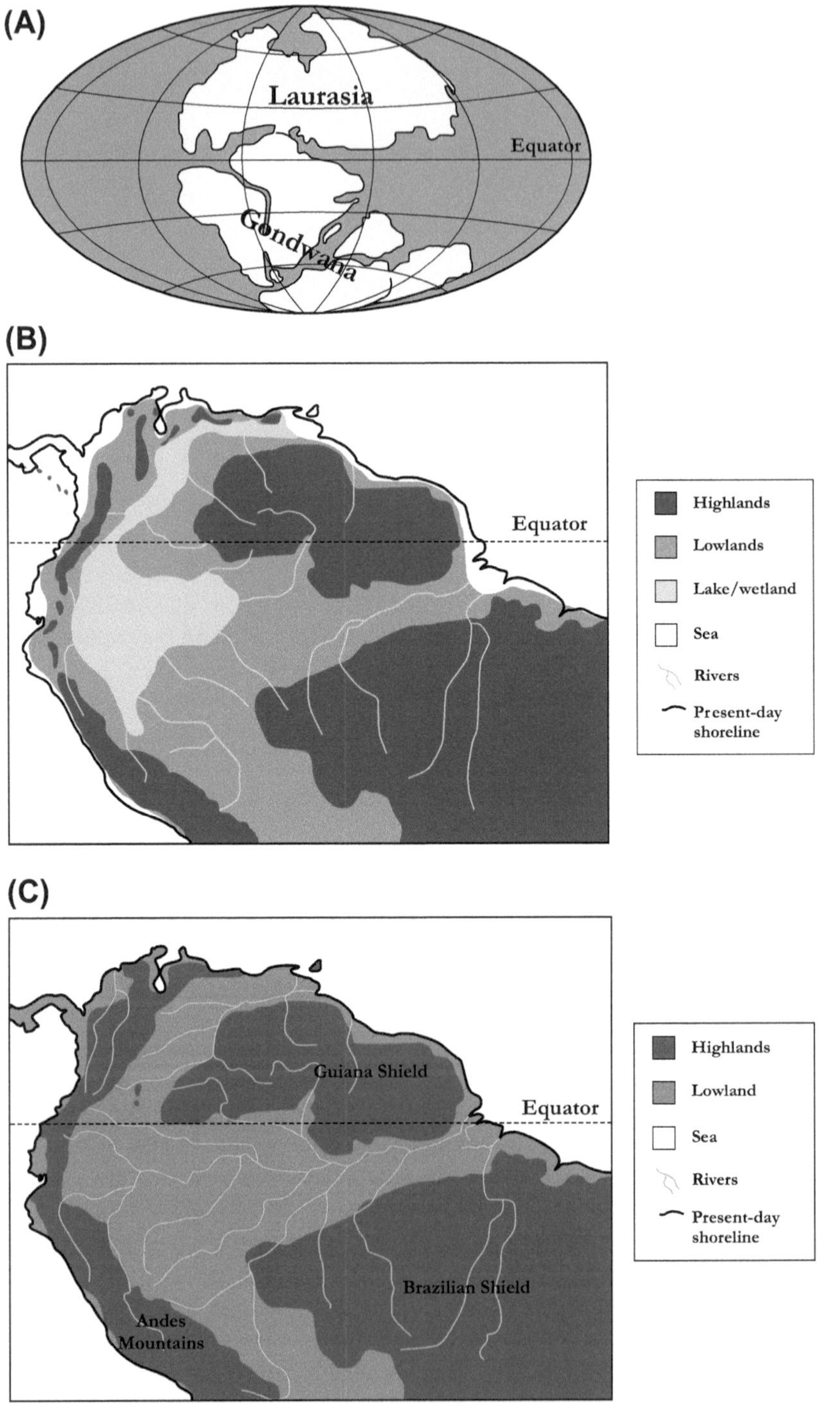

FIGURE 3.1 Phases of the dynamic environmental history of the Amazon Basin. (A) The breakup of the supercontinent Pangaea into its northern (Laurasia) and southern (Gondwana) parts some 200 million years ago. (B) The lacustrine and/or marginal marine settings at sea level with connection from the Caribbean during the Middle Miocene, c. 23–10 million years ago. (C) Present-day Amazonia is characterised by the Brazilian and Guiana shields, the Andes Mountains, and the vast sedimentary basin in the areas between them; B and C are modified from Hoorn et al. (2010).

(D)

FIGURE 3.1, cont'd (D) Amazonia today is shared by nine different countries. Most of the capital cities are located outside of the Amazonian rainforest area, which is roughly depicted in grey.

to subside and get filled with increasingly thick layers of sediments, while some other areas experienced relative uplift. In those areas that were above the level up to which the river floodwaters extended, soils were exposed to surface erosion. With time, these processes created hilly undulating landscapes with small creeks or rivers in the valleys. Erosion of the top sediments also re-exposed some of the underlying layers in parts of the basin. As a consequence of all these dynamic processes, the mosaics of soils made of different parent materials, so typical of western Amazonia today, began to be formed. In turn, in the northeast and southeast of the Amazonian region, where the old bedrock shields of the so-called Amazonian Craton occurred, the conditions influencing soil formation were more stable and gradual weathering processes slowly kept transforming the landscape.

Some three million years ago, yet another dramatic event took place when the Isthmus of Panama was formed, uniting the North and South American continents (Smith and Klicka, 2010). The formation of this corridor, which was first formed by a chain of islands, allowed an increasing exchange of species between these two land masses, which had been isolated from each other and developed distinct faunas and floras during the almost 200 million years since the breakup of the supercontinent Pangaea. When the isthmus of dry land was finally formed, a great interchange of terrestrial fauna followed, leading to new processes of diversification but also to the extinction of many species that did not adapt to the new conditions and competitors. For example, many of the mammal species unique to South America succumbed when having to struggle with the newcomers from the north.

Today, the drainage basin of the Amazon River constitutes the world's largest fluvial network, with an extension of over 6 million km^2 (Figure 3.1C,D). The general features characterising this area include low elevation, on average less than 100m above sea level, and an overall uniformity of the physical landscape. Relative elevation differences within the lowland basin are mainly within only a few tens of metres, yet in the Andean foothill margins and near some scattered mountain peaks in the Guiana Highlands also, greater differences with steep slopes occur. Most of the basin has a wet or humid tropical climate with relatively low seasonal variations, but towards the northern and southern boundaries of the region, the climate becomes increasingly seasonal, with at least a clearly distinguishable dry season for a few months per year.

When the current Amazonian landscape is seen from an airplane window it looks like a flat (or slightly undulating), featureless green carpet that never ends, only striped by rivers of various sizes and shapes (Figure 3.2). Where precipitation levels are very high (above 2000 mm/year), equatorial evergreen rainforests prevail, but at the margins of the basin there are also semi-evergreen seasonal forests and savannah vegetation. Also, wetlands abound in the floodplains of many large rivers and in the northern and south-western grasslands. Scientists working in the area have documented very high levels of biological diversity all over the basin, but as, for example, many tree species look quite alike, it has been difficult to detect clear spatial patterns in the distributions of the different species and habitats. This view changed only as research proceeded with the help of satellite imagery and other geographical data sources, which revealed many such spatial phenomena in the region that could not be perceived on the ground level (Salo et al., 1986; Tuomisto et al., 2005; Toivonen et al., 2007; Irion and Kalliola, 2009). The mechanisms behind the region's high species numbers are, however, still not fully understood. Undoubtedly, the very long-endured tropical conditions of the region have had a role to play, but also a number of distinctive processes and periods of landscape evolution have been, and still are, important factors in the generation and maintenance of high biological diversity (Salo et al., 1986; Wesselingh and Salo, 2006).

A closer look at the landscape-forming processes operating in Amazonia helps one realise that the apparent uniformity of the region is indeed misleading. Due to the forceful geological processes presented above, the physical geography of the Amasonian landscape has undergone and is still subject to complex dynamics of the abiotic environment (Sioli, 1984; Räsänen et al., 1987; Kalliola et al., 1993). Particularly, the western parts of the basin, which are the focus in many of the coming stories from the forest floor, have experienced interesting developmental dynamics. The Sub-Andean zone behaves like a funnel that brings together many rivers, ultimately forming the mighty Amazon, which then flows to the east as a relatively straight floodplain band between the Brazilian and Guiana shields and finally discharges its water masses into the Atlantic Ocean.

FIGURE 3.2 The seemingly endless and uniform green carpet of lowland rainforest extends to the horizon in north-eastern Peru.

The dominating processes within the Sub-Andean belt include not only the continuous accumulation of new sedimentary layers but also the recycling of the once-deposited sediments through continued erosional and depositional processes. Also, the neotectonics of the Andean orogeny keep on modifying the geological structures in this region under the seemingly flat sedimentary surface upon which the present-day rivers flow and forests grow (Latrubesse and Rancy, 2000; Räsänen et al., 1990). All these dynamic aspects, in turn, influence the edaphic conditions in different parts of the region, giving rise to distinctive vegetation types and habitats for animals.

As a consequence of all these processes, the Amazon Basin is a complex system comprising a diversity of different forms of landscape (Figure 3.3). Even though many areas may look quite homogeneous to an inexperienced observer, a closer examination starts to reveal differences in their biophysical and biological characteristics. Only through learning to understand the landscape-level developmental history of Amazonia is it possible to comprehend why different vegetation types, and consequently particular plant and animal species, occur in specific parts of the basin. This reality, in turn, strongly influences where human settlements occur and where different resource use activities have emerged and become established. And, as a consequence, it is increasingly so that not only do natural dynamics modify Amazonian forests but also that human activities have become a dominant force transforming the landscape (Davidson et al., 2012).

FIGURE 3.3 The natural landscape of Amazonia is highly variable across the region. (A) A rocky creek near the foot of the Andes Mountains in Ecuador; (B) a noninundated evergreen (*tierra firme*) forest in Ecuador; (C) Tiputini, a lowland river in Ecuador; (D) an evergreen rainforest in the Andean-Amazonian transition zone in Madre de Dios, Peru; (E) humid grasslands in the north of Amazonia in Casanare, Colombia; (F) a white sand forest near the city of Iquitos, Peru; (G) a floodplain forest in the Tahuayo River, Peru; and (H) mangrove vegetation in the Amazon estuary, Brazil.

EARLY HUMAN OCCUPATION

We will never know exactly when the first humans reached Amazonia, but what we do know for sure is that the Americas remained empty of humans much longer than the other large land masses on Earth, namely, Africa and Eurasia. This has had a profound effect on the human–nature interactions in these different parts of the planet. In Africa, the fauna and flora gradually co-evolved with various species of human-like hominids since at least 2 million years ago. A few hundred thousand years after the first hominids had started to roam in Africa; some of them had also spread to parts of Asia and Europe, and then finally, some 200,000 years ago, modern humans of the species *Homo sapiens* emerged in Africa, from where they soon started to disperse to other continents, first to Europe and Asia. Our species reached Australia about 50,000 years ago, and the conventional view is that it took still longer before they set foot on land in the Americas. This crossing took place via the Bering Strait when people migrated from North-east Asia to North America about 14,000 years before present (BP). In addition to this wave of migration, it has been suggested that the Americas were actually colonised by humans earlier than only 14,000 years ago (e.g. Neves et al., 2007; Roosevelt et al., 1996), and possibly later by people from other directions, perhaps including even Polynesians arriving to South America from the Pacific (Matisoo-Smith and Ramirez, 2010).

In any case, in Amazonia the presence of humans has been tracked back only a relatively short time. Remains of stone tools as well as of harvested plants and animals found in a cave near Monte Alegre in Brazil tell a story of an 11,200-year-old campsite used by people who were foragers relying much on fishing but also hunting a wide variety of animals and collecting molluscs and the fruits of palms and other wild trees (Roosevelt et al., 1996). A couple of millennia later, humans were already widely spread – although hardly very abundant – over Amazonia, and had, at least in some places, taken up some small-scale gardening to complement their hunter-gatherer subsistence practices. Remains at the Peña Roja site in Colombian Amazonia, dated to between 9250 and 8100 years BP, include stone axes (which are indicative of tree felling), the remains of primitive cultivars, as well as thousands of seeds, out of which the majority are palm seeds, indicating that palm fruits were an important component of people's diet (Piperno and Pearsall, 1998; Oliver, 2008).

Amazonian dwellers developed agricultural techniques gradually over thousands of years, including domesticating plant species (e.g. today's staple, the manioc, *Manihot esculenta*) and adopting crops from other regions (Clement et al., 2010). Among these crops was maize, which originated in Mesoamerica but was being cultivated in Amazonia already around 5300 BP (Piperno, 1990). By some 2000 years ago, many of the main Amazonian rivers were already surrounded by large human settlements with dense populations, and these persisted up to the sixteenth century. Highly organised societies with dense populations developed also in the forest–savannah mosaic landscape of the south-western fringes of the region, where large-scale geoglyphs are still visible, standing as

FIGURE 3.4 Oblique view of pre-Columbian earthworks in the Fazenda Colorada site in the Brazilian State of Acre. Deforestation has made these human-made structures clearly visible in the landscape. *Photo: Martti Pärssinen.* (Please see the colour plate at the back of the book.)

evidence of these ancient civilisations (Schaan et al., 2012; Pärssinen et al., 2009) (Figure 3.4). These cultures may have emerged by 5500 BP, and the earthworks construction began 3000 to 2000 BP (Schaan et al., 2012). In most of Amazonia, however, population densities seem to have remained low, such as in much of upland western Amazonia. Archaeological evidence from Ecuador suggests that settlements and agriculture there were confined to the crests of ridges, away from the main rivers. The reasons for this are not clear, but it may have to do with defence considerations, as well as with the greater ease of felling trees and weeding on the crests, in comparison with riverside habitats (e.g. Netherly, 1997; P. Netherly, personal communication; Sirén, 2004).

The prehistoric people did not arrive and proliferate without an impact. While the densely populated banks of the main rivers probably became subject to heavy deforestation, most of Amazonia remained with lower population densities, and these humans transformed the landscape in much subtler ways, although not always without long-lasting consequences. Where agriculture was practiced, the methods

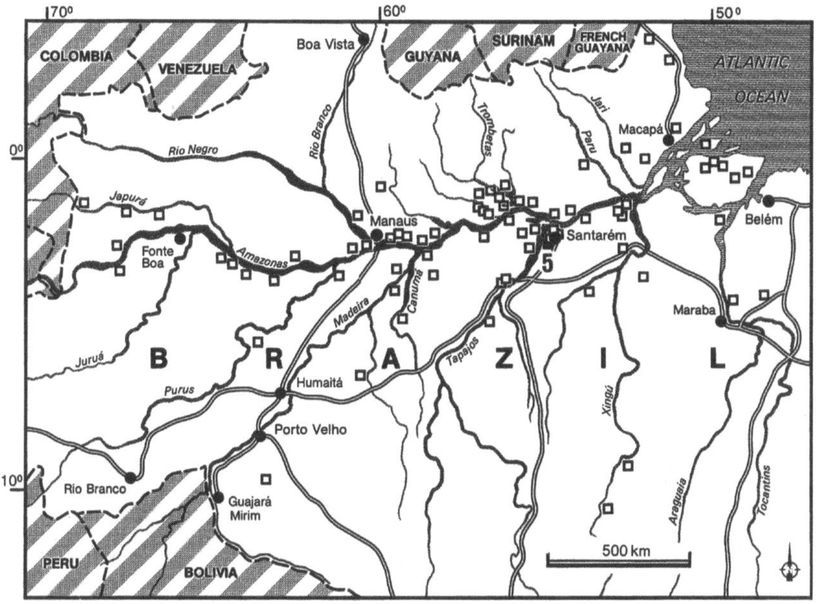

FIGURE 3.5 Location of known Amazonian dark earth sites (open boxes) in Brazilian Amazonia. *Source: Glaser and Birk (2012).*

used by the early Amazonian farmers were at least somewhat different from those of today, as there were no metal tools available. This implied that farming probably was mostly carried out on semipermanent or even permanently cleared plots, as clearing new land every few years would have been impractical. Crops were thus planted in the same plots over and over again for many years, and when forests were cleared, this was often done with the help of fire (Denevan, 1998).

Thus, probably only small areas were cultivated, but sometimes fires may have escaped, causing wildfires affecting larger areas, especially during the period 8000–4000 years BP, when the climate was drier (Betts et al., 2008). Some parts of Amazonia, however, particularly in the west where the climate is the wettest, seem to have remained unaffected by forest fires for at least the last 21,000 years (Power et al., 2010). If the direct impact on the forest cover caused by agriculture seems to have been modest in terms of the area under cultivation, discoveries of the so-called Amazonian dark earths – very fertile soils characterised by high concentrations of charcoal, other organic matter, and often pottery fragments – are another indication of major settlements with permanent agriculture in mostly the lower parts of the basin (Glaser & Birk, 2012; Levis et al., 2012) (Figure 3.5).

Nevertheless, hunting, fishing, and gathering remained major life-sustaining activities, and consequently some of the legacies of the prehistoric Amazonians are also related to wild species harvest. These impacts are however more difficult to perceive, study, and quantify. At least one of these subliminal effects of early humans has been studied empirically: in some areas, early hunter-gatherers seem to have contributed to the enrichment of forest with those trees and palms that bear edible fruits. For example, a study by Levis et al. (2012) in the interfluvial area of the Purús and Mamoré Rivers in central Amazonia shows how the density of individuals belonging to useful tree and palm species, such as the Brazil nut tree (*Bertholletia excelsa*; see also Shepard and Ramirez, 2011) and the *caiaué* palm (*Elaeis oleifera*), increase with decreasing distance from rivers, where supposedly human settlements were located.

The appearance of humans may also have had an impact on some of the Amazonian fauna. In particular, relatively large arboreal animals tend to be very vulnerable to hunting by humans, which is the only predator species that can fabricate missile-type weapons such as blowguns or bows and arrows. As the remains of hunted animals seldom are preserved in the rainforest environment, it is, however, hard to find firm evidence that prehistoric Amazonians hunted any species to extinction. An indirect indication that this may have happened is, however, the fact that in the Amazonian and Neotropical forests of today, there are much fewer large mammals present than in African forests (where the fauna evolved side by side with humans).

In African forests, 60 percent of the species of nonvolant mammals are greater than 1 kg, and 22 percent are greater than 10 kg, whereas the equivalent figures for the Neotropics are only 38 and 7 percent, respectively (Fa and Peres, 2001). The largest animal present in the African forests today is a mammal species, the African forest elephant (*Loxodonta cyclotis*), which can weigh up to 6 tonnes, whereas the largest animals in Neotropical forests are the tapirs, of which the species occurring in Amazonia (the South American tapir, *Tapirus terrestris*) weighs at most some 320 kg, and the species of central America (Baird's tapir, *Tapirus bairdii*) only slightly more.

This has not always been so. Many large-bodied species, so-called megafauna, that were adapted to semi-open landscapes most probably thrived in the forest–savannah mosaic that covered large parts of central and southern Amazonia between 25,000 and 10,000 years BP, when the climate was drier than it is today (van der Hammen, 1974). In the approximate times of the arrival of humans, 1.5-m tall and 1.5-tonne toxodons (*Toxodon*), similar to a rhinoceros but without a horn, roamed the forest floor in the south-western fringes of Amazonia, grazing the underbrush and eating fruits and leaves. Five-tonne ground sloths (e.g. *Eremotherium*) reached leaves and fruits from as high up as 5 m. Also, elephant-like mastodons (e.g. *Haplomastodon*) lived in central Amazonia (de Fátima Rossetti et al., 2004; Hubbe et al., 2013).

However, megafauna suffered a wave of extinctions globally, starting approximately 80,000 years ago in Northern Eurasia and then progressively occurring elsewhere, most notably in Australia and the Americas (Owen-Smith, 1989). The fossil record is sparse in Amazonian rainforests, and while there is no direct evidence of human-assisted megafaunal extinctions in the area, a number of discoveries suggest that elsewhere in South America, such extinctions at least coincided in time with the appearance of early humans (Hubbe et al., 2013). The role of humans in the disappearance of the now-extinct megafauna has been long debated, and there is still no full consensus about the relative importance of the different potential causal factors (Barnosky et al., 2004; Barnosky and Lindsey, 2010; Wolverton, 2010). It is most likely, however, that climate change had an important role, and in some cases this also coincided with new anthropogenic forces (Prescott et al., 2012).

Whether or not this occurred with human contributions involved, the loss of megafauna may have in itself also contributed to habitat changes. The browsing and trampling of large megafauna animals tend to have a profound impact on vegetation, keeping down the underbrush and creating gaps in the forest. It even has been speculated that such closed forests with dense underbrush that today are typical for parts of Amazonia may have been less common before the megafauna disappeared (Schüle, 1992). In any case, as little sunlight reaches the lower strata in many of the forests of today's Amazonia, there is little food for herbivores with terrestrial habits, and most animal life is found in the canopy. In addition, by eating fruit and then defecating somewhere else, the megafauna also dispersed seeds, and when these species went extinct, many tree species could have lost important dispersal agents. Some such plant species may even have gone extinct, but others survived, still today displaying characteristics, indicative of adaption to dispersal by megafauna, such as very large fruits surrounded by edible fruit flesh (Figure 3.6). Some of these plant species may have been aided by humans to survive, as humans also have eaten their fruits and contributed to their dispersal, whether accidentally or by purposeful planting (Guimarães et al., 2008).

In sum, while large tracts of the Amazonian rainforests certainly still form a great wilderness compared to many other parts of the globe, some parts of the basin that today seem pristine may actually have been shaped by humans, and

FIGURE 3.6 Megafaunal-dependent plant species as identified in Brazil by Guimarães Jr. et al. (2008): (A) *Attalea speciosa*, Arecaceae; (B) *Mouriri elliptica*, Melastomataceae; (C) *Hymenaea stigonocarpa*, Fabaceae; (D) *Genipa americana*, Rubiaceae; (E) *Salacia elliptica*, Celastraceae; and (F) *Annona dioica*, Annonaceae. *Photos from the original article.*

in many places the forest is more of a secondary nature than has been previously thought (Denevan, 2011; Pinedo-Vásquez et al., 2012). The impacts of early humans in Amazonia cannot, however, be understood using a general model to uniformly cover all of the vast and diverse basin, but rather the human occupation and its consequent impact are likely to have been highly variable along different environmental gradients across the region (Barlow et al., 2012).

THE CONQUEST AND THE COLONY

Although Europeans had reached the American continent by the late 10th century, when Vikings made expeditions to the eastern coast of North America, it was only after the second European 'discovery' of America by Christopher Columbus in 1492 that the presence of Europeans had any great impacts on the societies and environments of the Americas. Spain and Portugal were the most important colonial powers in South America, but also other Europeans, including the Dutch, French, and English, rushed to exploit the 'new' continent of the south (Hemming, 2008). Among the most significant first impacts of the newcomers was the introduction of diseases, typically spreading even far ahead of the Europeans themselves, and often severely affecting local indigenous populations and thus also their resource use patterns. Another early impact with equally devastating effects was slave raiding, which was commonplace throughout the sixteenth century and beyond. Although slave raiders themselves seldom left written records of their activities, contemporary observers have described how their incursions affected the indigenous population extensively well up the main rivers of the basin (Hemming, 2008). The mobility of people who either fled the slave raiders or were caught by them contributed further to the propagation of pathogens.

The Europeans increasingly explored Amazonia (Figure 3.7), and along with military excursions, various religious orders also became increasingly active in the region. In their need of 'souls' and labour, they hunted down and caught 'Indians' in the forest and gathered them in mission villages (called *misiones* or *reducciones* in Spanish and *missões* or *reduções* in Portuguese), further contributing to the proliferation of diseases. Missionaries were aware of and concerned about the fact that many of the native people died of diseases in these places, as they were helpless to stop the mortal epidemics that raged in Amazonia about every 20 years throughout the seventeenth century (Taylor, 1986). That many natives also fled from the mission villages perhaps just helped the disease pathogens even more, spreading them among those people still living dispersed in the forests away from the populated centres.

As a consequence of these changes, the densely populated chiefdoms and agricultural plantations, observed by the expedition of the Spaniard Francisco de Orellana (1511–46) in 1542 when making the first European descent down the Amazon River (Hemming, 2008), had soon disappeared. In their place soon grew jungles so dense that they were later thought to be pristine forests. Orellana was even thought to have lied, and only in the late twentieth century did

FIGURE 3.7 A map of South America from the year 1694. Note how Amazonia and other interior parts of South America mostly appear as detail-poor areas between major rivers flowing to the Atlantic Ocean.

archaeological excavations along the Amazon reveal that there had indeed been dense populations along the river in the times of European conquest and long before (Heckenberger and Neves, 2009; McMichael et al., 2012).

Where the reorganisation of the human landscape as a result of the conquest was most drastic, particularly along the lower Amazon River, it implied a change in the anthropogenic pressures towards the Amazonian ecosystems as many indigenous ways of life were destroyed and new habits were introduced by the newcomers, including cultivation of new crops. The indigenous population fiercely defended itself in many areas, most remarkably in many upper rivers, engaging in periodic skirmishes with the Europeans and local native groups allied with the intruders (Hemming, 2008). Some indigenous groups escaped from the margins of navigable rivers to increasingly remote headwater areas. This strategy has been repeatedly employed up until much more recent times, and even today there are a number of isolated groups of humans living in the few pockets of wilderness deep in the forest, still far enough from the aggressive dominant society (e.g. Huertas Castillo, 2004).

The times of war, disease, and slavery took a heavy toll on the population of Amazonia. At least in the western part of the region, the population is estimated to have reached its lowest point in the late eighteenth century, after which it began to slowly increase again (Taylor, 1986). This same phenomenon was general in most of Amazonia, although it may have had a slightly different timing elsewhere in the region. Pathogens had become less virulent, and the surviving people had become more resistant against the diseases, with an effect, for example, on the mission communities by making them no longer such deathly disease vectors as they had been before. Little by little, more people came to live in many of the mission communities, which meant that it was not always necessary for the missionaries to catch them in the forest and bring them to the 'civilisation' by force anymore. Thus, gradually the spatial distribution of human populations changed from very dispersed to more concentrated in larger settlements along the main rivers.

The missionaries and rural patrons provided a steady supply of metal tools, facilitating tree felling, weeding, and canoe making. This opened new options for both agriculture and extractive activities. Slash-and-burn agriculture, so typical for Amazonia today, was made a much more attractive alternative compared to more sedentary practices; instead of farming the same plot again and again over many years, it now became simpler to abandon a site, let the forest grow back after a couple of years of farming, and simply go and clear the forest in another place. Likewise, the increased facility of making canoes aided river travel and the exploitation of resources found in aquatic environments and further from the population centres. Metal tools also made it possible to cut down trees in order to harvest fruits or palm hearts, and it facilitated the elaboration of hunting arms. Making spears and blowguns using the hard wood of the peach palm (*Bactris gasipaes*), as was common in parts of Amazonia in the twentieth century, must have been very difficult before there were metal tools.

During the years of being a colony, Amazonia became, for the first time in history, connected to overseas markets. Without underestimating the pre-Columbian trade connecting Amazonian lowlands to densely populated regions such as the Andes (Korpisaari et al., in press) and the Caribbean coast, the flow of raw materials initiated by the Spanish and particularly the Portuguese was something completely new in the rainforest. Agricultural production and extraction of natural resources now served not only the American population but also the interests of the colonising powers thousands of kilometres away in Europe. The most important agricultural products exported from Amazonia included sugar (and rum made of it) and tobacco. Outside the agricultural colonies, products such as Amazonian manatee (*Trichechus inunguis*) oil, river turtle eggs and oil, fish and meat, seeds, and other parts from native plants not found in Europe started to be traded.

Also, during the eighteenth and nineteenth centuries, many European scientists travelled extensively in Amazonia to study its natural and human landscape. One of the first of these explorers was the Frenchman Charles Marie de la Condamine (1701–74), who travelled to Amazonia in the mid-eighteenth century. Although not a botanist or zoologist but rather an explorer widely educated in many fields ranging from military practice to humanities, geography, mathematics, and astronomy, he put particular emphasis on the Amazonian natural products that could have future use and thus economic value. Among the plant products he reported were wild rubber as well as the bark of trees in the genus *Cinchona*, which would later turn into the source of quinine, a much-needed remedy for malaria. In this way, de la Condamine laid foundations for future industrialisation of these biodiversity resources. Others followed, among them Alexander von Humboldt (1769–1859), Henry Walter Bates (1825–92), Richard Spruce (1817–93), and Alfred Russel Wallace (1823–1913).

DAWN OF INDEPENDENCE AND NATIONAL INTEGRATION

During the early nineteenth century, Spain and Portugal progressively lost control of their colonies in South America (although the three Guianas of Britain, the Netherlands, and France stayed under their colonial rulers). Many of the new independent nations possessed territories in the Amazon Basin and thus had high stakes when national borders and natural resources were contested and overall development of the region was envisioned. By the mid-nineteenth century, all of the six independent countries established in the region – Bolivia, Brazil, Colombia, Ecuador, Peru, and Venezuela – had initiated at least some sort of integration policies in order to ensure their presence in Amazonia. Governments tried to attract colonists (both because of sparse population as a result of disease and violence, and because people of European descent were believed to be more efficient than the indigenous people as developers of the region) by facilitating land acquisition in the rainforest, by developing river transport, and through tax relief, among other means. Despite all of these efforts, however, in the mid-nineteenth century most of Amazonia was, from a coloniser's perspective, still a great wilderness covered by lush and hostile forests, and populated by native groups many of whom were not subordinated to the white rulers confined to the few outposts of their 'civilised' world.

Nevertheless, there was an increasing number of wild species products whose commercialisation started to attract the attention of local merchants of European or *mestizo* origin, as well as newcomers to the region. As these new resources started to emerge in the rainforest area, also the race for territories intensified. One of these early products was the bark of trees in the genus *Cinchona*, which is used dried and powdered for the treatment of malaria. After the discovery of the quinine alkaloid by French scientists in 1820, the bark and its derivative soon became global commodities that increasingly started to draw expeditions to the western limits of Amazonia, where *Cinchona* trees grew in the forests of the Andean flanks. Growing global demand induced increasing prices and led to the progressive extirpation of the species from many areas, confining it to ever more isolated pockets of the mountain forests.

The trade on *Cinchona* bark helped merchants to accumulate capital and contributed to the reconstruction of many old and abandoned terrestrial and riverine trade routes in western Amazonia. Soon, however, it became obvious that there were two particular threats to the quinine-based economy. First, overharvest made it increasingly difficult to acquire raw material to be processed; and, second, production based on plantations, if achieved, could easily outcompete natural quinine production. Although at least the Peruvians were aware of this latter threat and tried to fight it through export restrictions, colonial powers of the time sponsored expeditions to acquire and export seeds and saplings of the *Cinchona* trees. And soon, European powers started to extract quinine from the bark of trees growing in plantations established in Southern and Southeastern Asia.

Another important natural product of the latter half of the nineteenth century in the same region was a hat. Panama hats are woven using fibres of a plant called Panama hat plant (*Carludovica palmata*), and they definitely came to have much importance in parts of the region; for example, these hats formed 90 percent of the value of exports from Peru through Brazil in 1855–71 (Coomes, 1995). The Panama hats got their name because they were sold in the Isthmus of Panama in mass numbers when East Coast North Americans, along with other would-be gold miners from Europe, headed for California, choosing this route in order to avoid the even more rough and dangerous alternative of travelling through the great North American plains. In western Amazonia, similarly to the *Cinchona* bark and quinine trade, the Panama hat trade helped the merchants to accumulate capital and develop trading networks. However, in contrast to the quinine boom, Panama hat plants did not suffer considerable overharvest as the plants grew almost as a weed in human-modified landscapes. Nevertheless, this boom also came to a gradual end, but for a different reason. Mostly, its decline can be linked to the fact that the value of the trade, after all, was not enough to overcome the difficulties related to maintaining the transport routes. Both of these examples, the fates of the quinine and the Panama hats trade, show how extractive economies depend on a multitude of factors, of which some depend on the decisions of the extractors while many more are beyond their direct control.

In the times of the trade in quinine and Panama hats, several other wild species resources also were extracted in different parts of Amazonia. For example, eggs of river turtles were collected and adult turtles hunted for their meat in great volumes along the Amazonian rivers. Whereas the indigenous population had used turtles and their eggs from time immemorial, during the colonial times and particularly in the eighteenth and nineteenth centuries, Europeans also started increasingly to engage in these activities, and it has been estimated that the number of eggs collected annually at the mid-eighteenth century was counted in tens of millions, and the adults butchered in millions of individuals. Even so, some parts of Amazonia still teemed with turtles by the time of the travels of the famous British naturalist Henry Walter Bates in the mid-nineteenth century. However, in other areas the turtle populations started to show clear symptoms of overharvest, and only when a new boom based on another set of wild species started to draw men from commercial turtle harvest to more profitable work did the turtles get a moment of relief. This new resource was wild rubber.

THE GREAT RUBBER BOOM OF AMAZONIA

When the French scientist Charles Marie de la Condamine described the use of rubber in Amazonia in the mid-eighteenth century, he found the product fascinating. The rubber latex was used by the native population in many ways, including as toys, torches, and waterproofing material. These uses, common also in Central America, had been known by Europeans long before de la Condamine's expedition in the 1730s, but because of the quality of natural rubber, the products' durability and persistence in changing conditions of temperature and humidity were low. However, de la Condamine was the first European scientist to describe the use of this material in the Amazonian context, and although he was not precise on the different species and types of wild rubber used by the native Amazonians (Whaley, 1948) his role was central in making wild rubber a well-known natural product in Europe of that time.

However, it was not until the development of the modern rubber vulcanisation process (a chemical process in which raw rubber gets improved elasticity and durability properties) that the industrial applicability of rubber was radically improved. This revolutionary innovation was made by Charles Goodyear in the United States and Thomas Hancock in England almost simultaneously in the mid-1840s, and it drastically changed perspectives on the uses for this natural raw material and soon led to a global demand with unprecedented magnitude for any Amazonian resource so far. The global markets grew fast as new applications of industrial rubber kept emerging alongside developments in other fields of industry from the 1860s. The story that followed is one of the most referred-to socio-economic phenomena in the use of wild species worldwide and through all times. Not only did rubber serve for manufacturing waterproof footwear and clothing but also, by the 1890s, it was used for cable insulation and, in particular, for pneumatic tires. This happened after the Scottish inventor John Dunlop came up with the idea of an inflatable tyre in the late 1880s, and soon after, the market was opened first for bicycles and then quickly for automobiles. Expanding demand made the price of raw rubber soar, and Amazonia with its natural monopoly of wild rubber tree populations entered an era of unprecedented economic boom.

The growing value of the rubber trade from the 1870s transformed many parts of Amazonia into vibrant hubs of economic activity. Amazonian countries, most of all Brazil (Barham and Coomes, 1996), Peru (Santos-Granero and Barclay, 2002), and Bolivia (Gamarra, 2007; Justiniano, 2010), were quick to start harnessing this new source of wealth, and also overseas powers, through private companies, steamed into the rainforest in order to make the best out of the expanding global market. Practically all of rubber production was exported, and myriad small and large exporting houses bloomed along with lucrative imports businesses in the major rubber centres such as Belém, Santarém, and Manaus in Brazil and Iquitos in Peru (Figure 3.8). For half a century, Amazonia held a virtual natural monopoly of the world's wild rubber market, and the wide distribution of the rubber tree stands, as well as the diversity of exporting

FIGURE 3.8 A rubber boom-era building today functions as a primary school in Iquitos, Peru. This and many other nearby buildings are finely decorated and, at the time of their construction, were an indication of considerable wealth.

houses, made it impossible for just one power to control the trade at the level of the Amazon Basin. At the local level, however, the picture was quite different.

Because rubber tapping is highly labour intensive, the availability of a disciplined workforce was of utmost importance for the local patrons supplying the exporting houses. The great tragedy of the rubber boom was in the making when local native populations were recruited into the rubber economy. In many cases, indigenous people were subject to outright slavery, and when showing resistance or exhaustion, they were killed or tortured. Also, this era saw the consolidation of the famous Amazonian arrangement of tying rural workers to local patrons through a vicious circle of debt. The patronage system, in Spanish called *habilitación* (meaning, literally, "enabling") and in Portuguese *aviamento*, is based on advance payments, most commonly in overvalued provisions, that were then repaid by the labourers with undervalued extractive resources or agricultural products. This relationship between the patrons and their workers commonly created an endless debt trap and dependency, so-called debt bondage, which is not always very different from slavery either. This kind of patronage and 'enabling' are, however, not only about greed but also an adaptation to the high risks inherently present in extractive economies in dynamic, unpredictable, and low-governance environments such as Amazonia (Sears and Pinedo-Vásquez, 2011). In some places, however, the rubber traders could not establish that degree of control over the indigenous people, who worked collecting rubber only to the extent that they wanted to do so themselves.

Rubber can be extracted from a variety of tree species, of which the most important belong to the genera *Hevea* and *Castilloa*. The best quality rubber, called Pará Rubber, was obtained from *Hevea brasiliensis* rubber trees. The Pará rubber from *H. brasiliensis* latex was usually extracted by cutting a series of slashes in the tree trunks and letting them bleed the rubber, which was then collected by the tapper. *Castilloa* trees, in turn, were most often felled in order to extract all of the latex at once (Santos-Granero and Barclay, 2002). These different methods of extraction had different effects both on the tree populations and on the rubber extractors. Bleeding the latex was sedentary and bound to a permanent network of paths connecting the trees that, if dealt with carefully, kept producing from year to year. Felling the trees for latex, instead, moved the extraction frontier further as the populations got exterminated (Santos-Granero and Barclay, 2002). From the harvester's viewpoint, bleeding the trees, as a sedentary activity, often led to strict control of the tappers and even to slavery. This kind of control was much less feasible in those cases when rubber was extracted in a moving frontier. Felling rubber trees for latex was easier to carry out on occasional expeditions that did not form full-time employment throughout the year and thus did not interfere too much with other rural work (Santos-Granero and Barclay, 2002). As the value of rubber increased, ever more distant locations of Amazonia were integrated into the rubber economy.

As the rubber extraction expanded and new frontiers were opened, the rubber patrons often moved their labourers with them (Santos-Granero and Barclay, 2002). This contributed greatly to the ethnic mixing of different indigenous peoples, both among the different native groups and with people of European, African, or Asian ancestry, which largely has created the human landscape of Amazonia today. Rubber estates also contributed, however, to the sedentarisation of the population and to increased access to and dependence on metal tools and other manufactured products provided by the patrons, although the opposite also took place. Particularly where rubber tappers were forcefully and violently recruited and controlled by patrons and rubber company foremen (Figure 3.9), and where there still was wilderness of very difficult access, some human groups decided to flee and stay away from the dominant society. Some of these groups are still present in the remotest corners of Amazonia, many times in transboundary forest areas, where they are often, somewhat erroneously, referred to as uncontacted peoples (Huertas Castillo, 2004).

The rubber boom produced such richness that it changed Amazonia in many ways, most visibly in the urban centres of trade, such as Iquitos in Peru and Manaus and Belém in Brazil. The boom was at its peak in the first decade of the twentieth century, but the seeds of the collapse had been sown much earlier. In 1876, the British explorer Henry Wickham (1846–1928) had managed to export tens of thousands of rubber tree seeds from Brazil to England, from where they were then taken to the Asian colonies of the British Empire. In contrast to the story of the *Cinchona* trees smuggled from Peru by another Briton, Clements Markham (1830–1916), just over a decade earlier, there was no law that would

FIGURE 3.9 Indigenous Witotos enslaved by the foremen working for the Anglo-Peruvian Rubber Company that was run by the famous Peruvian rubber baron Julio César Arana. *Photo taken by Walter Hardenburg in 1908.*

have prohibited what Wickham did. Thus, contrary to what is commonly said, this was not a case of bio-piracy. In any case, as the value of rubber soared, filling the wallets of the Amazonian rubber barons, the British were busy at their plantations in Asia, and soon their work started to pay off. As the Asian plantation rubber inundated the global markets, the prices plummeted. This led many of the rubber exporters to ruin, and although the production held out in some areas (Figure 3.10), the profits had melted to next to nothing and the boom had ended.

POSTRUBBER WILD SPECIES HARVEST

The decades after the collapse of the great rubber boom saw a number of other, smaller export cycles that took place in different parts of Amazonia. This was nothing new, and some of them were already in the making during the height of the rubber trade. It is also important to note that these other booms did not form a sequence over the Amazon Basin but rather partly overlapped in time and took place in different times in different parts of the region. Likewise, there are different reasons for the development of these booms in different places and for their collapse or gradual decline – the boom did not always lead to a bust.

An example of an export cycle is the case of the plants first extracted and then later mainly cultivated to produce a natural insecticide called Rotenone (Coomes, 1995). Many plants in the genus *Lonchocarpus* of the legume family

FIGURE 3.10 Wild rubber is still extracted in many places in Amazonia. These chunks of rubber are from Pando, Bolivia *Photo: Natalia Aravena.*

produce chemical compounds of this substance, which faced massive demand for export during the 1940s. Importantly, the same plants are also used locally to produce a poison used in fishing. Another boom developed, for instance, in the Department of Loreto in north-eastern Peru, where many local patrons organised their workers in the extraction of the hard seeds of the ivory-nut palm (*Phytelephas macrocarpa*, in Peru called *tagua*), which were used principally as raw material for the manufacturing of buttons for clothing. This activity had already started in the last years of the rubber boom in the early 1900s, and it continued for almost 50 years (Santos-Granero and Barclay, 2002). Its end in Loreto was related to the emergence of plastic buttons that were much cheaper to produce, but outside Amazonia, such as in the coastal areas of Ecuador, the production was more competitive and the exports of 'plant-ivory' buttons and figurine carving continued (Runk, 1998). Recently, the extraction of plant-ivory also has been promoted in Peruvian Amazonia as an alternative income-generating activity (Figure 3.11).

Soon after the plant-ivory exports had started to grow, Peruvians also began to extract precious timber for the export market. As an example of how distant Amazonia was from many of the national capitals, it is noteworthy that in the first half of the twentieth century, almost all valuable timber logged in Loreto in north-eastern Peru was exported; even the big-leaf mahogany (*Swietenia*

FIGURE 3.11 (A) The hard seeds of the ivory-nut palm (*Phytelephas macrocarpa*) are used for carving figurines sold as souvenirs in the Pacaya Samiria National Reserve in Peruvian Amazonia. (B) In the same protected area, almost all standing trees of big-leaf mahogany (*Swietenia macrophylla*) have been extracted by illegal loggers. Occasionally the logs are seized by the reserve authorities, as has been the case with this one. (C) Hunting black caimans (*Malanosuchus niger*) has been illegal in Peru since the 1970s, but large skulls, such as this one in the village of San Lorenzo in Peruvian Amazonia, still tell a story of killing these animals. *Photo: Ilari E. Sääksjärvi.*

macrophylla) that was needed in the country's own capital, Lima, was transported first down the Amazon to the Atlantic, then north to the Caribbean and through the Panama Canal to the Pacific, and finally down the coast back to Lima (Santos-Granero and Barclay, 2002). Mahogany was targeted in the accessible riverside forests, and the natural mahogany stands that could be easily accessed were logged from the 1910s onward; by the 1940s, the stocks had been exhausted to such a degree that there was no more mahogany to be logged in these areas. Improved technology and the development of infrastructure, however, have helped loggers to expand the frontier, and mahogany has been exported from Loreto almost until the present, extirpating the species to such a degree that it has been declared commercially extinct in the region.

In the early twentieth century, other tree species were targeted to extract latex used for various purposes. For instance, the latex of manilkara trees (*Manilkara bidentata*, in Peru called *balata*) was used to cover electric cables and golf balls, and a manilkara boom continued until the late 1940s. Another species, *leche caspi* (*Couma macrocarpa*), produces a latex that serves various purposes and for which this species has been targeted by extractors. None of these latex products, however, came even close to the importance of the wild rubber that, again during World War II, lived through another short but very important export boom. This was caused by the advance of the Japanese troops in Southern Asia, leading to problems of supply to the Allied countries. Provision of this raw material had become so essential, not only for civilian industries but also to the machines of war, that, in order to avoid being left without rubber, the United States (and, to a lesser degree, other Allied powers) turned their eyes to the old source of wild rubber, Amazonia.

Although there were plans to begin production in plantations, the still relatively recent experience of wild rubber stands, combined with the urgency of the matter, led back to this latter option. Thus, the United States promised to buy Brazilian rubber with a guaranteed high price in exchange for getting all of the production. The needed influx of workers was to be generated by turning the recruiting towards the poverty-stricken north-east of Brazil and offering free transport and healthcare for new tappers, in addition to a pension. Needless to say, the *soldados da borracha* (soldiers of rubber) recruited this way mostly saw only the transport to materialise. However, all this did not yield the volumes of rubber expected by the US government, and moreover, because the boom era organisation of the rubber-tapping economy, including the role of middlemen, had persisted in many places even when the profits had become meagre at best, it was difficult to change it all to benefit the new buyers (Revkin, 2004). The new boom nevertheless was an important episode that also reinforced rubber tapper extractivism as a resilient form of livelihood in some parts of Amazonia.

After the war, more export cycles succeeded. For example, rosewood (*Aniba rosaedora*) oil was distilled from the tree's heartwood, in order to supply this natural fragrance component to the international perfume industry. Although the boom's peak was short lived, it efficiently extirpated rosewood populations in many parts of the Amazon Basin. From the nineteenth and turn of the twentieth centuries onward, a boom of bird hunting was experienced in the great grasslands north of the Amazon Basin, where countless egrets, herons, and ibises were killed for their feathers to supply the expanding fashion industry in Europe and North America.

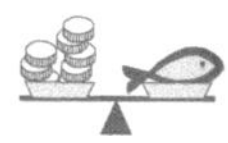

This same trend was reinforced in the mid-twentieth century, when animal trade increased drastically as pelts and high-quality leather were exported from all parts of Amazonia. Animal hides and pelts weighed relatively little but yielded a high price, and they could be preserved for quite prolonged times. Thus, lured by the high prices paid for animal hides and pelts, hunters penetrated into some of the most remote and inaccessible parts of Amazonia, setting up hunting camps where they could spend months in a row (Álvarez Alonso, 1997). In terms of number of animals hunted, this trade mostly focused on peccaries and deer, out of which, in particular, the former had a skin that was held in high esteem in Europe and the Unites States for making luxury items such as gloves, shoes, and purses (Redford, 1992). Although hunted in much smaller numbers, jaguars (*Panthera onca*) and giant otters (*Pteronura brasiliensis*) were less resistant to hunting pressure and suffered a severe impact (Figure 3.12).

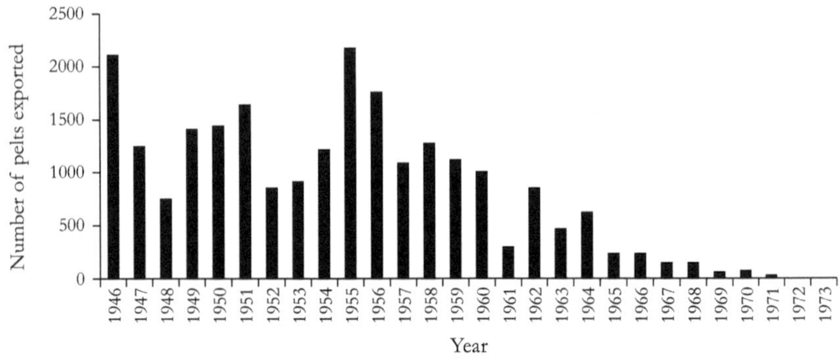

FIGURE 3.12 The number of skins of giant otter (*Pteronura brasiliensis*) exported from Peruvian Amazonia in 1946–73 (Recharte Uscamaita and Bodmer, 2010).

Moreover, a new market was also established for live monkeys as test animals in the biomedical industry.

During the rubber boom, indigenous population had in many places decreased, and different native ethnic groups had increasingly become mixed, both with other indigenous groups and with people originating outside Amazonia. As a consequence, a growing number of nonindigenous rural groups of people employed traditional Amerindian practices to make their living from the forest (Coomes and Barham, 1997). This is the main rural population in many parts of Amazonia still today, called *ribereños* in Spanish and *caboclos* in Portuguese. The existence of *ribereños* and *caboclos* owes a lot to the rubber boom, but they did not emerge out of nowhere after the boom, nor did their development take place equally and simultaneously all over Amazonia. Rather, they have become independent small producers at different times and because of different reasons in distinct places across the Amazon Basin. An important component of this process has been the political emancipation linked to improved transport infrastructure and education. As a consequence, the ties of rural peons controlled by patrons engaged in agricultural and other production increasingly begun to break from the 1960s onward (Santos-Granero and Barclay, 2002).

Yet other growing extractive economies developed around animal products, and in particular around the trade of live animals for pets and ornamental fish. Some long-distance trade in animals and animal products existed in Amazonia already in pre-Columbian times, as caimans and anacondas were brought from the Amazon Basin to the Inca Empire's administrative centre, Cuzco, where they were kept in menageries (Gilmore, 1950). However, the effect of this trade was minimal compared to the new hunting pressures triggered by the demand that led to the legal exportation of almost 2 million live animals from Peruvian Amazonia alone during the years 1965–73, after which the trade was banned. Particularly large parrots and macaws were captured for trade (González, 2003). The total figures of animals killed and those exported alive were undoubtedly multifold compared to the numbers reported (Álvarez Alonso, 2012).

FIGURE 3.13 The introduction of hunting dogs to Amazonia facilitated hunting of several species of ungulates and large rodents.

Many dynamic aspects of human culture induced increasing pressure on the Amazonian forests and their biological diversity. Metal tools such as knives and axes that had become available basically everywhere were increasingly accompanied by other, yet more efficient technological innovations that helped harvesters to be really efficient in the field and facilitate product transport. Even indigenous groups living in a kind of 'self-chosen' isolation (so-called uncontacted groups, whose isolation is often far from voluntary because it is forced by threats of disease and/or violence) typically acquire metal tools by seizing them from other indigenous people, the campsites of loggers or hunters, or even tourist lodges. Access to outboard motors and big canoes cut the relative distances and people were able to transport extracted fresh products with improved efficiency to cities and other consumption centres. Particularly this concerned fish, but also game meat and many palm fruits. Remote villages therefore became increasingly integrated into regional economies.

Towards the end of the twentieth century, shotguns had already replaced the much less effective 'traditional' weapons such as blowguns, bows and arrows, and spears almost everywhere. Also, hunting dogs had become almost universally available, as increased contact with urban centres made it possible to continuously bring in new dogs to replace those that succumbed due to disease, parasites, snake bites, or other calamities (Figure 3.13) (cf. Koster, 2009). Also, torchlights became available, making it possible to hunt at night. New fishing technology brought to Amazonia during the twentieth century included not only nylon lines, diver's masks, and gill nets but also dynamite and synthetic pesticides. Recently, in addition to improved transportation, fishing and hunting technologies, also chainsaws have become increasingly common possessions even in many remote indigenous communities. All this made life a lot easier, but the downside was that it also made it so much easier to exhaust resources.

Another important development during this epoch, however, was that conservationist concerns became incorporated, at least to some degree, into national and international policies. For instance, as the catastrophic impact of commercial hunting on animal populations started to become obvious, governments began taking action. Brazil banned all commercial hunting by 1967, but the law could not be efficiently enforced (Da Silveira and Thurbjarnarson, 1999). The Unites States banned imports of products from giant otters in 1969 through the Endangered Species Conservation Act, but by then these magnificent animals had already been eliminated from large parts of Amazonia, and the trade had collapsed. Peru banned commercial hunting in 1973, and in the same year the governments of 80 countries agreed on the Convention on International Trade in Endangered Species of Wild Fauna and Flora (CITES), which came into force in 1975. This implied that importing products from many threatened species listed in the annexes of the convention became prohibited or restricted, and, as overall demand for these products started to fall on a par with changing consumer attitudes, much of the commercial hunting ceased. Since then, populations of some species, including giant otters, have slowly recovered.

Whereas most commercial wildlife hunting nowadays is prohibited in the Amazonian countries, there are notable exceptions. Many species of birds, for example, can be commercially hunted or captured. One curious case is also that of hunting peccaries in Peru. The two existing peccary species in Peru are hunted, in the first place for subsistence purposes. But whereas their meat can be legally consumed by the hunters and their families only, their hides can be sold commercially. For example, in 2011 the Peruvian government's Forest and Wildlife Service authorised the commercialisation of 70,784 hides of collared peccary (*Pecari tayacu*) and 48,268 hides of white-lipped peccary (*Tayassu pecari*). Also, part of the commercialised hides can be exported under CITES permits if this is with added value through processing and, for example, through the elaboration of leather products.

Towards the end of the twentieth century, trade of wild meat towards urban centres increased in various parts of Amazonia (Suárez et al., 2009, see also Sirén, 2012). This differs from the commercial hunting for the fur trade in the past, in that it focuses on the same set of species that are hunted for subsistence purposes, and not at all on such species as jaguars (*Panthera onca*) and giant otters, previously so highly esteemed. Overall, subsistence hunting removes a larger amount of animal biomass from the forest than commercial hunting does. Nevertheless, in certain locations commercial hunting can be very significant, such as in areas that are uninhabited but yet relatively well connected with major urban centres. For example, in Peruvian Amazonia large cities such as Iquitos and Pucallpa consume large quantities of game meat, turtle eggs, and other wildlife products. This hunting pressure has decimated the populations of large animals such as ungulates, tapirs, river turtles, manatees, as well as large monkeys and birds in many areas around these cities (José Álvarez, personal communication, April 2013).

In sum, the economic downturn after the end of the Amazonian rubber boom was only temporary, and its impact may be overemphasised by the dramatics of the fall. More broadly, most of the region witnessed accelerating economic development and environmental change throughout the twentieth century and, of course, up until today. Although Amazonia, just as any other part of the world, has experienced all kinds of development, it is fair to say that it first became part of the global economy through the trade of wild species products. Whereas all changed with one emblematic wild species product (as natural rubber was), the economic role of wild species harvest in contemporary Amazonia is somewhat different, and no one product has a comparable role. Although the times of the rubber barons and other all-mighty patrons are mostly gone, different forms of forced labour are still present in forestry, agriculture, and mining, in some parts of Amazonia (Youatt and Cmar, 2009; Phillips and Sakamoto, 2012).

SMOKE AND NEW WINDS OVER AMAZONIA

Today, the most valuable natural resources in Amazonia are increasingly those that are produced in agriculture and through the extraction of nonrenewable resources such as minerals, gas, and oil. Likewise, deepening urbanisation processes and associated growth of consumption also shape the demand for resources. Many small outposts of a handful of thatched-roof houses have grown into towns, and some such towns into cities with traffic lights, concrete houses, internet cafés, and all of the other services imaginable in any modern urban centre. Not only have many newcomers to Amazonia directly settled in some town or city, but also the young generation has increasingly left their indigenous communities or the farms that their settler parents or grandparents once hacked out from the rainforest, in order to find a better life in the city (Figure 3.14).

However, this process has also led to increasingly blurred lines separating urban and rural dwellers as many families have retained links between houses in rural areas and family members in towns and cities; moreover, this is not only about maintaining pre-existing links but also about forming new ones (Padoch et al., 2008). This is also because wild species harvest, such as logging and fishing, keeps employing people living in urban centres. Also, many indigenous persons have participated in these kinds of processes often but not always leading to assimilation into the majority culture. Many people from Amazonia, settlers as well as indigenous, also have participated in a wave of migration to national capitals and other metropolises as well as North America and Europe.

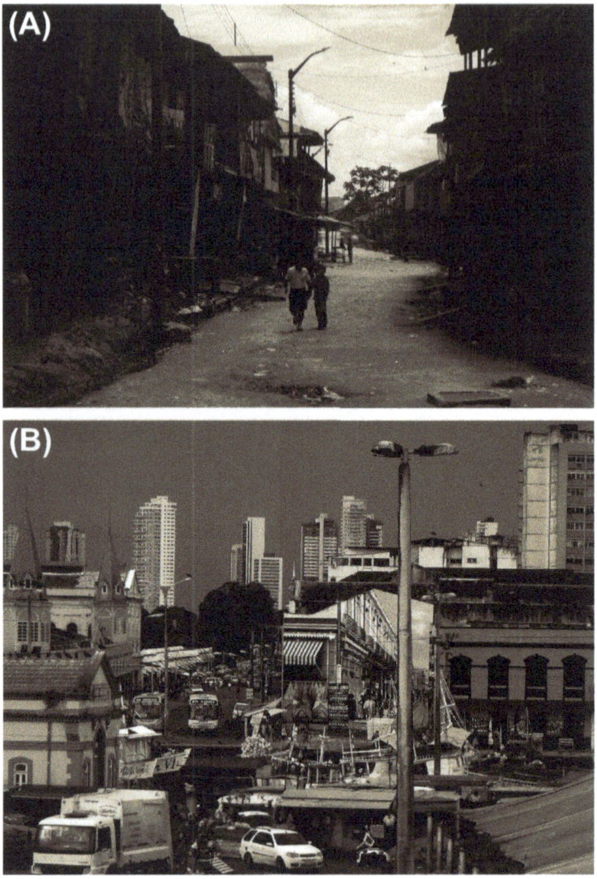

FIGURE 3.14 Urbanisation of Amazonia in (A) the Belén neighbourhood in Iquitos, Peru; and (B) Belém, State of Pará, Brazil.

During the twentieth century, many indigenous people in Amazonia experienced rapid population growth as death rates decreased and birth rates remained as previous levels, and the same happened also to the population of mixed ethnic origin (McSweeney and Arps, 2005). The most significant factor of demographic change in contemporary Amazonia, however, has been the influx of settlers as people have left their homes in southern and north-eastern Brazil, or in the highlands or the Pacific coast of the Andean-Amazonian countries, in order to begin a new life in the rainforest. This colonisation of Amazonia by nonindigenous peoples often has been promoted by governments, in part pushed by the perceived overpopulation and lack of land elsewhere, in part pulled by desires to start to develop the region more forcefully. The colonisation wave began much earlier but became dramatically accelerated in the 1960s and 1970s, when new sources of funding, including international lending, and in Ecuador also oil revenues made it possible to construct penetration roads into Amazonia on a much larger scale than before.

The settlers cut down the forest along the roads and planted pasture or cash crops such as rice, coffee, or sugar cane. Indeed, deforesting the land was often a legal prerequisite for obtaining a land title, which in turn made it possible to get access to credit and subsidies. Particularly in south-western Brazilian Amazonia, in places like the States of Rondonia and Acre, the land conversion and concentration of property in the hands of powerful landowners were of such magnitude that towards the end of the 1980s, its environmental and social effects became increasingly visible both locally and internationally. This development also increasingly started to corner more traditional lifestyles and livelihoods, such as extractivism. Countless colonisers and peasant workers annually took advantage of the dry seasons in the region, burning down felled forest vegetation. But as the thick clouds originating in the raging fires soared into the Amazonian skies, also global conscience and new forms of resistance started to arise.

This reality was also linked to important changes in the political character of this region during the latter part of the twentieth century. While many of the Amazonian countries left the military rule of past decades behind to increasingly embrace representative democracy, also the world was experiencing profound political change. The Cold War coming to an end was about to liberate much financial resources and creative energy in the field of civil society. This was seen particularly clearly in the way in which indigenous and other local organisations at regional and national levels started to emerge as international actors, often seen as potential allies by international environmental nongovernmental organisations. Growing public concern in the Amazonian countries, as well as internationally, triggered processes in favour of environmental conservation and promotion of indigenous and local rights. These processes also placed focus on extractive economies as alternative livelihood strategies.

As part of these discussions and processes, extractive reserves of different types were increasingly established in Amazonia, including the famous Reserva Extrativista Chico Mendes, which was created in the State of Acre in Brazilian Amazonia in 1990 shortly after rubber tapper activist Francisco 'Chico' Mendes (1944–88) was assassinated by contract killers (Revkin, 2004). In Peru, areas set aside for similar purposes were called 'community reserves'. This is an indication of the strong role that community-based conservation and natural resources management achieved in the environmental movement towards the end of the twentieth century. No doubt, the extractive reserves in many cases managed to halt deforestation, but they have also been criticised by conservationists who argue that although increased levels of extraction and associated economic development can lead to improved living standards for the extractors, it also implies increased impacts on the natural ecosystems. Others criticise the concept for other reasons: extractivism fails to bring extractivist families out of poverty, and living conditions in the reserves remain precarious.

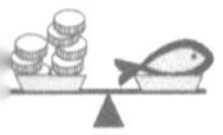

It is also true that at the local level, another problem is insufficiently developed processing technology that hinders the commercialisation of many extractive products, leaving the harvesters in poverty. In any case, if the alternative to extractive use of forest resources is land conversion, there is little doubt that extractive reserves serve the conservation of biodiversity. Meanwhile, the extension of formally protected areas in Amazonia also increased sharply in the late twentieth century (Schulman et al., 2007), as did also the areas set aside for indigenous peoples, whether as commonly owned by indigenous nations or communities or as exclusive usufruct rights granted to indigenous peoples on land formally owned by the state.

Wild species are constantly harvested for export across the region. Export here means products that are sold outside of Amazonia; many of them have not only international demand but also important national markets in the non-Amazonian regions of the Amazonian countries. Among the most important of these products are an ever-expanding list of timber species, Brazil nuts *(Bertholletia excelsa)*, açaí fruit *(Euterpe oleracea)*, wild rubber, and ornamental fish. The harvest of many of these products has never ceased since they first started to be exported, but they were rediscovered as something with conservation value during the boom of 'productive conservation' in the late 1980s and the 1990s. However, many traditional nontimber forest products' uses are in danger of getting forgotten. For example, in an ethnobotanical study made in Eastern Amazonia, Shanley and Rosa (2004) reported declining plant uses in the form of frequent statements such as 'My grandmother used this' and 'We do not use it anymore'. When cheap industrial substitutes become available, people use wild species more seldom, and finally also the knowledge about them may disappear.

Those wild species uses that found their way to modern market economies are among those that have often persisted or even expanded, compared to subsistence-related species or products that are not appreciated by dominant cultures. Some wild species products with at least centuries of local human use history in Amazonia, such as Brazil nuts have only recently become global commodities, while others, including many of those described in this chapter, have lost their relative importance. Economies based on wild species harvest have evolved along with the overall societal change, and different parts of Amazonia have seen very different development processes depending on national policies, geographical accessibility, and the surge of new transportation routes, among other factors.

Despite the achievements discussed here regarding the modernisation of wild species-based economies as alternatives to land conversion, deforestation and industrial-scale selective logging continue to change the Amazonian landscape at high speed. In places, they seem to have escaped beyond control, yet there is also some evidence that recent measures of land use policy are serving to protect some forests (Oliveira et al., 2007). Although secondary growth takes place in recovering forests, short rotation times mean that temporary agriculture has changed the forest ecosystem permanently in many areas. Human disturbance is increasingly transforming the Amazonian forest biome, potentially taking it towards a transition to a disturbance-dominated regime with changes in water and energy dynamics, at least in the southern and eastern parts of the basin (Davidson et al., 2012). There is also a risk of coupled influences of deforestation and the stress caused by climate change (Malhi et al., 2008).

Wild species harvest explains little of Amazonian deforestation and forest degradation, but there is increasing evidence of extensive impacts of selective logging of valuable timber (Asner et al., 2009). However, although the single most important economic activity behind land conversion in Amazonia as a whole is cattle ranching (Nepstad et al., 2009), the main causes of deforestation vary greatly across the region. For example, many areas have experienced large cumulative deforestation as a consequence of cash crop cultivation, including plantations of illegal crops such as coca (*Erythroxylon coca* and *E. novogranatense*) (Salisbury and Fagan, 2013). In general, migratory farming (called slash-and-burn, shifting, or swidden agriculture) is responsible for large accumulated deforestation in Amazonia, but while this kind of deforestation often leads to relatively rapid secondary growth of forest, industrial large-scale agribusiness also has been introduced in many areas, particularly in Brazil and Bolivia, with soya recently being the most extensively cultivated crop, and with much more long-lasting consequences (Figure 3.15).

All of the important land use patterns are related to infrastructure, and changes are often directly triggered by improvements in transport infrastructure, in particular by road construction (Mäki et al. 2001). A continent-wide process called Initiative for the Integration of the Regional Infrastructure of South America (IIRSA) is currently enhancing regional integration, and this is particularly visible in the construction of physical infrastructure such as road building, improvement of waterways, and establishment of hydroelectric power plants (Killeen, 2007). In sum, this will have very significant impacts on the natural habits in Amazonia, and thus also on wild species harvest.

As a general trend, in the 1990s and 2000s, the governments of the whole Amazonian region have put increasing emphasis on environmental issues. The opening for signature of the Convention on Biological Diversity in 1992 and the rapid ratification of the convention by the Amazonian countries have been accompanied not only by legislative reforms better aligning environmental law with international standards but also by substantial changes in the associated science–policy interface. Whereas the particularities of how biodiversity issues are dealt with vary from country to country, and money, politics, and power keep running much of the environmental decision making in Amazonia just as elsewhere in the world, the CBD nevertheless has also had real effects on the baseline worldview of biodiversity policy makers and managers.

In Table 3.1, we present a general outline of major periods of Amazonian history with emphasis on wild species harvest and the context in which it has occurred. The five distinguished periods are of different durations, and also the criteria upon which they are highlighted in this table may vary. Despite these

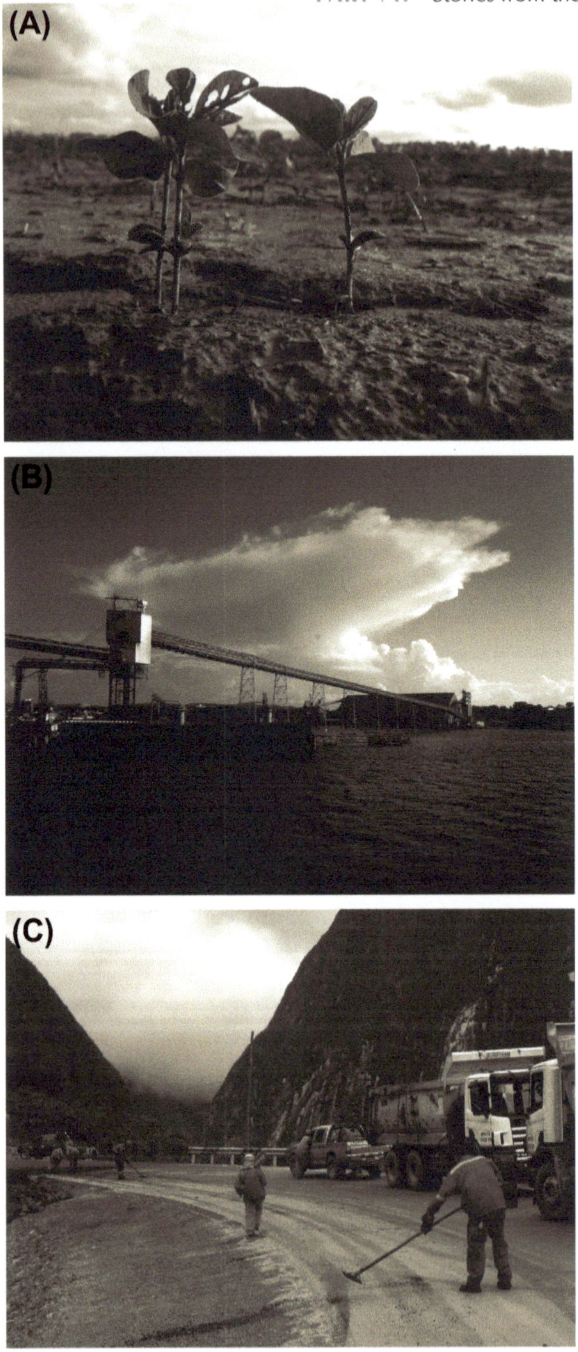

FIGURE 3.15 (A) In the Department of Santa Cruz in Bolivia, soya is cultivated by both large agro-industrial actors and colonist farmers. (B) A soya terminal in Santarém, Brazil, receives and sends further the production from large areas in central Amazonia. (C) The Interoceanic Highway cuts through the Andes mountains in Cusco in order to facilitate the transport of people and commodities between the Pacific coast of Peru, Amazonia, and the Atlantic coast of Brazil.

TABLE 3.1 General Outline of the Most Distinctive Periods of Wild Species Harvest in Amazonia

Period	Global Perspective	Amazonian Perspective	Wild Species Harvest in Amazonia
Before human arrival (until 13,000 BP)	Evolutionary processes in nature; Presence of early humans on other continents and maybe in some other parts of the Americas	Amazonian landscape and life forms evolving without human interference	None
Indigenous cultures (until the sixteenth century)	Expansion of human population; development of local cultures; increasing intercontinental human interactions in the old world	Arrival, adaptation, and distribution of humans in Amazonia	Subsistence hunting and gathering; discovery of useful wild species as well as ways to use them; trade with Andean highland peoples and inhabitants of coastal regions
Demographic collapse and recovery (sixteenth to eighteenth centuries)	Overseas expeditions from Europe; conquest and colonies established	European exploration and conquest; indigenous populations decimated by diseases	Decreasing subsistence harvest; increasing harvest for overseas trade
Booms and busts (late eighteenth to mid-twentieth centuries)	Worldwide political turbulence but also scientific, technological, and economic advancement	Forceful overseas dominance giving way to national integration and planned regional development	Subsistence harvest increases again; massive extraction of selected wild species for exportation
Integration and globalisation (from the mid-twentieth century onward)	Postcolonial modernisation with economic growth; globalisation and international agreements	Infrastructure development and mass immigration; growth and diversification of regional economies; road construction and deforestation; increasing claims for indigenous rights and conservation	Increased extraction of timber, fish, and wild game; traditional knowhow of wild species use starting to vanish; bioprospecting

limitations, we hope that this summary overview helps readers not only to identify some key aspects of Amazonian history in a nutshell but also to realise how the present state of affairs has evolved. The table also presents some linkages that join the development in Amazonia with simultaneous processes occurring elsewhere in the world. Although expressed by a few words only, these linkages help to make the particular points that we consider important.

The more integrated the world becomes, the less isolated any wild species harvest can be in terms of broader scale developments all around the world. This is so even if the harvesters themselves ignore or prefer to overlook this reality. As an example, after the moment when a combination of domestic development pressures and technological advancement initiated European overseas expeditions and conquest, their legacies became such that they can be seen even in present-day Amazonia. This explains much of the languages used, the ethnic identities of people, the distribution of major population centres, the particular ways in which biological matter has been dealt with as a resource, the harvest technologies, and the forests emptied of some of their natural species, to list a few examples of these legacies. Wild species harvest still is an important and diverse field of Amazonia's regional cultures and economies, but the current state of affairs is not static, and situations constantly evolve as part of overall global developments.

Chapter 4

On a Winding Trail towards Sustainable Hunting

THE *MUNDITI* CALL

In the dry season, the *munditi* sings at night, and that is when you can hunt them. You get up early when the night is at its quietest. After drinking a few bowls of warm manioc brew, you take your gun and your jungle knife and walk out to the forest. Perhaps you already hear one singing, perhaps several, or perhaps you just hope to hear one as you get further out into the woods. You may light a torch made of the splinted resin-loaded wood of the *inayu* palm *(Attalea maripa)* to light your path, as they did in your grandfather's time, but more likely you take along a torchlight and a pair of alkaline batteries. Nowadays, you may have a long way to walk until you get near a *munditi*, so you must take care not to run out of light before you have reached your destination.

The *munditi* is what the nocturnal curassow *(Nothocrax urumutum)* is often locally called in Ecuador. It is a distant relative to the hen, typically weighing almost 2 kg, and its meat is very tasty. The song of the male is mystical, even magical – very unlike the song of almost every other bird. It consists of two deep, muffled tones – the second one longer than the first – followed by an instant of silence. Then, three slightly rising tones in a row, with the last one longer and just slightly (not even a halftone) higher than the second. Again, there is silence, and finally a weaker, but sharp, short high-pitch rasping yell. The males sing to attract females. Towards the end of the mating season, you can sometimes hear the females respond to their suitors with a song that follows a similar rhythm as that of the males, but it is less melodic and goes down in pitch instead of up.

Getting to the tree where the *munditi* sings is not easy. The terrain is rugged. If you try to go straight to where the song sounds, you will end up stuck on some steep slope where you can progress only so slowly that you will never get to the bird in time. You have to follow the crests of the ridges as long as you can, abandoning the crest only as you make the final approach to the tree where the bird is sitting. Several *munditi* may be singing from different trees, so you must choose one. Suddenly, one of them may become quiet, or another one that was quiet may start to sing. Only as you approach the tree where the *munditi* is do you realise how strong the song actually is, although it may have been just barely audible when you first heard it.

Diagnosing Wild Species Harvest. http://dx.doi.org/10.1016/B978-0-12-397204-0.00004-8

You must be in place about half an hour before dawn starts to break, because that is when the bird stops singing. You should find a suitable place to sit and wait near the tree. Hopefully you have brought with you a dark long-sleeve shirt to protect your arms from the mosquitoes while waiting and something else to cover your head and face. Then, you just wait. Every now and then, you should uncover your eyes and check whether the sun begins to dawn. Once it starts to get light, quietly load your gun and look steadily to the canopy. After some time, a *munditi*, or perhaps even two, will descend from the canopy to a lower branch or even down to the ground. If you are lucky, you can return home with enough meat to provide some delicious meals for your family (Figure 4.1).

A MAN AHEAD OF HIS TIME

'We are exterminating the *munditis*', exclaimed Don Raúl, a man in his upper 40s living in a thatched-roof house. 'Every dry season, when the *munditis* sing, we go and kill them. If we continue like this, there will be none left. We always talk about the indigenous people protecting the nature, but it is not true. We are ourselves exterminating the animals!' It was 1990, and Don Raúl was ahead of his time. Others had noted that game populations were dwindling, but typically they blamed somebody else for it: Others hunted too much, others entered grounds they had no right to, or some evil and powerful shaman had of pure envy hidden the animals

FIGURE 4.1 Sarayaku woman with her husband's catch. To the right is a *munditi*, a nocturnal curassow *(Nothocorax urumutum). Photo: Franklin Gualinga*

inside the ridges so that nobody should not be able to hunt them. Don Raúl, on the other hand, acknowledged the consequences of his own actions – that each time he hunted, he contributed to the process of depletion of the wildlife. This was true not only of the *munditis*. He was well aware that some other species were even worse off. He wanted to find a solution to this problem but did not know what to do, and he felt that he was talking to deaf ears when he tried to discuss the matter over and over again with other community members. A few years later, Don Rául died in a motorboat accident, still having not found a solution to this concern.

Twenty-two years later, Don Raúl's son Tupak, number seven out of eight children, is the vice president of the community of Sarayaku. This community is located in the middle stretches of the Bobonaza River in Ecuadorian Amazonia (Figure 4.2). Sarayaku consists of five hamlets within an hour walking distance from each other, with over 1000 inhabitants and over 1000 km^2 of hunting grounds. Tupak is today heading a meeting in the administrative building of the community, which is a palm-leaf hut with tile floor and equipped with solar panels, computers, and modern furniture of the cheaper brands. A young boy sits quietly with his head down. He is just 16 years old and has recently done what every teenager in the village, not long ago, would have been very proud of – not to mention their parents. He had found tapir *(Tapirus terrestris)* tracks under a fruit tree in the forest, and in the night he had returned with a torchlight

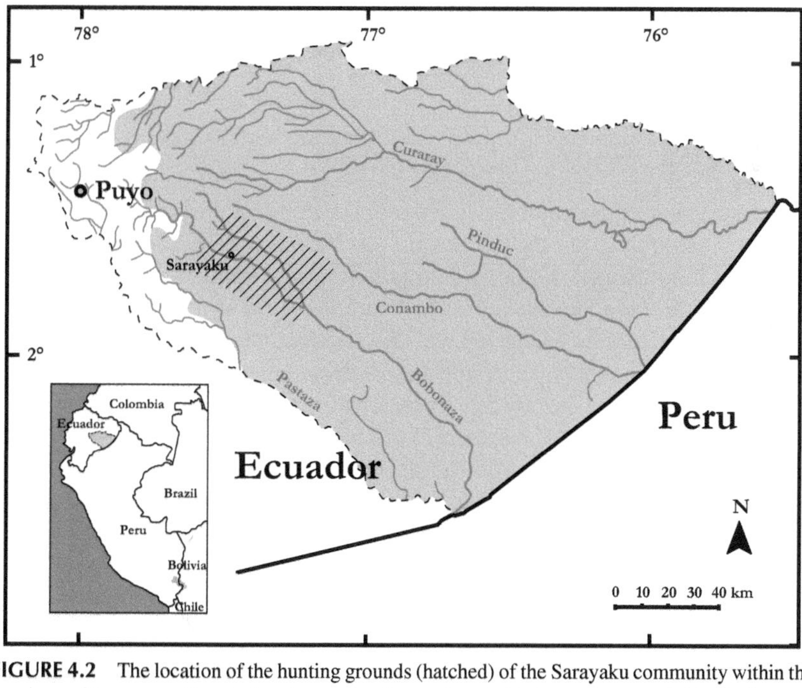

FIGURE 4.2 The location of the hunting grounds (hatched) of the Sarayaku community within the province of Pastaza in Ecuadorian Amazonia. Grey colour indicates areas occupied by indigenous peoples, and white indicates areas primarily occupied by non-indigenous settlers.

and a shotgun to wait. When the tapir came back to eat the fallen fruit again, the boy shot it dead and then cut the body into pieces to carry the meat home.

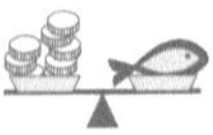

The tapir is, by far, the largest mammal in the Amazonian forests, often weighing some 200 kg, sometimes more. Killing such a large animal would have meant, just a couple of years ago, becoming a family hero for some time. The boy would have given away large chunks of meat to friends and relatives living nearby. Then, he would have loaded a couple of baskets of tapir meat into his canoe to go a few kilometres downriver to the central square, where other community members would have swarmed around in order to buy a piece of meat before it was gone. In a couple of hours, he would have earned money equal to a month's salary in this remote place where wage work is scarce. And his own family would still have had enough meat left so to feast for days, until they laughingly would have started to complain about the very particular stomach gases you get from eating dry-smoked tapir meat day after day.

But this day the boy sits here being accused by the elected leaders of his own community of having broken the local "law", which states that one must not kill tapirs. He faces a $500 fine, which is a small fortune in a place where the daily wage for unskilled labour is $10 – if paid work is found at all. His mother and father accompany him, but neither of them says much. The one who speaks for him instead is his much older brother-in-law. 'He will pay it. We are all aware that the tapir is an endangered species that needs protection', the brother-in-law says. 'All that we demand', he continues, 'is that all the others who have killed tapirs also get fined'.

BLOWGUNS, DOGS, AND SHOTGUNS

In the late eighteenth century, there were no permanent human settlements at the site of what is today the community of Sarayaku. Diseases brought from Europe to the Americas had caused repeated mortal epidemics during almost three centuries, reducing the native human population to a fraction of what it once was. Small mobile bands of people lived dispersed in the forest, away from the main rivers. The raids, *correrías*, for catching indigenous people and taking them to the mission villages only occasionally reached this distant area, and sometimes people fleeing the hardships of life in mission communities ended up here, where the risk of getting caught again was relatively small.

Around 1830, however, a mission community was established at the site where today's Sarayaku is located. People from different ethnic groups now voluntarily came to settle here. Metal tools provided by the missionaries facilitated forest clearing and weeding on the fertile soils of the alluvial plains. They also served to make canoes, further increasing the benefits of settling near the river instead of on the crests of the ridges, which previously, for reasons not fully understood, had been the preferred places for habitation and farming. Gradually, the mission community grew as increasingly more people came from the surrounding forests to settle. The previously disastrous epidemics had also become less mortal, as 300 years of exposure now had given people some immunity.

Not all people joined the mission community, and the surrounding forests continued to be inhabited by mobile bands of hunter-gatherers who heavily relied on wild fruits, as well as on hunting and fishing, and who practiced farming only to a limited extent. Gradually, however, more people came to settle in the mission community. Although hunting pressure previously had been fairly evenly distributed in space, it now began to be more concentrated to the immediate surroundings of the mission village. Socially, the community remained fragmented along lines related to ethnicity and extended families. The mission had no permanent presence of a priest, so when a priest left, the people dispersed out to their distant alternate homes, locally called *purinas*.

Hunting was carried out with spears as well as with bow and arrow, and probably also with blowguns, but most likely hunting dogs were not used. Dogs suffer high mortality in the tropical rainforest environment, and therefore hunting dogs became common only after steady contact was established with settlements outside the tropical rainforest that could provide a steady supply of dogs to replace those that succumbed to diseases or accidents. By the mid-nineteenth century, hunting dogs seem to have become common in the mission communities in the region. Dogs facilitated the hunting of many ground-living species, in particular collared peccaries (*Pecari tajacu*), pacas *(Cuniculus paca)*, and tapirs.

The rubber boom reached Sarayaku in the late 1880s. Although it brought slavery and death to many other indigenous peoples, the Sarayaku experience was different. Basically, all adult men and many women worked collecting *Hevea* rubber in the forest, but they did it of their own choice. Instead of bleeding the trees, the people simply felled them. Through this work, the people in Sarayaku could acquire industrial goods such as pots, axes, and clothes. Some even came to know city life as they travelled all the way downriver to the Peruvian city of Iquitos together with the rubber traders.

The missionaries had ruled the community with firm hand ever since its foundation, but since the rubber boom, it had to share the power first with the military when a small garrison was set up in Sarayaku and later with a civil representative of the national government, the *teniente político*. The native inhabitants of the community, however, did not have any of their own social organisation at a higher level than extended families. Although there were a handful of officials, *varayuks*, in charge of keeping order in the community, these were appointed by the missionaries rather than by the community members. In spite of this, the Sarayaku people retained quite a degree of freedom. Contemporary observers noted, sometimes with surprise and consternation, that the Sarayaku people were free to choose for whom to work, and they also could choose simply not to work if they did not want to. This was in stark contrast to other areas, where the indigenous people were owned by 'white' masters through debt bondage.

During the first half of the twentieth century, hunting was still carried out mainly using blowguns with poison darts. The areas in the immediate vicinity of the village were already hunted out of many species. However, some 4 km away from the village centre, wildlife was abundant, including even vulnerable species such as spider monkeys (*Ateles belzebuth*) and woolly monkeys (*Lagothrix* spp.). By the 1950s, however, the blowguns were gradually replaced by muzzle-loaded shotguns. Those who had a shotgun typically carried both the shotgun and a blowgun while hunting, using the blowgun for small and the shotgun for big prey. Ammunition was expensive, so people used to load just 3–5 shots and carefully retrieved the shots from the dead animals in order to reuse them. Another new hunting technology introduced in those years was the use of flashlights, used for waiting for pacas in the dark in the manioc fields or under fruit trees, as well as searching for the glowing eyes of caimans (*Caiman crocodilus*) along the river.

From the 1940s and on, travelling merchants came to buy furs from spotted cats, neotropical otters (*Lontra longicaudis*), giant otters (*Pternonura brasiliensis*), and collared peccaries, as well as gold, toucan feathers, the cinnamon-like spicy bark of *canela* (*Ocotea quixos*), pottery, and live monkeys. The fur trade continued up to the 1970s, when it ceased because international trade restrictions caused the demand to collapse. In contrast to many other areas, however, hunting for furs in Sarayaku did not cause any perceivable decline in the abundance of the species targeted by the hunters.

In the 1970s, the Ecuadorian government finally had achieved the financial means to make reality of old dreams. They promoted the building of penetration roads into Amazonia and sent in nonindigenous settlers to bring 'civilisation and progress' to the region. This was seen by many indigenous people as a serious threat. In Sarayaku, however, also other important social changes took place during this decade. Up to then, the governmental representative in the community, the *teniente político*, had always been a nonindigenous outsider, but now a Sarayaku native was appointed to this position. A few years later, the Sarayaku people formed a community organisation with an elected board of directors.

The impetus to form this organisation came from a number of young men who had gone to school in town, where they learned about how other indigenous peoples in the country were organising themselves and trying to defend their rights to their land. In addition, people who had been working on agricultural estates in the coastal region of Ecuador learned how the workers organised themselves in labour unions. Although primarily formed in order to cope with the external threat of colonisation, the organisation also became a forum for regulating internal affairs and resolving internal disputes. After having been ruled by outsiders for one and a half centuries, the Sarayaku people began to take charge of their own destiny. Finally in 1992, after a long struggle, the government granted communal land titles to Sarayaku as well as many other indigenous communities in the province. The territory of Sarayaku was then still intact, in contrast to many other communities closer to roads and towns that had lost much land to settlers.

About 1990 was also when the shift from blowguns to shotguns was fully completed. Most hunters now used more effective cartridge shotguns instead of the old style muzzle-loaders. Hunting with blowguns would now be a futile quest because wildlife was not as abundant as before. The forests adjacent to the village were emptied of many of the most valued game species. Woolly monkeys, the most prestigious prey of them all, maintained breeding populations only in the remotest corners of the community's hunting grounds, and spider monkeys were only occasionally observed – typically young individuals who left the group in which they were born and were now travelling to search for a new group to join. Some of these spider monkey youngsters, however, ended up in somebody's cooking pot before having fulfilled the objective of their journey.

Also the Salvin's curassow (*Mitu salvini*, locally called *pawshi*), a bird with similar habits as the nocturnal curassow but with an even deeper and more enchanting song, was extirpated from much of the community's hunting grounds, as was the tapir. The white-lipped peccary, the only mammal in the region that is a long-range migrant, had disappeared completely, leaving behind just the stories about how once upon a time herds of hundreds of peccaries

FIGURE 4.3 A Sarayaku hunter with a collared peccary.

used to enter right into the village itself. Collared peccary (Figure 4.3) and nocturnal curassow had fared only somewhat better. Many small species, however, remained minimally affected by hunting because they were hunted almost exclusively in the very near surroundings of the village. A growing share of the prey hunted was made up of relatively fast-breeding species, such as large rodents and armadillos, which, in addition, are tolerant to, or even benefit from, forest clearing and farming (Based on Sirén, 2004, Chapter 5).

EMERGING ENVIRONMENTAL AWARENESS

'You shouldn't shoot toucans here near the house', grunted Benjamín, a man in his 50s, to the young boys who had accompanied him on a long, but not so successful, hunting expedition. Although they had been camping out in the most remote parts of the community's hunting grounds where wildlife is abundant, their hunting luck had not been as good as expected. While returning, almost back home again, the boys had encountered a band of toucans fluttering from branch to branch through the canopy in search of fruits, and they had shot a couple of them. Toucans (*Ramphastidae*) have huge curved bills (Figure 4.4), and the larger species weigh about 600 g on average, just barely enough to make the bird worth the cost of a shotgun shell. The taste of soup made with smoked toucan is lovely, however, and as a bonus, its

FIGURE 4.4 A white-throated toucan *(Ramphastos tucanus). Photo: Eriberto Gualinga.*

plumage in black, white, red, and yellow was the preferred raw material for making feather crowns. Benjamín, who had walked a couple of hundred metres before the boys, nevertheless disapproved. 'I also saw the toucans, but I didn't shoot them,' he said, 'because I like to hear their song near my house. The song of toucans make you feel happy. Toucans you should hunt far away, not near the house!'

During the course of the 1990s, many people in Sarayaku became aware of the fact that wildlife resources were dwindling. However, doing something about it was anything but easy. Avoiding shooting small birds with a beautiful song near your own house was, of course, a commendable initiative, but effectively tackling the whole problem was much more challenging.

Many people actually saw wildlife depletion as something inevitable. The only alternative, many thought, would be to stop eating wild meat and instead eat domestic animals or industrial food. And this was not appealing. Although they realised that one day there may really not be any other alternative, the people wanted to continue hunting freely as long as possible. As one man put it: 'Let us first hunt all the animals, and then afterwards, when there are no animals left, we can start to raise animals instead'. The idea to restrict hunting, but not altogether prohibit it, in order to manage the resource base in a sustainable manner, was alien to most people – although not to all. Marlon, for example, a teenage boy studying the last year in the newly founded upper secondary school of Sarayaku, made his graduation thesis about hunting. Marlon proposed that the community could set aside a wildlife reserve where hunting would be prohibited altogether. This, he thought, would secure the survival of wildlife populations in spite of hunting continuing outside the reserve. Marlon would later become part of a new generation of community leaders that emerged in the late 1990s who had much more than basic literacy. They often were fluent Spanish speakers who were well informed about current issues on the national and international political agenda.

Having reached the primordial objective of securing the community's land rights, the focus of the community organisation changed from 1992 and onwards. It became more oriented towards regulating internal affairs and promoting community development. In practice, it was becoming a legitimate local government. All important decisions were made in large community assemblies where all community members had the right, as well as a social duty, to participate. A four-day assembly was organised to formulate through consensus a set of rules to regulate life in the community. The participants were divided into various working groups; the group in charge of natural resources proposed a long list of drastic measures, including restrictions on hunting – most of which received applause and approval in the final plenary session. The next day, however, life continued as before. The children were hungry and the food supplies were empty after four days of sitting still, so many went out to hunt. The community lacked still capacity for enforcement of the rules it had set itself.

DEBATE AND INQUIRY

In 1998, the community embarked on a research effort in cooperation with one of the authors of this book (Sirén, 2004). During the following four years, local hunters provided data to the researcher about the trajectories they walked while hunting, as well as the hunting methods, the location, and the species of all hunting kills. Parallel to this, there were frequent meetings and workshops to discuss the research process and research results, as well as the problem issue itself and how to solve it. It was clear from the beginning that there was no lack of ideas about how to solve the problem of overhunting (Box 4.1).

Almost everybody had an opinion on the issue, but opinions differed and not all ideas were realistic. Some people kept thinking that the solution would be to switch to eating domestic animals and industrial food (but had only vague ideas about how to bring about that change), whereas others firmly rejected this proposition because they thought that eating wild meat was an important part

Box 4.1 Community members' proposals on how to solve the problem of wildlife depletion

- Voluntarily refrain from killing too much: When you encounter three Spix's guans (*Penelope jacquacu*), kill only one. Do not kill pregnant animals.
- There should be a law establishing how much you may kill. Permit killing a maximum of one collared peccary per month per person.
  - *Contrasting view: Here, things are not like in the outside world. Here we are free to do what we want.*
- Reduce the hunting for festivals.
- Quit having festivals altogether.
- Celebrate the festival every 2–3 years instead of every year.
  - *Contrasting view: The festivals are good, they are part of our culture.*
- We should make money and buy food instead of hunting.
  - *Contrasting view: We must not become like the mestizos.*
- Create a reserve where hunting is prohibited.
- Everybody must respect the hunting grounds of others.
  - *Contrasting view: It is not good to exclude others. What will happen to those who do not have any own hunting grounds?*
- Ban firearms.
- Make fish ponds, raise pigs, cows, chickens, and forest animals. Ask for money in order to finance it.
- Establish new settlements in order to ease population pressure.
  - *Contrasting view: If we make new settlements in remote areas, there will be no remote areas left.*
- We should not reproduce too much. Have just one or two children.
- We should talk with each other, especially among young people. (The old people already have their habits and won't change).

Source: Sirén (2004).

FIGURE 4.5 The native Amazonian fauna in the form that you often see it: cut in pieces and dry-smoked over open fire.

of indigenous culture (Figure 4.5). Some people thought that the problem was that the village had grown too large and populated; their solution was to split up the village, creating new villages dispersed all over the community lands, instead of living concentrated in just a small area. Still others thought that this would exacerbate the problem because then there would remain no forest areas at all that were distant from settlements, and thus with low hunting pressure and animal populations almost unaffected by hunting.

The quantitative research results pretty much confirmed what everybody in the village already knew. Some species were extirpated from the vicinity of the village. The yield, in terms of kilograms hunted per kilometre walked near the village was just one sixth of the hunting yield in remote parts of the community lands. In one sense, however, the results seemed to contradict a view put forward by many of the locals. Many blamed the annual community festival for the ongoing depletion of game resources (Figure 4.6). This old tradition consists of four days of celebrations preceded by almost two weeks of preparations, including hunting for food, animal hides, and feathers. Many thought that the intense hunting annually carried out for the festival was a main reason for the ongoing depletion of game resources. The research results, however, showed that the biomass hunted for the festival constituted only 4 percent of the biomass hunted annually. The impression of the festival hunting causing such huge mortality among the wild animals was just an illusion

FIGURE 4.6 Festival chiefs displaying the hides of the hunted prey. *Photo: Eriberto Gualinga* (Please see the colour plate at the back of the book.)

FIGURE 4.7 Everyday hunting often involves hunting relatively small prey species – in this case, nocturnal curassows (*Nothocrax urumutum*) and Spix's guans (*Penelope jacquacu*), birds usually weighing 1–2 kg.

created by festival hunters putting all they had hunted together during a week and a half in one big pile, instead of just consuming it separately in each individual home as is the case with normal everyday hunting outside of the festival hunting (Figure 4.7).

TAKING ACTION

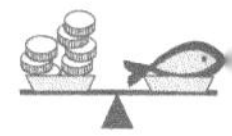

A new phenomenon occurred around the turn of the millennium. Some people from Sarayaku occasionally hunted wild meat for 'export' to other indigenous communities closer to town that had lost much land to settlers and had hunted out whatever forest was left, but still wanted wild meat to celebrate weddings or other traditional festivities. Many other people, however, thought that if even just the consumption of wild meat within the community was enough to threaten wildlife populations, the last thing one should do was send away meat to other communities.

Therefore, the community, in a public assembly, voted for banning this export. Such hunting for export, however, did not cease immediately. Some people either had not been present at the assembly or just did not care about this ban, thinking that it was something that people said but that nobody would enforce. However, this time the community board took effective action. When they heard that somebody was about to go hunting in order to send the meat away, they went to talk to the people in question in their own house, explaining about the reasons for the ban – and this approach worked. Everybody accepted the explanations and the hunting for "export" ceased. Since then, it has sometimes reemerged; sometimes the community board has taken action to stop it, but sometimes not. Because the community board members are changed every two years, the level of enforcement very much depends on the personal capabilities and ambitions of the persons who happen to be on the board at any given moment.

The community also decided to celebrate the community festival biannually instead of annually. Because the biomass hunted during the festival hunting actually is not very large, this may, as mentioned previously, seem to be an ineffective measure. However, looking more closely at the issue, it turns out that festival hunting typically takes place in the most remote forest areas of the community's hunting grounds and has a particular focus on hunting woolly monkeys, which is one of the species most severely affected by hunting (Box 4.2). To specifically protect woolly monkeys, as well as a few other rare species, restricting the festival hunting could potentially be a very effective measure.

Apparently, the concern of the people here was not so much the maintenance of the resource base for hunting overall, but rather the specific conservation of one or a few very threatened species. So far, this measure has not proven very effective. Although festival hunting is hard, it is also great fun, and many young men just love to participate. Therefore, as the frequency of festivals was halved, the number of festival hunters at each festival has instead more than doubled – and so has also the number of Woolly monkeys hunted for each festival (Sirén, 2012). The debate about the community festivals continues.

The community also set aside a wildlife reserve in 2003, as the young boy Marlon first had suggested a decade earlier. This reserve covered some

Box 4.2 Woolly Monkey Hunting

Hunting in Amazonia is usually a lonesome activity. The more people you have with you, the more noise you make. Plus, you usually find no more than one animal at a time anyway, with just a short moment to shoot before the animal flees, so there is not much to be gained from hunting with other people. Of course, there are some exceptions – one of them being the hunting of woolly monkeys, which is not only a way to acquire food but also an appreciated social activity (Figure 4.8).

The meat of woolly monkeys is highly esteemed, especially during the fruit season (March–April) when the monkeys build up their fat reserves. Hunting woolly monkeys also carries some prestige because it is considered to be difficult. In reality, finding woolly monkeys is much easier than finding other animals. Ground-living animals can be difficult to detect through the dense underbrush, but woolly monkeys are easy to spot up in the canopy. They live in large groups that travel around in the canopy, so the sound of branches bending under their heavy weight and the vocalisations they make to keep the group together can be heard from far away. When you hear a group of woolly monkeys, you should quietly approach them, spreading out such that each one in the group tries to find a different monkey to aim at (preferably a large male).

Once the first shot is fired, the monkeys flee. The hunters then each try to follow one monkey. The monkeys usually slow down soon enough, such that they usually can be shot. The difficulty of hunting woolly monkeys lies, for one part, in getting to where the woolly monkeys are; nowadays, they are found only in the remotest corners of the community's hunting grounds. Also, the woolly monkeys usually are so high up in the canopy that they are close to the limit of the reach of the shotguns employed for hunting. Unless you shoot a monkey in its face or chest, a gunshot does not kill it; it just leaves the monkey wounded. Despite all of the prestige that surrounds woolly monkey hunting, some hunters today have mixed feelings about it – not only because they have realised that the woolly monkeys are threatened, but also because of their human-like appearance and behaviour.

50 square kilometres. One of the hamlets, Chuntayaku, was in charge of the practical management, whereas the community board took care of the administrative issues and funding was provided by a German nongovernmental organisation. The project funds were used to pay for permanent surveillance by two reserve guards. One more reserve was founded in 2005, by the people in another hamlet, Shiwakucha. This reserve, however, has been laden with conflicts and problems since the beginning. The area that became a reserve had been used as hunting grounds by people from two different hamlets. When one of the hamlets created the reserve, many people in the other hamlet felt that they had captured both the land and the project funds for themselves in a way that was unfair to those in the other hamlet.

Opinions about the effects of the reserves have been mixed. Rumours about encroachment into the reserves have abounded. Sometimes, the reserve

FIGURE 4.8 Sarayaku hunters with Woolly monkeys. *Photo: Eriberto Gualinga.*

guards have expelled hunters from the reserve, but at other times the guards themselves have been accused for clandestinely hunting while on duty. On the other hand, there are also reports about a remarkable recovery of wildlife populations within the reserves. In particular, the abundance of collared peccaries and tapirs was reported to have increased rapidly. Groups of up to 30 woolly monkeys also have been reported in these areas, where they previously were absent.

The people of the Chuntayaku hamlet then made yet another brave move. They decided, unilaterally, to ban all hunting of tapirs. The ban concerned all inhabitants of Chuntayaku, although not the inhabitants of the four other hamlets even though their respective hunting territories partly overlap. Considering the high value of the meat of an entire tapir, it was indeed impressive that they had the courage to make this decision.

Perhaps this decision had to do with an experience a few years before in another part of the community (although nobody really knows for sure). A tame female tapir, caught in the forest as a baby, was tagged in the ear. She was gradually reintroduced into the wild, in a forest area near the village where all tapirs long ago had been hunted out. Seeing the tag in the ear, no hunter shot at this tapir. After some time, she began to take long hikes away to forest areas where she could meet wild male tapirs, but then she always returned to the area where she had first been released and received some feed until she had adapted completely to the wild. Every two years, she gave birth to a baby tapir. As long as they stayed in the same area, people were reluctant to shoot them, so gradually the tapir population in the area increased again.

After the Chuntayaku people had set aside their reserve and banned tapir hunting altogether, sightings of tapir tracks near the hamlet became more and more common. One day, a tapir appeared right by the house of a man in his 60s, who did what he always would have done – he shot it. His joy came to an abrupt end, however, when the rest of the people in the hamlet learned about what he had done, got angry at him, and collectively decided that the meat must be shared, in equal pieces, among everybody in the hamlet. After that, nobody in Chuntayaku hunted any tapir for years. However, they sometimes grunted that it was not fair that they abstained from hunting tapirs, while the people from other hamlets continued to hunt them.

A couple of years later, the community got funding for an environmental management project. As part of this endeavour, there was again a series of workshops and a meeting that ended in a series of regulations concerning natural resource use. Among these was a total ban on hunting tapirs in the whole community territory. Again, however, enforcement was lacking, and most of the regulations remained as paper products without much impact in real life.

Somehow, however, the ban on hunting tapirs had been remembered by many. People hunted just as many tapirs as before, but this did not result in admiration anymore. Hunting tapirs resulted in nasty comments behind the hunter's back and sometimes direct complaints in community assemblies. Complaints without exception were met by counterarguments: 'It was not within the reserve', 'It was in my own hunting grounds', 'I hunt whatever I want, nobody has the right to prohibit that'.

This cycle continued, year after year, until 2011. The people now obviously wanted discipline and order. They elected a community board consisting of people determined to bring order and who were not afraid of confrontation. During their two years in office, the new board fined over 60 persons for different offences, including not only unlawful hunting, but also stealing, spreading rumours, fighting, and selling alcohol. Initially, this angered many of the affected, who were not used to such strict application of disciplinary measures. The community president, José, however, would then just respond, 'Criticise me if I do this alone', alluding to the fact that he only enforced what the community members had decided jointly in the community assemblies held every month. He felt that he had firm support from the great majority of the community members.

Among those who were fined were four persons who each had killed one tapir during the nine months before this chapter was written, constituting a sharp decrease from the 20–40 tapirs that previously used to be hunted annually in the community. Three of the hunters were fined $500 each. Because they had no money to pay, they paid instead by working for the community in various types of construction and maintenance work, such as keeping trails and bridges in shape. The young boy mentioned in the beginning of this chapter, however, was given a fine of only $200, considering that he was the first person ever who had been fined for this offence and previous offenders had been left in impunity.

FIGURE 4.9 Hunters and a white-lipped peccary.

Another interesting development was the return of the white-lipped peccaries (Figure 4.9). This species had been virtually absent in the area in the mid-1990s but then started to recover. In 1999–2001, it constituted four percent of the total hunted biomass in the community (Sirén et al., 2004), and by 2008–2009, the harvest of this species had increased sixfold, making it the most hunted species of all, contributing to no less than 23 percent of the total hunted biomass (Sirén, 2012). This occurred even though there were no specific conservation measures specifically targeted at this species. Because it is migratory and roams far distances, the relatively small wildlife reserves in the Sarayaku territory hardly could have much effect on it. Nobody really knows for sure why this species disappeared and then returned. Nature is unpredictable.

Chapter 5

Fishing in and Fishing out the Amazon

A BOUNTY IN THE RIVER

'There is plenty of fish for everyone out there', says Ricardo, gazing into the distant treeline that draws a sharp green stroke between the brown water and the blue sky in the horizon. He rests his hand on a metallic rudder and stays quiet. In his 40s, Ricardo is in charge of *Mi marido* ('My husband'), a 15-m wooden vessel carrying an eight-man crew (Figure 5.1). The men are fishers hired from the poor riverine neighbourhood of Belén in the outskirts of the vibrant Amazonian city of Iquitos, capital of the Loreto region (Figure 5.2). The fishermen work for one of the city's ship owners, entrepreneurs that equip boats and crews to supply the urban markets of the region's population centres. The owner of *Mi marido* has several similar boats operating not only from Iquitos but also from Pucallpa, another Peruvian rainforest city a few days up the Ucayali River.

Now *Mi marido* has left behind the fishing port teeming with boats and fishermen and is heading to the main channel of the Amazon River. The boat's final destination is in the mouth of the Shishita, a small tributary of the mighty Amazon some 200 km, or almost 20 hours of journey, down towards the Brazilian and Colombian borders (Figure 5.2). The crew has plenty of time to fix broken meshes and to place lead weights and floaters into net linings (Figure 5.3). Preparing the gear is of utmost importance – it is the time of the *mijano*, migration of fish, during which great catches are the rule. These famous migrations are triggered by seasonal changes in water levels; as the fluvial dynamics blur the line between the terrestrial and aquatic worlds, fish also move from small tributaries to the main rivers to breed, and then from the river channels into the inundating forest in search of food and shelter, and then again from the forest back to the rivers as the floods retreat. Different species and different areas have distinct migration patterns and timings, but most of the important species do migrate; some large Pimelodidae catfish species even migrate thousands of kilometres (Cañas and Pine, 2011).

Although the schools of some of the fish species hide among the submerged vegetation across the floodplain and are thus difficult to catch, others migrate in great numbers and fill the river channels, stuffing the professional fishermen's coolers. Now the waters are rising, and large areas on both sides of the Amazon and its tributaries are flooding. This is a signal for many important characid

Diagnosing Wild Species Harvest. http://dx.doi.org/10.1016/B978-0-12-397204-0.00005-X

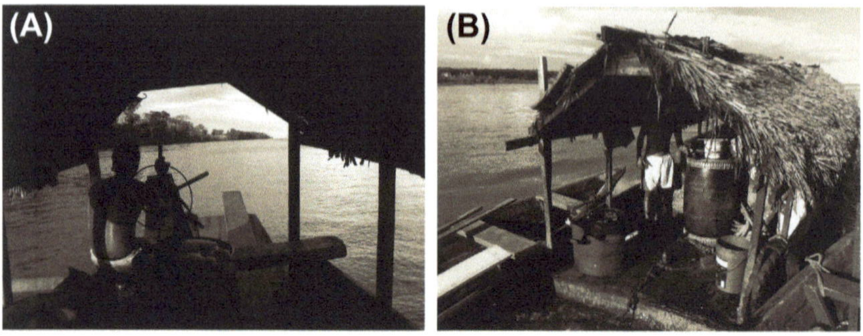

FIGURE 5.1 (A) *Mi marido* on its way down the Amazon River. (B) Palm-thatch roofs and simple open galleys characterise the Iquitos commercial fishing fleet. The auxiliary canoes on both sides of the boat decrease its velocity but are essential for the fishing itself.

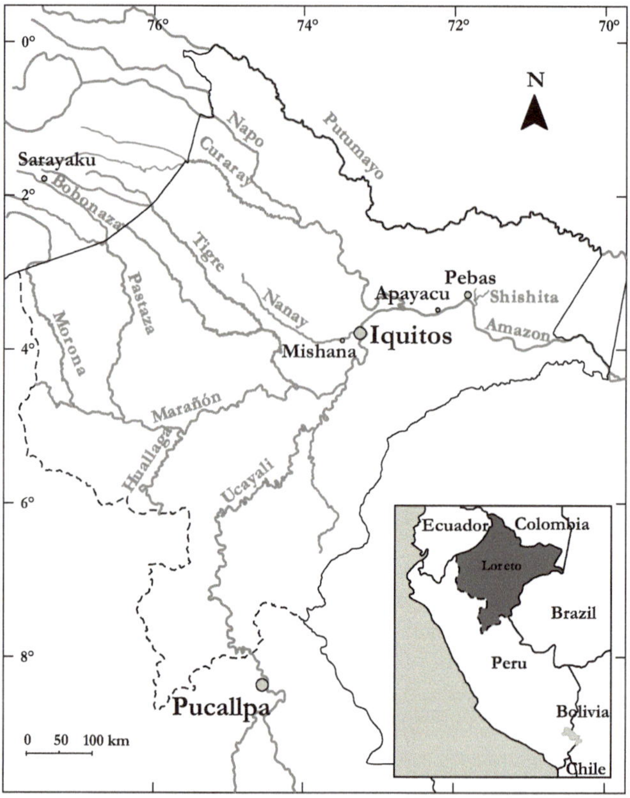

FIGURE 5.2 Map of northern Peruvian Amazonia and adjacent parts of Ecuador, with places mentioned in this chapter indicated. Fish is transported to Iquitos and other population centres in the region over long distances. Sometimes, the fishing teams spend two to three weeks out in the river. During that time, fish is stored with ice.

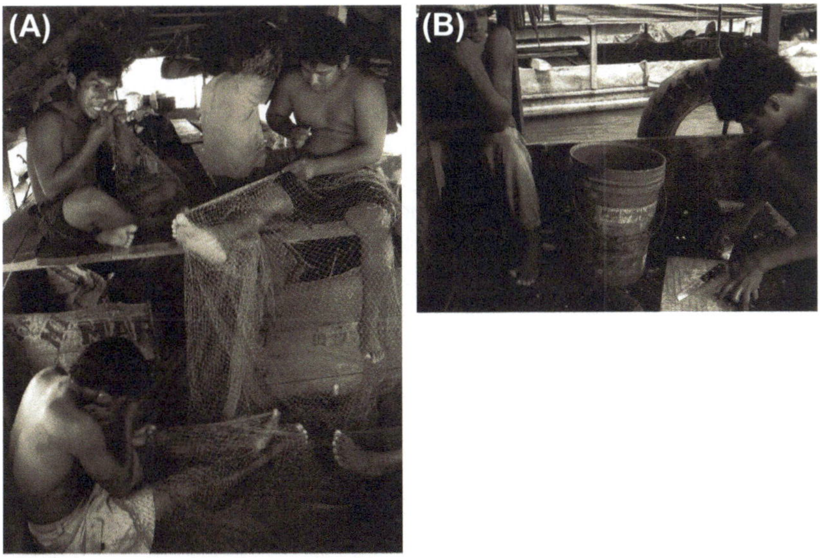

FIGURE 5.3 (A) Purse seine nets must be in good shape when fishing starts; sometimes hundreds of kilos of fish are caught in one single try. (B) Life on board is sometimes monotonous, fishing for work and fish for food. Please see the colour plate at the back of the book for (A).

species (Characidae), such as the black prochilodus *(Prochilodus nigricans)* – in Peruvian Amazonia known as the *boquichico* (Anderson et al., 2009). When the rivers rise, large schools of these fish swarm from the black water tributaries into the sediment-laden turbid main rivers to breed (de Jesús and Kohler, 2004). Many fish, just like the black prochilodus, lay eggs during the rising water levels in order to make the most of the flow of water into the inundating forest that teems with feed and provides safe habitats for the larvae.

Although the seasonal floodplains cover only a minor portion of the whole Amazon Basin, roughly 12 percent of Peruvian Amazonia experiences floods every year. In many parts of the lowlands, the inundated areas extend dozens of kilometres from the main rivers, and dry land seems to disappear altogether for months. The resulting floodplain forests are extremely important habitats for many fish species, which feed on invertebrates, leaves, fruits, and seeds that are found in abundance in the forest invaded by the rivers. Where fish are plentiful, people also take their share of the abundance; young and old, men and women alike are seen angling or fishing with nets in their dugout canoes below the branches of the trees and shrubs that drop their fruits straight into the mouths of the fish waiting in the river.

In Loreto as elsewhere in Amazonia, the majority of the total fish catch is consumed directly in rural areas. In the region, 90 percent of the rural population depends upon fish as the main source of protein intake (IIAP, 2009). However, the region's rivers also are a vital warehouse for the nearly half million inhabitants of Iquitos, where the urban demand is attended by commercial fishers such as those on board *Mi marido*. The commercial fishing fleet contributes about a quarter of all fish landings in Loreto (García et al., 2009). In terms of landings volume, the black prochilodus, or *boquichico*, is the most important species. It is sold in abundance in the markets of Iquitos, a city consuming up to 40 tonnes of a great variety of fish every day (López Ríos, 2010).

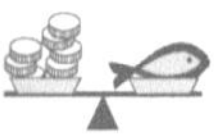

Despite the high demand and consumption, in Peruvian Amazonia all commercial fishers operate on a rather small scale and use simple technology. As an activity, however, fishing forms one of the most visible extractive economies, not only in Peru but all over Amazonia. Rustic coolers stuffed with ice and frozen fish are a common sight in every port along the Amazonian waterways. Being widespread and vitally important for a great proportion of the total human population of the region, fishing also sustains a series of other economic activities linked to transport, boat building, fish handling and processing, as well as all kinds of commerce and services.

Indeed, fishing has formed a cornerstone of Amazonian economies for millennia. Old chronicles mention how the native Amazonians were blessed with rivers teeming with all kinds of fish and other wildlife for food, while the Europeans were often helpless and did not know how to capture these creatures (de Carvajal, 1543; Hemming, 2008). Compared to earlier times, the demand for fish today keeps growing at an unprecedented pace, along with ever more efficient and less selective fishing methods. This means that many of the most important commercial fish species are commonly captured before they have reached reproductive maturity (García et al., 2009). On the other hand, minimum catch-size limits, if very near the size of reproductive maturity and strictly followed, may also have an unexpected effect: if only the mature and efficiently reproducing fish are caught, this may reduce the spawning result and create a selection bias that favours smaller fish.

Overfishing changes the structure of the fish fauna in Amazonian rivers. Currently, there are visible signs of the infamous phenomenon of 'fishing down the food web', meaning that the largest species from the top of the food web are depleted, thus shifting the fishing pressure successively down the web to smaller species (García et al., 2009; IIAP, 2009). This is a tragic development because, as a whole, many Amazonian fisheries seem to be underutilised (IIAP, 2009): while more fish of many lower-value species could be caught, the most wanted species are driven toward local extinction. In Peruvian Amazonia, particularly the highly valued large species such as *gamitana (Colossoma macropomun), sábalo* (*Brycon* spp.), and the red-bellied pacu (*Piaractus brachypomus*) have suffered from excessive fishing during the reproductive migrations. In addition, the habitats of these fish are often threatened as floodplain forests are logged or converted to seasonal pasture or agriculture. Less trees offer less fruit for the fish and, vice versa, as a result of overfishing, there are less fish to disperse the seeds of the fruiting trees left standing.

THE COSTS AND BENEFITS OF LIFE ON BOARD

The owner has equipped *Mi marido* with fishing gear, fuel, food, ice, and insulant material, hoping to see his crew back in the harbour after two weeks. The boat's cargo hold should then be filled with eight tonnes of a mix of frozen fish and ice. In Peruvian Amazonia, few boats exceed the storing capacity of 10 tonnes and practically none use advanced fishing or cold storage technology. Some boats have proper isothermal boxes, but many others, like *Mi marido*, just use their plain cargo holds to store the fish. The boat is packed with purse seine and gill nets, but most of its hold is filled with sacks of ice-bars. This essential product has been brought into the boat directly from the boat owner's ice factory; the 250 twenty-five-kilogram sacks make 6.25 tonnes of ice, and the load is completed by 50 sacks of rice husk used as insulant between the precious white bars and the hot tropical air. In Amazonia, rice husk is commonly used as insulant, instead of the sawdust that is common in many other parts of the world: 'This is because most commercial timber species in Amazonia contain aromatic oils that would affect the smell of the fish', Ricardo explains.

Mi marido pushes slowly down the river. When the boat approaches the riverbank, the drowsy put–put of its 24-hp motor reflects from the lush vegetation. Children play on inundated football fields and many houses can only be reached by canoe. *Mi marido* drags two large wooden canoes tied to its rear. These so-called auxiliary boats are used in the fishing itself, while the main vessel stores the catch and forms a base camp for the crew. The galley of the boat is found in the rear; equipped only with a single shelf loaded with salt and onions, it is just an open space near the water level, accommodating the bottom of an oil barrel used as a fireplace. Towards the bow, after an open space where the motor roars, the boat's cargo hold is covered with wooden boards forming a floor. A ridge roof made of palm thatch protects the crew from the burning sun and the violent tropical rains, but it is so low that below it a grown-up person can only move well when bent down. The floating dwelling is home to the crew for the next 10–15 days, and the men accommodate their few belongings in the edges of the floor or hang them from the ceiling. Sometimes, they stay out in the river for almost three weeks.

By nightfall, the boat reaches the village of Apayacu. In dim moonlight, *Mi marido* approaches the riverbank, where it is tied to a branch of a small tree. The fishers extend blankets over the boards of the floor and hang mosquito nets to cover their hard berths. A simple wire running below the apex of the ceiling connects a car battery to a couple of light bulbs, creating a cosy atmosphere. Soon the lights are turned off and everyone holes up in their mosquito nets. One of the younger men listens to music from his cell phone, but soon the rhythms of *cumbia* cease, leaving the stage free for the choir of insects, nocturnal birds, and frogs that surround the boat in the darkness.

A couple of showers fall during the night but the dawn finds the boat already heading down the Amazon River, with the first rays of sunlight penetrating the forest foliage and raising tatters of mist over the canopy. There is still a few hours' journey ahead before the boat reaches the mouth of Shishita, and the young brothers Jaime and Rither kill time trying to capture a dark hairy tarantula they have found inside a car tyre functioning as a fender in the rear of the boat. They manage to catch the spider into a bucket and boast it will bring them good money back in Iquitos. Whether or not that will happen, the boys need extra income because the fishers' pays are generally low, and the youngest and least experienced crew members are in the low end of the wage scale.

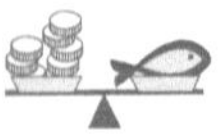

The boat's owner has to take into account several other costs apart from the fishers' salaries. Depending on several factors including seasonality, boat and motor size, distance, and flow of river, up to half of the operational costs of a fishing trip consist of fuel (IIAP, 2009). Normally, ice makes up another large share, commonly around a third of the total costs of a trip. In the case of *Mi marido*, this investment is under the ship owner's personal control because he gets the ice from his own factory. Overall, it is common wisdom that although any job might be better than no job at all, commercial fishing hardly manages to improve the living standards of the majority of the fishermen themselves in the long term; the commodity chain suffers from informality and intermediary merchants many times take a lion's share of the profit made (IIAP, 2009). However, other paid work is often scarce and fishing employs people throughout the year.

CONFLICTS AND COMPETITION

While contemplating the river and its edges, it seems that everybody is fishing here. Pink Amazon River dolphins *(Inia geoffrensis)* surfacing not far from the boat share the river not only with yellow-billed *(Sternula superciliaris)* and large-billed terns *(Phaetusa simplex)*, ospreys *(Pandion haliaetus)*, ringed kingfishers *(Megaceryle torquata)*, snowy egrets *(Egretta thula)*, and great egrets (*Ardea alba*), but also with local fishers sitting in their small dugout canoes with nets or lines and hooks. The riverside people, *ribereños*, are the champions of this semiaquatic landscape. Their culture is a mixture of indigenous and colonist influence, adapted to the drastic changes that take place every year in their environment. Fishing is a safety net and fish are a staple for them. Therefore, potential for conflict exists when commercial fishers, such as the crew of *Mi marido*, meet local dwellers who depend on subsistence fishing.

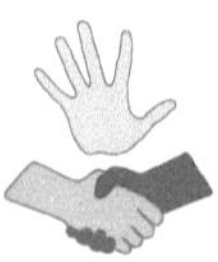

Laws and regulations concerning fishing in Peruvian Amazonia are often difficult to enforce in practice. To make things more complicated, the basis of the fisheries legislation to be followed in Peruvian Amazonia is to a large extent based on the nationally important ocean fisheries (IIAP, 2009). This means that the particularities of the Amazonian environments and fisheries are only superficially addressed. As all natural resources, in Peru fish resources constitutionally belong to the state. They can be exploited upon temporal licences, authorisations, permits, and concessions granted by the state, with associated fees to be paid. All commercial fishers should thus operate under a degree of formality and monitoring. This is only in theory, however, because authorities lack the needed resources to enforce the formal regime and collect and verify information about the fleet, effort, and catch (de Jesús and Kohler, 2004).

The role of active community-based measures has increasingly stood out in relation to the deficient role of fishery authorities (García et al., 2009). *Mi marido* is a 'freezer', a commercial fishing boat of which some are infamous for their habit of entering small rivers and lakes in community lands and, according to many, leaving behind empty waters and hungry stomachs. The problems have been exacerbated by the fact that some of the commercial fishers use all imaginable means to fill the coolers of their boats with fish: in addition to efficient purse seine nets and gill nets, they sometimes also use explosives and poisons that indiscriminately kill all kinds of fish. These practices are illegal, but their control is difficult (Box 5.1). The crew of *Mi marido* unanimously affirms that they stick to legal methods, and according to them nets are efficient enough for them to accomplish the mission and get back home in two weeks with the cargo hold filled. Many local communities, however, have encountered different attitudes and found the only hope was to close access to the most important water bodies in their control. This strategy has been increasingly successful (García et al., 2009), but it has also involved a level of risk to the physical safety or occasionally even to the lives of local community members involved in control activities.

The boat makes a short stop in the small riverside town of Pebas, where Ricardo buys nails and a knife. The port of Pebas teems with boats and canoes, many transporting frozen fish. After two more hours down the river, *Mi marido* finally reaches the mouth of the Shishita River. The boat is anchored next to a half a dozen similar vessels that are already feasting on the offerings of the river. The place is dominated by a small sawmill; during the low water season, it processes timber logged in the forest concessions up the Shishita. Now it is out of function; but although the rising Amazon has flooded the mill and its circular saws lie idle, the air is thick with a strong timber scent. The barefooted boys find a killed pit viper *(Bothrops sp.)* upon a pile of sawmill waste. The place seems like a good base for fishing: at least firewood is abundant for the boat galleys.

The fishers do not want to waste time; the eight-tonne goal means full days of hard work. So Rither, Ricardo, and Milton board one of the two auxiliary canoes with a 75-m long and 2-m wide four-inch-mesh gill net, and they head to the river. When they reach a place just a couple of hundred metres upriver from the camp and another 100 m from the riverbank, Milton leaves an empty plastic bottle floating in the river, with one end of the net tied to its neck. Rither then uses the canoe's 15-hp outboard motor to steer it back towards the riverbank and again downriver while Milton keeps releasing the net into the river in such a way that it finally forms something like a gentle letter J (Figure 5.4).

There are several other canoes bordering the riverbank, all with fishers using their gill nets along the shoreline. The nets and the canoes do not get entangled

Box 5.1 Fear of Hostile Fishermen

Mishana is a small settlement located on the banks of the Nanay River, a few hours from the river mouth near the city of Iquitos (Figure 5.2). People in Mishana make their living by subsistence agriculture, extraction of forest products, fishing, and hunting. The latter activities also generate cash income, yet the scale of these economies is low. Some Mishana dwellers are also temporarily employed by the local authorities or tourism companies. Near the community there is an oxbow lake formed by an old bend of the Nanay River, one of the traditional fishing areas of the community.

In 1999, a protected area, the Allpahuayo Mishana National Reserve, was created including the lands of Mishana. Consequently, the early 2000s saw increasing collaboration between the local communities and researchers and resource management professionals from Iquitos and other parts of the world. Local resource use was a particular topic of interest for all parties. During one of the meetings in Mishana, an elderly woman described a severe problem the community members had suffered increasingly often – causing even fears of threat to the community lifestyle.

The problem she described was the commercial fishers who suddenly and frequently appeared in large boats and entered the lake. The armed invaders behaved in a hostile manner and refused to speak with anyone. They fished with explosives, poisonous barbasco root, or pesticide. When they collected the killed or fainted fish of commercial value, they just stuffed them into the cargo hold coolers and sailed away. Only after they had gone, people from the village dared to return to the lake to see what had happened. Dead fish of the smaller species were floating all over. As a consequence, the local fish populations were heavily reduced by this method of fishing and the people of Mishana were left with mixed feelings of anger, fear, and helplessness.

However, there were ways to fight this kind of exploitation. After obtaining formal land titles, Mishana and a number of other communities in the Nanay River started to gain more control over many of their lakes, in part due to internal self-organisation and in part with the backing of the authorities.

because the flow of the river gives them all the same speed. When two river dolphins approach the net, Rither angrily yells at them, saying they should leave the nets alone. He claims that the dolphins do not get entangled in the nets but that they compete with the fishers over the same prey. The dolphins could not care less about a man shouting in a canoe, and maybe Rither's anger is also just a piece of theatre; traditionally Amazonians have left the river dolphins alone, out of cultural respect for the species. Recently, however, there are increasing reports from Brazil (Loch et al., 2009), as well as from Bolivia, Colombia, and Peru (Aliaga-Rossel, 2002), of dolphins that have been entangled in nets being killed and their meat being used as fish bait.

In practice, however, the most serious competitors of fishermen are other fishermen. In addition to direct competition for the fish resources, theft of fishing gear is a common problem. This time, the danger nevertheless turns out to be something

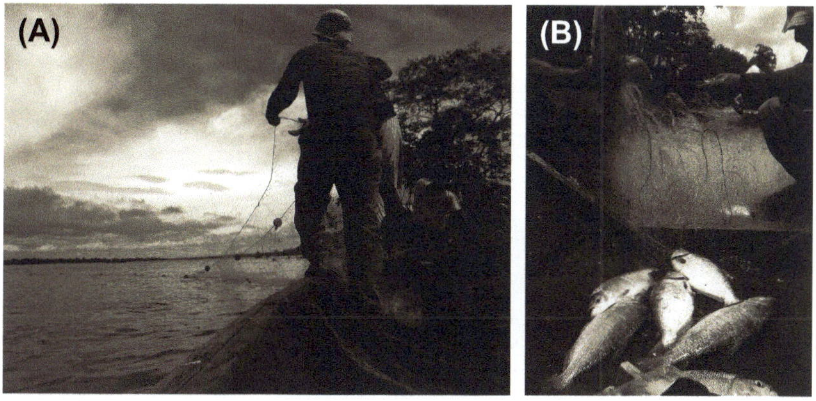

FIGURE 5.4 (A) Gill nets can be efficient in catching larger fish. Sometimes, however, fish seems to be scarce. (B) *Sábalos (Brycon* spp.*)* are high up in the list of the most preferred catch.

else. The net has been in the water only for about 10 minutes when another canoe pushes up the river and approaches the upper line of the net marked by floaters. First, the fishers do not pay much attention to the other canoe. These crossings happen all the time. The normal way to avoid getting the propeller entangled in the net is just to lift the outboard motor at the right moment. The typical Amazonian small motors called *peque–peque* are particularly easy to handle because their propeller is at the end of a long shaft that can be lifted with little effort.

This time, however, the net's lead weights have not been well placed. The net does not hang down from the floaters but rather lies almost horizontally near the surface – something that has gone unnoticed. Thus, when the man in the passing canoe puts the propeller of his motor back into the water after crossing the floaters, it immediately catches the net and entwines it around the motor's shaft. Ricardo bottles up his bitterness in an admirable way when dealing with the two men that have destroyed his gear. After a short conversation, three small hands of plantain change owner as compensation. In sum, the first draw has yielded in 15 kilos of plantains and two small *boquichicos* – the first fish, and quite a modest catch considering the eight-tonne goal.

While Milton and Rither try to recover what can be saved from the damaged part of the net, the others remove the lead weights and floaters from the intact part of it and stick them into another larger gill net. In the boat next to *Mi marido*, the crew is depositing a catch in the boat's hold (Figure 5.5). Two fishermen sit in a large canoe and select the fish to be taken in from those that are too small. The bigger fish are left on the bottom of the canoe while the smaller ones are tossed back into the river. Jaime watches the operation while gutting a few *palometas* (*Mylossoma* sp.) for lunch. As the youngest crew member, he is confined to the galley with the task of preparing a simple fisherman plate of boiled manioc and plantain accompanied by fish, boiled as well – all seasoned with abundant salt. Commenting on the small fish returned to the

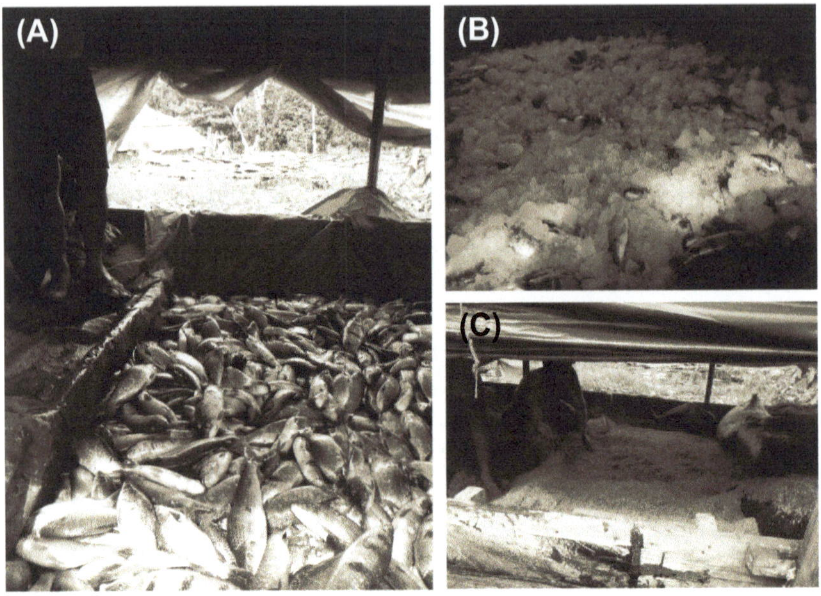

FIGURE 5.5 (A) This vessel is steadily approaching its catch goal of five tonnes. The fish is stored in layers: (A) fish, (B) ice, and (C) rice husk.

river, Jaime says that they are released live so that they can keep growing. Many, however, stay on the surface with their white bellies up.

DIVERSITY OF SPECIES AND TECHNIQUES

In Peruvian Amazonia, it has been estimated that up to 200 fish species have commercial potential for human food, but only some 65 are currently landed in the main ports by commercial fishers; of these, just a handful of species make up more than 90 percent of the catch (García et al., 2009; IIAP, 2009). The total number of fish species in the Amazon Basin is unknown, but the number might be up to 3000. *Mi marido*'s neighbours' catch includes many small *lisas* (*Leporinus* sp., *Schizodon* sp.), *yulillas* (*Anodus* sp.), *palometas,* and some *panshinas* (*Pellona* sp.) that are mostly thrown back to the river (Figure 5.6). The bigger fish that are carefully separated on the bottom of the canoe are mostly *boquichicos*, *sábalos (Brycon* sp.*),* and *tucunarés (Cichla monoculus)*. These fish are then deposited in the boat's cargo hold in layers among crushed ice, which comes from large bars that one of the fishers keeps pounding with a heavy wooden club in the rear of the boat. The fish and the ice layers are then veiled by blue plastic, which is then covered with rice husk; this again is lined with plastic, more fish, and so on. The fishers say that in five days the crew has caught two tonnes. Another three tonnes are still to be caught before the boat can head back to Iquitos.

FIGURE 5.6 Some of the common catch species. Top left: *palometa*, Middle left: *lisa* (*Schizodon* sp.), bottom left: *panshina* (*Pellona* sp.); right: *yulilla* (*Anodus* sp.).

Fishers' days are monotonous, and they stay out in the river for long periods of time. This means that a large part of the catch is stored in ice for up to two weeks or even more before the fish is sold in Iquitos or other regional centres. This obviously affects the quality of the product, not to mention that the ice itself is seldom produced using purified water, further increasing the risk of bacterial contamination. Because of deficiencies in the commodity chain, fish from Peruvian Amazonia rarely qualifies for external markets but rather is consumed within the region. This situation is somewhat different when moving down the Amazon through Brazil towards its estuary, where much larger boats are working with more advanced capture and storing technologies.

These conditions, though, are just dreams in the mouth of Shishita. Now it is Jorginho, Félix and Víctor who board the small canoe with the larger gill net. The 75-m net's width is 4 m, but the doubled effort does not lead to an improved catch – not a single fish is caught during the 20-min attempt. The third time looks as bad as the second until the last few metres of the net are dragged out of water, revealing seven beautiful *sábalos* wriggling in the air. This species is popular in Iquitos restaurants and is apparently high up in the list of the most wanted catch. However, most fish forming this *mijano* are *boquichicos* concentrated in the mouth of the Shishita River to spawn so that the offspring then will enter the floodplain forest along the rising waters. The fishers say that San Pedro de Shishita is a good place to fish both during the rising-water and the descending-water seasons. Between the peaks of fish here, they work elsewhere where other migrations bring abundant fish, in places like the Curaray River west of the Napo or on the lower Ucayali River. There is always somewhere to go and fishing is one of the prime employers in Amazonia, both in the rural areas and in the cities, where many young men are unemployed or suffer from serious subemployment. It is easy to hire them as commercial fishers; although the wages are low, no formal education is needed and there is work practically all year round.

Discouraged by the minimal catch, Ricardo decides to change the method (Figure 5.7). Jorginho, Víctor, Milton, and Ricardo now board a canoe, steering it into the black waters just inside the mouth of the Shishita. A purse seine net lies as a huge green pile on the bottom of the canoe as the four fishers enter the river, slowly observing the two other groups of fishermen in canoes floating near the riverbank and cleaning their nets. As the canoe slides through the calm black water, it passes near a gill net left there by one of the other groups of fishers, and Ricardo cannot resist checking whether his colleagues have been lucky or not. He lifts the net and takes a closer look at the handsome *sábalos* entangled in its meshes before letting the trap with its fish back into the water.

The men sit in their boat for several minutes just waiting for the right moment and the right spot to make their move. Then, all of a sudden, Jorginho takes one end of the net in his hand and jumps into the water with a great splash. At the same moment, Milton starts the motor and the canoe launches forward with great noise. Víctor keeps passing the net into the water as fast as he can while the canoe draws a large circle in the surface of the river. Meanwhile, Ricardo repeatedly splashes the water with a paddle to frighten the schools of fish into the open water rapidly encircled by the net. The circle closes in a minute, and Jorginho, still swimming in the river, hands the end of the net to Víctor. Now, the seine net forms a complete circle in the river, with the upper line of the net kept on the surface with floaters and the lower line drawn down by lead weights.

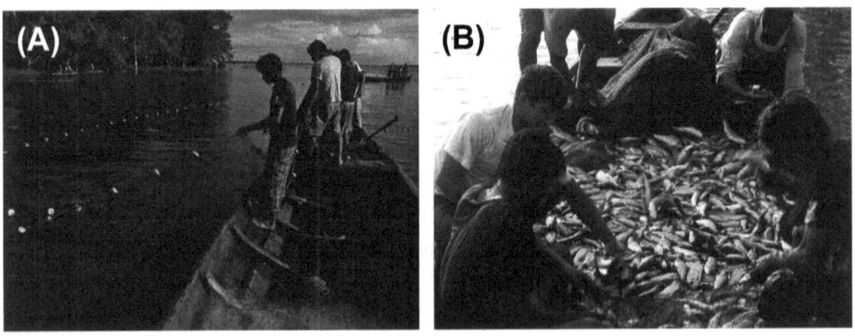

FIGURE 5.7 (A–B) Just one attempt of the purse seine net in the mouth of the Shishita River resulted in an abundant catch of *boquichicos* and other common fish.

While Jorginho climbs agilely into the canoe, Víctor and Milton start pulling the net back on board. They first draw in only the lower line, tightening the bottom of the net and ultimately closing it altogether. Thus, the net now forms a huge purse that prevents fish from fleeing. Only the smallest ones escape either through the mesh of the net or by jumping out of the water and crossing the net in the air. The purse is now closed from below, and the men start to draw the upper line into the canoe, too. While the purse gets smaller and smaller, the catch concentrates into one glittery tumult of fish. Finally, Jorginho, Víctor, and Milton manage to drag the bag full of fish into the canoe's bottom.

The hundred-kilo mass of thrashing and flopping fish must then be sorted, separating the larger fish that qualify for sale from the majority of them that are too small to be sold and thus are worthless. Most of the catch consists of *boquichicos* and *palometas,* but the men also pick up some *sábalos* and *tucunarés*. When the valuable fish lie in a pile in one end of the bottom of the canoe, the men are all covered in shiny scales. They now overturn the net back to the river, liberating a great volume of small fish – some live, some dead. While the net is cleaned of floating plants, pieces of wood, and other litter, many of the small fish floating belly up in the river start moving and finally plunge into the relative safety offered by the black water.

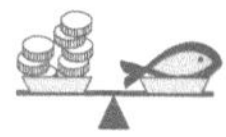

The Sun is about to set, and two large macaws fly over the campsite heading east. The beautiful nightfall is in contrast with the looks of the sawmill. On an inundated volleyball field, black vultures *(Coragyps atratus)* fight over fish guts and whatever they can find while a lonely dog wanders on a mountain of saw timber waste looking for something to eat. The full moon rises over the sawmill and distant thuds of ice being crushed in a nearby boat mix with a loudly crackling radio playing *cumbia*. There is still a long way to go and long days of work to get *Mi marido* back home.

A MIGRATORY RESOURCE UNDER INCREASING PRESSURE

In Ecuador, some 500 km upriver from where Ricardo navigates *Mi marido*, Emerson stands in the Bobonaza River, watching with amazement a strange fish he has just harpooned. Although he has been fishing on an almost daily basis, even before he reached the age to start school, he has never seen anything like this fish. The fish weighs several kilograms, and unlike most large fish species he is used to, it is disc-shaped rather than elongated. Emerson understands that this is the fish that the elders call *ruyak gamitana* in their native language, kichwa, but he does not first realise that it is actually the same species recently brought as fry to the community of Sarayaku by aquaculture extensionists who call them *cachama blanca* in Spanish. Emerson is in his 20s, and few in his generation have ever caught this kind of fish in the wild. The elders, however, tell that this fish, the red-bellied pacu *(Piaractus brachypomus)*, was very common still in the 1930s, as was its relative the black pacu *(Colossoma macroporum)*. But they both decreased little by little and finally disappeared completely except for very occasional observations, such as this catch by Emerson (Figure 5.8).

In daily speech, the people in Sarayaku refer to the Bobonaza River simply as *Hatun Yaku*, meaning 'Big River', as it is the biggest river flowing through the community's lands, reaching up to 40 m width in some places. In the context of the Amazon Basin, however, it is just a minor tributary, far up in the headwaters, and people living further downriver consider it to be just a creek. Some fish can be caught here any time during the year, but the real fish season starts more or less in September, a month or two after the onset of the dry season, and lasts for about half a year (Box 5.2). The people living along the Bobonaza, mostly Kichwas, but also some Achuars near the mouth at the border to Peru and some Spanish-speaking *mestizos* further upriver, catch the fish using hook and line as well as gill nets and the poisonous roots of the *barbasco* plant (*Lonchocarpus nicou*). Some also fish with dynamite, and occasionally somebody fishes with pesticides.

When Emerson caught his fish, it was the month of February. He had gone with a group of almost 20 men to hunt and fish to provide food for the annual community festival to be celebrated soon. Each one had brought some barbasco roots, which they all plant in their fields. Then, standing in the water of one of the largest tributaries to the Bobonaza, they crushed the roots with a club to get the poison out. The poison paralyses the respiratory

FIGURE 5.8 Emerson standing in the Bobonaza River with the strange fish he had caught. *Photo: Eriberto Gualinga.*

Box 5.2 Fish Migrations in Amazonia

The general pattern of fish migrations (simplifying it quite a bit) starts with the downriver drift of newly hatched fish larvae from certain localised spawning grounds to large nursery grounds, where the larvae develop into fry and continue growing. These young often thrive in wetlands and the shallow waters of the seasonally inundated forests that are typical for the larger rivers of the Amazon Basin. Juvenile fish then slowly travel back upriver in order to reach the spawning grounds again at reproductive maturity. There is also a seasonal rhythm of migrations. When the water level in the large rivers is high, the fish spread laterally in order to feed in the inundated forests of the floodplain. In the so-called dry season, on the other hand, the water withdraws from the inundated forests; then, the fish instead migrate upriver towards the headwaters, returning back downriver after several months, just in time to reach the spawning grounds on schedule.

Although this is the general picture, the migratory patterns vary between species and between rivers. One species of large catfish, *Brachyplatystoma rousseauxii*, known as *dourada* in Brazil and *zúngaro dorado* in Peru, is known to migrate between spawning grounds in western Amazonia and nursery grounds in the estuary of the Amazon River, 5500 km away (Figure 5.9). Most species migrate somewhat shorter distances, but the migratory patterns are not well known. These migrations of the fish imply that fishermen living hundreds, or even thousands, of kilometres away from each other depend on the same resource pool for their livelihood; thus, they are affected by each other's actions but know very little about each other and rarely, if ever, have a chance to communicate with each other. Moreover, it is not only fishing that interferes with migrating fisheries: land-use changes and infrastructure construction affect river hydrology and form barriers for fish movements in different parts of the Amazon Basin (Castello et al., 2013).

system of the fish, and as they come to the surface to gasp for air, the men harpoon them.

Barbasco fishing is often carried out by small parties of people in forest creeks, but several times per year the people in the communities along the Bobonaza River also arrange huge communal fishing events, where hundreds of people – men and women, children, adults, and elders – participate by throwing barbasco into the Bobonaza River itself (Figure 5.10). Although this may appear to be a traditional and natural way of fishing – the barbasco is a native plant to the Amazonian forest – it is in fact not necessarily so environmentally friendly, given the massive fish deaths it causes. After a stretch of the river has been fished with barbasco, it is almost void of fish and may remain so for weeks. Fishing with barbasco in such a large river as the Bobonaza is probably also a quite recent phenomenon.

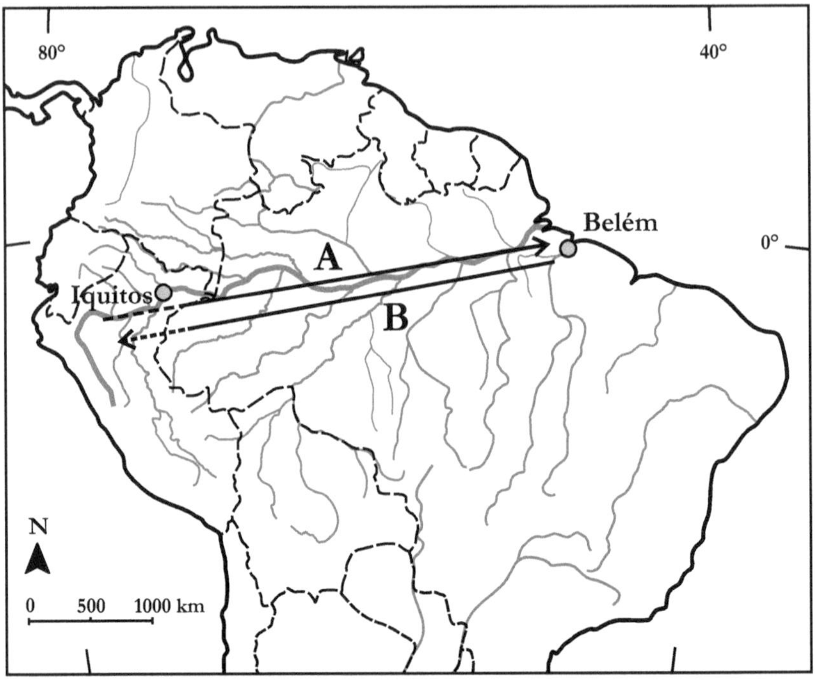

FIGURE 5.9 The larvae of the catfish *Brachyplatystoma rousseauxii* drift downriver from the spawning grounds to the nursery and juvenile growth area in the estuary of the Amazon River, a journey that may span up to 5500 km and take some 13–20 days (A). After three years, the fish start to migrate upriver again towards the spawning grounds in the headwaters, a journey that may take a year and a half (B). *Modified from Batista and Alves-Gomez (2006).*

Long ago, Emerson's ancestors used to fish with hooks and lines made of natural materials, as well as with barbasco, spears and harpoons, and sometimes simply by stirring up mud in small forest creeks to clog up the gills of the fish, making them suffocate. But, there were not too many fishers and the levels of catch were moderate compared to the area's fish populations. When more and more people came from the surrounding forests to settle by the riverside, however, fishing in the 'Big River' became an increasingly important source of food. New technologies were gradually adopted to make fishing more efficient, such as the technique to make fish nets from palm fibres, introduced in the late nineteenth century by refugees who had fled the cruel treatments to which rubber traders subjected the indigenous peoples in the Napo region further north.

Similarly, the fishing effort increased as people started to, more frequently, use barbasco to fish in the 'Big River'. This practice was prohibited by the government representative (*teniente político*) in the 1940s, but when the representative, a *mestizo* having married a native Sarayaku woman, got old and feeble in the 1960s, the people again took up the practice to fish with barbasco in the 'Big River'. Also several other new fishing methods were introduced around 1960.

FIGURE 5.10 Barbasco fishing in the Bobonaza River. *Photo: Eriberto Gualinga.*

Travelling salesmen brought dynamite, thus introducing fishing with explosives, whereas protestant missionaries brought diving masks, particularly useful for catching the nonmigratory small armoured catfish (*Loricariidae* spp.). All this was complemented by those Sarayaku natives who had travelled to do wage work on banana plantations in coastal Ecuador and brought back the technique of fishing with pesticides.

In the long run, all of these new fishing methods have not brought prosperity to the Sarayaku fisheries. The catch of large fish, particularly large catfish, has on the contrary decreased during the last few decades, and little by little the people have become aware of the negative impacts of some of these seemingly efficient fishing methods. Around the turn of the millennium, many people started to criticise those who fished with dynamite. This was so serious that the practice actually soon diminished and became very uncommon in Sarayaku, although it continues being the principal fishing method used in some communities further upriver. Fishing with pesticides, on the other hand, has become absolutely socially unacceptable. Some people have vowed never to fish with pesticides again after a joyful fishing tour to a solitary forest creek turned into a nightmare, with fish dying en masse as if the massacre would never end. In recent years, there have indeed been cases of some creek suddenly becoming void of fish, shrimps, and snails, leading to suspicions that someone has fished with pesticides. This, in turn, has led to investigations, interrogations, accusations, and threats of expulsion from the community, although in the end it has not been possible to identify the guilty.

The community of Sarayaku has, in fact, also tried to reduce barbasco fishing in the 'Big River'. Many people prefer to keep the fish in the river so that the young boys can catch some with hook and line every day, rather than killing them all with barbasco and then being without fresh fish for weeks. Thus, barbasco fishing of the 'Big River' is not done as frequently as before, not so much because of any formal restrictions, but more because it is not viewed quite as favourably as it used to be. Meanwhile, the use of nylon gill nets has also increased the efficiency of fishing. This can potentially be harmful to the fish stocks; however, fishing with gill nets is probably less wasteful than fishing with barbasco, pesticides, or dynamite because it does not kill small fish or so many more fish in addition to what the fishers actually retrieve.

Similar changes of values also have taken place in some other communities along the river. In others, however, the people still continue fishing with barbasco just as before, without hesitation. The question is then whether it pays for one single community to put restrictions on fishing when other communities do not and when much of the fish are migratory. No one really knows how far downriver the fish come from, or the real impact of the commercial fishery downriver on the subsistence fishery practiced by the communities along the Bobonaza and other headwater rivers. Managing and governing fisheries based on migratory species is a huge challenge.

Chapter 6

River Turtles – They Have Come Back

A THATCHED-ROOF BIOLOGICAL STATION

A tall, very thin man was on his knees on a lonely riverside beach lined by luxuriant green forest. He had drawn his wooden canoe partly up from the black water and placed a pile of plastic bowls on the sandy ground. The man was digging the sand by hand, removing it with great care and with a concentrated look on his weather-beaten face. Soon, he picked up a small white egg and contemplated it in the flat of his hand. He then placed it carefully in one of the bowls, with a good layer of sand lining its bottom, and continued working. He picked up more eggs from the sand and placed them in the bowl just as carefully as he had with the first one. He kept doing this for a good while, making sure that all the eggs were placed in the bowl in exactly the same position as the one they had in the sand.

Finally, more than 30 eggs later, the man stood up, stretched his legs, grabbed the bowl, and started to slowly walk towards the edge of the beach. After a while he stopped, placed the bowl on the ground, and studied the sand with great interest. Then he kneeled down, burrowed his hands in the sand, and started digging again. He placed more white eggs in the bowl, carefully separating the second bunch from the first with a layer of sand. He repeated this several times using new bowls until he seemed to be happy with the results.

This is how it might have all happened, although we cannot know for sure, because the man was utterly alone, although maybe accompanied by distant yells from a band of howler monkeys *(Alouatta seniculus)*. What we do know is that the man was Pekka Soini, a Finnish-born self-taught biologist, and the beach was not far from a small hut where he lived. It was a rustic construction on a high riverbank deep inside a vast stretch of floodplain rainforest that extends between the great Amazon tributaries Ucayali and Marañón (Figure 6.1). It was also in the Pacaya Samiria National Reserve (INRENA, 2000) that had been established by the Peruvian government in 1982. The reserve was large even by Amazonian standards, covering a total of 20,800 square kilometres – approximately the size of the country of Slovenia. Importantly, the designation of the area as a national reserve – instead of a more strictly protected national park – gave it a degree of freedom for local resource use.

Soini had already worked some years in the area prior to its declaration as a national reserve, and he closely witnessed its creation. Originally, Soini was not

Diagnosing Wild Species Harvest. http://dx.doi.org/10.1016/B978-0-12-397204-0.00006-1

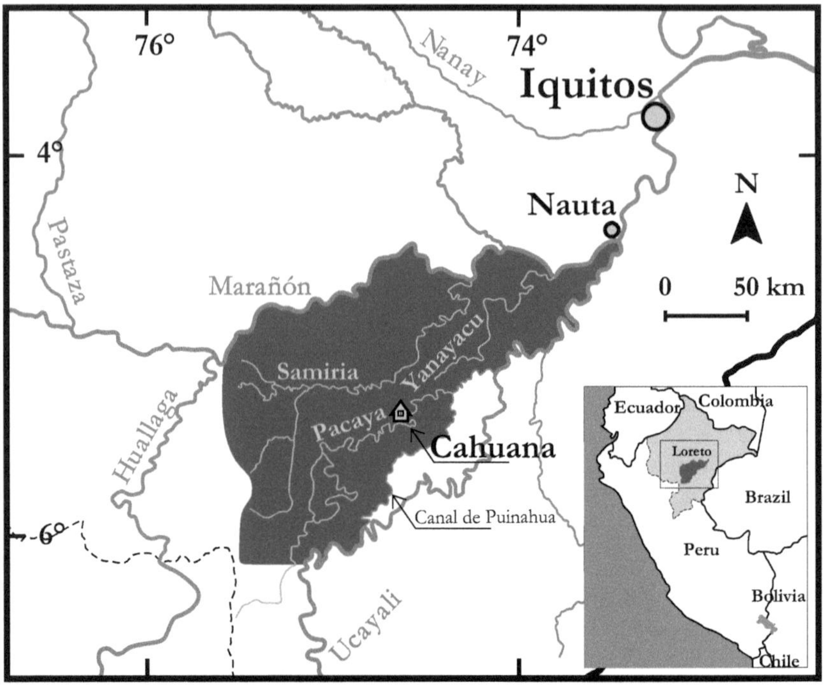

FIGURE 6.1 Map of the Pacaya Samiria National Reserve in northeastern Peruvian Amazonia. The 20,800 square kilometres reserve lies in a vast floodplain forest area between the large Amazon tributaries Marañón and Ucayali.

a man of science but rather was a sailor and navigation officer. First, he sailed the seas all over the globe in search of what he considered to be the real jungle of Tarzan, inspired by the reading of books by Edgar Rice Burroughs. However, he did not find a rainforest that satisfied his imagination until he reached Peruvian Amazonia in the 1960s. Then, Soini settled down in the northeast of the country, in the city of Iquitos surrounded by forest in all directions. The call of the forest was irresistible and soon drew him out of the city. In 1979, he moved to the Pacaya River, and by the early 1980s, his work was becoming increasingly well known in small but dedicated circles within the scientific community, both in Peru and abroad. His home also was visited by a number of these researchers.

In the open yard next to Soini's precarious palm thatch-roofed hut, there was a peculiar sandbox surrounded by a low fence. It was an experiment he had been developing for some time, and he was becoming increasingly convinced that it actually worked. It was all about turtles. These animals are an essential component of the aquatic ecosystems in Amazonia, and they also have been a central part of the local diet since prehistoric times. First, the native Amazonian cultures took advantage of the existence of such a magnificent resource, and then the European invaders could not but marvel at the abundance of turtles that served the local people. They were extracted from the wild for meat and eggs and also kept in captivity for consumption. The river turtles were increasingly harvested (Schneider et al., 2011), but their numbers still remained remarkably high in the nineteenth century when many Europeans travelled in Amazonia. However, the risks of overharvest were already noticed by some observers (e.g. Herndon and Gibbon, 1854, p. 242). More recently, the growing *ribereño* population had extracted turtle meat as well as eggs for trade up to volumes that had progressively grown beyond the limits of natural population regeneration – in many places to the point that the turtles virtually disappeared, leaving the local population without this very valuable resource.

The area protected as a national reserve was well known for its turtles. After arriving in the Pacaya River, Soini could not help but take notice that the river's turtle populations suffered from serious overharvest. One of the most harmful extractive activities was the hunting of females during the nesting period; this both affected the number of offspring and targeted the most critical segment of the populations – the breeding females. Furthermore, when the already heavily harvested turtle nests were not being pillaged by humans, they often suffered a similar fate because of natural predators or flooding. As a consequence, the turtle populations were increasingly made up of only the adult turtles that had escaped hunting but could not produce sufficient offspring to replace the increasing losses. The protection of adult females and their nests from humans and other predators would have been an effective measure in many cases, but it was excessively costly along the uninhabited stretches of the river. Even if control could have been established, it would not have helped against flooding, which was a significant cause of egg mortality.

What was the alternative, then, if the nests could not be protected in their natural environment? Maybe the solution was to remove the eggs from the beaches and relocate them to places where protection would be feasible? These ideas had already surfaced in Brazil, where the government was involved in turtle conservation, with varying success, including projects for collection and translocation of turtle eggs (Alho, 1985; Bonach et al., 2003; Schneider et al., 2011). Yet in the early 1970s, adult turtles had also been translocated in Brazil; a small local population of reproducing female arrau turtles *(Podocnemis expansa)* in Monte Cristo was reinforced with the introduction of 130 females from the Trombetas River. The population that had achieved such an impulse grew in less than two decades from a little over 200 to 1600 females (Soini, 1999).

Soini was aware of these efforts, and having observed the situation in Pacaya he believed that vigilance – and more specifically, human dwellings in the immediate vicinity – was the key to guarantee the safety of the turtle nests. An experiment of such protection was favoured by the fact that now he lived in the middle of an important turtle population. Soon Soini's 'Biological Station of Cahuana', which he shared with his wife and a few animals on the bank of the Pacaya River (Figure 6.2), became one of the main hubs for his local experiment

FIGURE 6.2 The place of Pekka Soini's home, the Biological Station of Cahuana, on the bank of the Pacaya River, today accommodates a control post of the Pacaya Samiria reserve authorities. (For colour version of this figure, the reader is referred to the online version of this book.)

on turtle egg translocation. It was, however, important to adapt any methods to the local reality of the Pacaya Samiria reserve and carefully monitor their success or failure. For this, Soini needed local help.

WATCHING AND LEARNING

The reptile fauna of Peruvian Amazonia is amazingly rich; just within the group of aquatic turtles, the region is home to 10 species. Of these, Soini was especially interested in three that had considerable economic and cultural importance for the riverine people in Pacaya Samiria. The largest of the three is the stunning arrau turtle (*Podocnemis expansa*), locally called *charapa*. Its particular importance is shown in that even the people of the region of Loreto are colloquially called *charapas* (although some find it defamatory). The sheer size and the large volume of eggs that arrau turtle nests commonly contain make them a considerable resource for the local population, both for subsistence and for trade. Large female arrau turtles can reach almost one metre in length and weigh over 60 kg, while their nests can provide up to five kilos of eggs.

Another species, the six-tubercled river turtle (*Podocnemis sextuberculata*), locally called *cupiso*, is much smaller. The largest females are hardly 35 cm long and weigh just around four kilos. Nevertheless, this turtle is also subject to collecting pressure. The third species, larger than *cupiso* but smaller than *charapa*, is the yellow-spotted river turtle (*Podocnemis unifilis*), the *taricaya*. With females reaching up to 50 cm in length and 12 kg in weight, this species compensates for their smaller individual size with abundance. When Soini started working in the Pacaya River, *taricayas*, although heavily harvested both for their meat and their eggs, were still quite common in many parts of the basin.

The main feed of all these turtles is aquatic vegetation, including plants such as pistia (*Pistia stratiotes*) and water hyacinth (*Eichornia crassipes*). Other important sources of food for the turtles are fruits and seeds, but they also eat invertebrates and fish and occasionally scavenge on dead fish and other larger animals. When river turtles are abundant, they can form a very large part of the vertebrate biomass in aquatic ecosystems; because of this fact, added to their diverse diets, river turtles can play very important roles in the food webs and energy flows of their habitats. It is thus not wrong to say that river turtles often perform key functions in rivers and lakes (Moll and Moll, 2004, pp. 64–76), although their exact roles and what happens in their absence remain questions of speculation.

River turtles are predominantly aquatic, as their name suggests. They live in the rivers and lakes of the Amazonian lowlands and only come out of water to enjoy the sunlight on beaches, branches, and tree trunks or to lay eggs. This is when they most easily fall prey to the few enemies that the strong-shelled adult turtles have apart from humans; the jaguar (*Panthera onca*) is practically the only wild animal species that has the physical strength needed to break the shell of a large adult turtle. Thus, adult river turtles have few enemies where they are not hunted. Juveniles instead are predated by various other animals and even more feed on turtle eggs, including ants and several bird and reptile species. The juveniles are not safe even in the water; many large fish species as well as caimans feed on them. However, if turtles manage to avoid falling victim to these dangers, they are long-lived. There is no exact information on turtle longevity in the wild, but most probably they can live for decades. Because they reach reproductive maturity between their fourth and ninth year of age, depending on the species, they are able to produce considerable numbers of offspring during their lifespan.

Amazonian river turtles lay eggs during the period when the low-water season reaches its extreme; in Pacaya, this means approximately the period between late July and early October. The females ascend from the river to the sandy beaches that are exposed by the decreasing water levels. Then they dig a hole in the sand using their hind limbs, after which they lay a series of eggs in this cavity. Finally, the turtles cover the nest again with sand and leave the eggs to incubate alone. The nests can be found by looking for tracks that the females leave in the sand or by walking on promising beaches using a light stick to search for spots of loose sand, which indicate the locations of the nests. Depending on the species and the individual, female turtles can lay from about 10 to more than 150 eggs at a time. Arrau turtles lay the most eggs, whereas six-tubercled river turtle nests are considerably smaller.

The nests also tell about the state of river turtle populations. One way to use the nests as a proxy for the population size is to count them along a stretch of river, checking all of the potential nesting beaches several times during the reproduction period and then extrapolating the results to cover all similar areas. *Taricayas* normally lay about 35 eggs per nest. Although the other species' females lay only one set of eggs during one breeding season, *taricayas* can do so up to three times. This is why the number of *taricaya* nests cannot be used as a direct indicator of the number of adult females living in a given area, but rather the number of nests has to be multiplied by a factor of 0.67 if it is assumed that half of the females lay just one set of eggs while the other half lays two sets. Another way to estimate the numbers of *taricayas*, which like to sunbathe on branches or beaches, is to directly count the turtles from a canoe or a boat.

These were among the countless observations that Soini made and meticulously wrote down in his notebooks while living in his Pacaya hut. He then typed the texts on letter sheets and sent them through his contacts in Iquitos to be further sent to collaborators all over the world. Many of the papers were later published by Soini alone or in collaboration with other scientists. Soini himself was not ambitious about being a famous scientist, but rather he wanted to keep a low profile and continue his fieldwork – and that is what he did.

THE START OF THE RECOVERY

One thing Soini did was to build artificial beaches, consisting of a kind of sandbox to accommodate the turtle nests that were carefully collected from their original locations on the riverside beaches. Soini had constructed his Cahuana Station artificial beach in a sunny place from which he had first removed vegetation, including roots. He made the sides of the box from palm trunk boards and then he filled the frame with a layer of clean sand up to a thickness of about 40 cm. The box was surrounded by a protective fence. Taking the plastic bowls with turtle nests covered by sand, one after another, he placed and reburied the eggs in the sandbox. After careful observation and improvement of his technique, he knew that the eggs had to be reburied by imitating the original form and size of the nest and, again, maintaining the original position of each egg to interfere as little as possible in their natural incubation. Soini proceeded to do this with each nest he had collected. He carefully wrote down in a notebook the basic information about each nest, including where and when it was collected and how many eggs it contained.

Local knowledge about the reproduction and use of turtles was applicable to their conservation in the Amazonian socioeconomic context, but nevertheless there was still a great need for knowledge on basic river turtle ecology. On the practical side, it was known that the incubation of *taricaya* eggs takes around two months, after which the hatchlings remain in the nest for a week or more before they are ready to burst into the open air, trying to get into the water as rapidly as possible to avoid predators. Soini and others had noticed that hatchlings usually headed for the water during or after rainfall and practically always at night. He wrote, for example, that during the incubation period and the time the hatchlings remained in the nest preparing to dig their way out of the sand, it was important to keep the sand free from any weeds that would start to grow rapidly (Soini, 1999).

Soini wrote that it is important to notice when the hatching happens. If the hatchlings, leaving the nest, cannot find their way into the water because they are blocked by the protective fence, they may easily perish or be eaten by animals such as the yellow-headed caracaras (*Milvago chimachima*). His observations also led to the recommendation to liberate the hatchlings immediately near the bank of a river or lake if they have left the nest at night. If they are found in the morning, they should be stored in bowls to be liberated the next nightfall in order to protect them from predators (Soini, 1999).

Lessons learned continued to accumulate. For example, the position and the properties of the sandbox often were of critical importance for the hatching outcome. Like many other reptiles, river turtle eggs develop into male or female turtles according to the incubation conditions, of which the most important is temperature. Due to hunting that particularly targets adult females on nesting beaches, many overharvested turtle populations already have a biased sex ratio. Because one male turtle can mate with more than one female during the reproduction season, it is important to make sure that incubation in sand boxes does not reinforce the excessive proportion of males in the population. Too many males could result if the incubation temperatures are too low. Thus, Soini (1999) wrote, it is important to manage the temperature by means of controlling exposure to the sun.

As time passed, the work was extending with the participation of new local collaborators from the reserve staff, local communities, and other institutions. In the late 1990s, Soini and his collaborators had accumulated an important body of experience on the methods and approaches that were useful for improving the situation of the turtle populations. Soini was still worried about the future of river turtles, but the concern was increasingly mixed with optimism; there clearly was potential for the turtle populations to recover, which would also benefit the local users of this resource. In particular, signs of success could be seen in the abundance of yellow-spotted river turtles. What then needed to be done was to disseminate the lessons learned from these experiences in the national reserve.

SPREADING THE WORD OF SUCCESS

Today, the human population of the Pacaya Samiria reserve is about 24,000, and up to 68,000 more people live in its buffer zone. Most of the population is engaged in a typical *ribereño* life, consisting of some combination of different economic activities such as farming, fishing, hunting, and gathering of forest products. Within the limits of the reserve, the population density is only a little more than one person per square kilometre. This is not much if you compare the size of the reserve to, for example, Slovenia, which has the same area but a population of two million people (100 persons per square kilometre). The pressure on natural resources in the reserve, however, is also affected by the demand from nearby cities, including Iquitos, which has almost half a million inhabitants.

As of the year 2011, there were 32 turtle management groups in Pacaya Samiria, managing *taricayas* with officially approved management plans (Harju, 2012). Each of these groups used a geographically defined management area, and they had a total of some 350 active members. The reserve authorities and local nongovernmental organisations (NGOs) provided training to the management groups and covered the costs of some of the materials needed. All work done by the management groups was, however, voluntary and unpaid. What motivated the people to participate was simply the increased supply of turtle eggs and hatchlings: the results of management could be felt and seen as tangible benefits. A central component of the Pacaya Samiria management scheme is the so called quota system that establishes a target number of eggs that should be collected. A defined percentage of these must be reburied in artificial beaches, and another proportion can be used for local consumption or be commercialised.

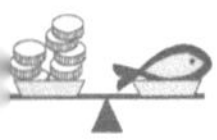

However, once the management groups have reached the set quota for reburied eggs, they are allowed to collect even more eggs from the nests and use them freely. Thus, the so-called quota system in practice does not define the harvest quota, but rather a minimum absolute number of eggs that the management groups must protect in order to ensure the conservation of the resource base. Once the eggs hatch, again the management groups are also allowed to take a determined percentage of the hatchlings to be sold to aquarium companies in Iquitos. These hatchlings are mostly exported to Asia (Harju, 2012). Because the yellow-spotted river turtle is listed in Appendix II of the Convention on International Trade in Endangered Species of Wild Fauna and Flora, all export shipments must be accompanied by documents that certify their legal origin and demonstrate that the harvest does not threaten the species. The management groups and the reserve staff record data, such as the total number of eggs and nests collected, the number of eggs collected per nest, the number of eggs used (sold or consumed) by the groups, and the number of hatchlings born alive. Based on this information, the reserve administration sets the following year's quota (Harju, 2012). In the case of the Pacaya Samiria hatchlings, the system seems to work and the export documents are continuously granted.

The reserve authorities and the NGOs involved have had an important role in promoting turtle management, but the first-hand observations of the success of turtle management by the local community members also have been very important. The word has spread among the local people, from community to community, based on perceptions on how people have been able to increase the turtle harvest through a relatively simple means. This has increased the interest in those communities and areas still outside the scheme. People who previously collected turtle eggs illegally have now also become turtle managers, and the harvest keeps increasing, year by year (Figure 6.3) (Harju, 2012).

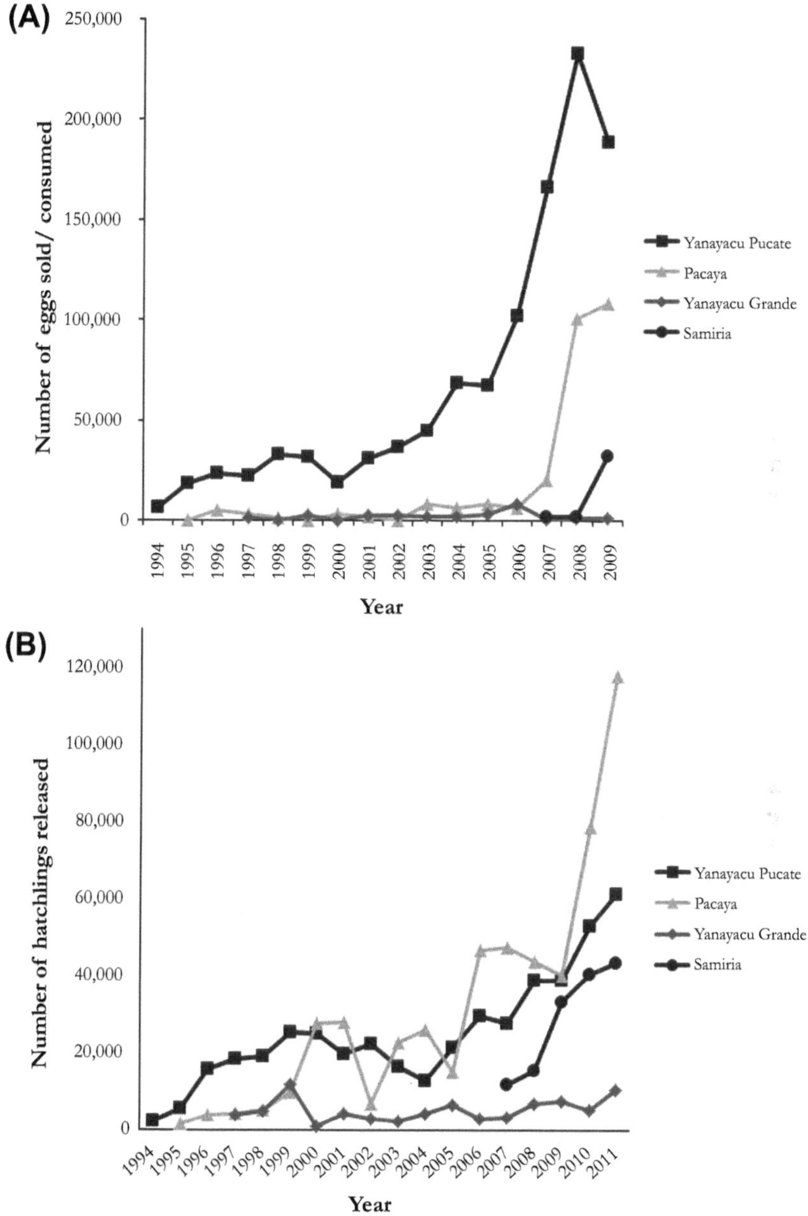

FIGURE 6.3 Records of management activities in the Pacaya Samiria reserve since 1994 indicate recovery of populations of yellow-spotted river turtles. (A) The number of eggs harvested by four management groups, not counting those that have been reburied. (B) The number of released hatchlings in the same areas. *From Harju (2012).*

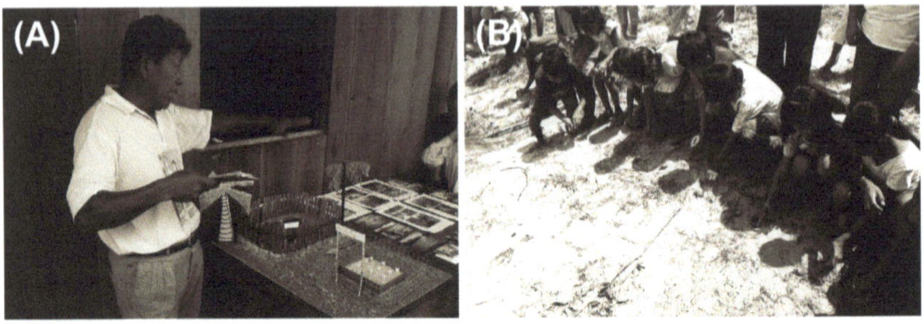

FIGURE 6.4 Turtle management has been promoted in many communities outside the Pacaya Samiria reserve. Teachers, pupils and parents participate in management activities near Iquitos. (A) Building models of the artificial beaches is one way to prepare the children for the work in practice. (B) The liberation of the hatchlings is one important step.

Turtle management in the Pacaya Samiria style has also spread outside the reserve, largely due to the efforts of different institutions such as the Research Institute of Peruvian Amazonia (*Instituto de Investigaciones de la Amazonía Peruana,* IIAP), the Peruvian-Finnish cooperation project BIODAMAZ (*Diversidad Biológical de la Amazonía Peruana)*, and a number of NGOs, particularly the Peruvian ProNaturaleza. Of the above institutions, IIAP has been actively involved in turtle management since the 1990s; in addition to the research activities of its staff, the institution has also actively promoted the cause of the turtles through other means. For example, a joint effort with BIODAMAZ in the 2000s introduced turtle management as a form of environmental education in many primary schools in and around another reserve, the Allpahuayo Mishana National Reserve near the city of Iquitos. In the background, however, there is always the work of hundreds of members of the management groups in Pacaya Samiria; without their successful efforts there would be no turtle eggs to be donated to those schools that participate in turtle management outside the reserve. Turtles have not only been visible in rural areas but also in the city, where the *Festival de la taricaya* has brought together school children and teachers to celebrate the recovery of the river turtles (Figures 6.4 and 6.5).

Another strategy that has been and will remain vital for the fate of the river turtles is the control of transport and trade of turtle products (Schneider et al., 2011). This can be carried out both at the rivers and at focal points such as in urban markets and control posts. There is also a need for monitoring protected areas and the creation of no-take-zones that help to conserve populations that remain under the threat of unauthorised harvest (Pineda-Catalan et al., 2012). When Pekka Soini started to work with turtles, the only protected area with important turtle populations was Pacaya Samiria. Since then, however, the picture has changed dramatically. The creation of other protected areas has helped to diminish harmful human interventions, such as hunting and egg collection, in some other basins and made the nest relocation

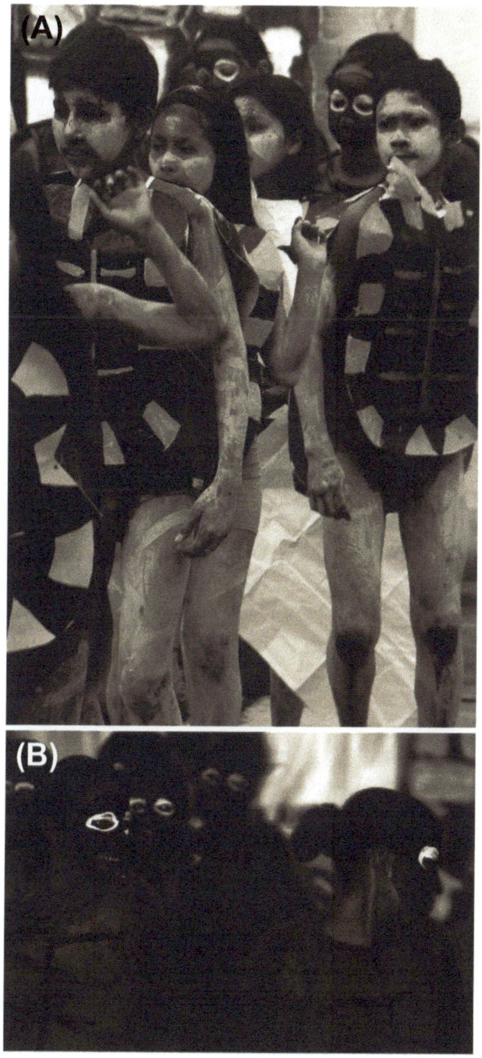

FIGURE 6.5 (A–B) A festival for the *taricaya*, the yellow-spotted river turtle, is organised in Iquitos, Peru, to promote the management scheme and environmental conscience in general.

scheme more widespread. Meanwhile, along with the strategy of active protection and propagation of focal populations in places such as the Pacaya River, the controlled use and trade of turtle eggs and hatchlings have become real options because of growing populations. In general, community-based management and use of river turtles has been shown to have potential elsewhere in Amazonia (e.g. Caputo et al., 2005).

FIGURE 6.6 Sunbathing yellow-spotted river turtles carefully follow the movements of humans and easily flee from them from a long distance.

Although more information about the effects of nest relocation and hatchling release schemes is still needed, local people in Pacaya Samiria seem to have no doubt that turtle management works, even if no systematic quantitative field inventory data were available to support that claim. Anyone travelling the rivers of the Pacaya Samiria reserve today can easily see *taricayas* sunbathing on beaches exposed by falling river levels or on branches sticking out from the water (Figure 6.6), and more recently even the huge arrau turtles have started to come back. In the times of Soini's arrival, the arrau turtle was a real rarity. They do not have the same habit of sunbathing that exposes the yellow-spotted river turtles to the human eye, but nevertheless Soini was sure there were few of them in the area. In fact, he estimated that for the Pacaya and Samiria River basins combined, the number of breeding female *charapas* at those times was approximately 400–600 individuals, of which 40 or so were killed annually either by jaguars or illegal hunters (José Álvarez, personal communication).

Today, the yellow-spotted river turtles have become a common sight after years of management, including the release of over a million hatchlings in Pacaya Samiria, and – as a result of management and turtle migration – they are increasingly visible elsewhere in the region as well (Figure 6.7). Remarkably, arrau turtles have also begun to recover, and they are now counted in a few thousands in the reserve. In fact, during the year 2012,

FIGURE 6.7 (A–B) Yellow-spotted river turtle hatchlings are both released into rivers and lakes and sold to aquarium companies to be commercialised as pets. For (A) please see the colour plate at the back of the book.

they were seen in a concentration of hundreds of nesting females in just one place, laying eggs on beaches in full daylight – something that has not been witnessed in Peru for over a century (José Álvarez, personal communication). Unfortunately, Pekka Soini is not here to see all of this; he died in 2004 (more about Pekka Soini's life and work, see e.g. Salo and Pyhälä, 2004).

The legacy of the people involved in turtle management in Pacaya Samiria lives on, however. When Finnish M.Sc. student Elina Harju did her fieldwork in Pacaya Samiria in 2011, many people told her about their satisfaction with the outcome of the management activities (Harju, 2012), as exemplified by the following quotes:

"We did not know what conservation is; we did not understand that a resource could run out" (member of recently founded Yanayacu Pucate Management Group)

"Before we just used everything as much as possible but now we know we should wait" (member of management group COMAPA 20 de Enero in Yanayacu Pucate)

"When you are an illegal you don't think that it (a resource) will run out, when you are organised you think". (President of management group Los Tucanes in the Pacaya River)

These quotes, which are just a few examples from a large amount of similar statements made by the interviewees, suggest the existence of a mutually enforcing link between management success and people's own experiences of learning by doing – leading to a positive feedback cycle of ever-improved management of the turtle populations in the reserve. Although continued conservation success should not be taken for granted, this is very encouraging.

Chapter 7

Palm Leaves, Sustainability, and Dignity

LEAVES FOR GOOD ROOFS

Among the most characteristic features of the communities of the middle and lower Bobonaza River in Ecuadorian Amazonia are the oval, thatched-roof houses (Figure 7.1). The oval shape does have a functional purpose; the otherwise thick layer of thatch made of palm leaves gets very thin at sharp corners, and therefore it deteriorates rapidly. Consequently, oval houses, without sharp corners, last longer. The aesthetical aspect is, however, also important for the indigenous Kichwa people living here. Having a good house for receiving visitors is a matter of pride, and this means that the house should not only be large but also be aesthetically appealing and provide a cool microclimate (Figure 7.2).

In a small clearing surrounded by forest stood the three houses of José, a man in his 40s. The largest house was oval shaped and almost 15 m long, and it had a thatched roof made of the leaves of a palm locally known as *wayuri*. This building had a dirt floor and no walls, and it functioned as the family's living room. Another smaller house, also with a thatched roof, was the kitchen, and there was still another small house that had a tin roof and walls made of boards, and this house contained the dormitories. The third house had been provided for free by a nationwide government-sponsored housing project. José was, however, not all that satisfied with that particular house. 'It is too hot', he said, 'only at nine o'clock in the evening it gets cool. And when it rains, it is very noisy'.

The *wayuri* (*Pholidostachys synanthera*) is a small palm tree whose leaves constitute the preferred material for thatched roofs among the people in this area. In Amazonia, its stem can reach a maximum of about 6 m in height, but usually it is much smaller, such that one can harvest its leaves without felling the tree, just by standing on the ground and using a machete. The popularity of *wayuri* roofs depends not only on the coolness they provide in the hot tropical climate; these roofs are also relatively durable, often lasting for over 20 years, whereas tin roofs often rust and start to leak much earlier than that. Also several other palm species have similar characteristics, but the *wayuri* has the additional advantage that it grows in very dense stands, something that reduces the time required for searching for the resource to a minimum, and

Diagnosing Wild Species Harvest. http://dx.doi.org/10.1016/B978-0-12-397204-0.00007-3

FIGURE 7.1 Thatched-roof houses in the Bobonaza River valley, Ecuador, seen from above *Photo: Eriberto Gualinga*. (Please see the colour plate at the back of the book)

FIGURE 7.2 Inside of a house with a *wayuri* roof.

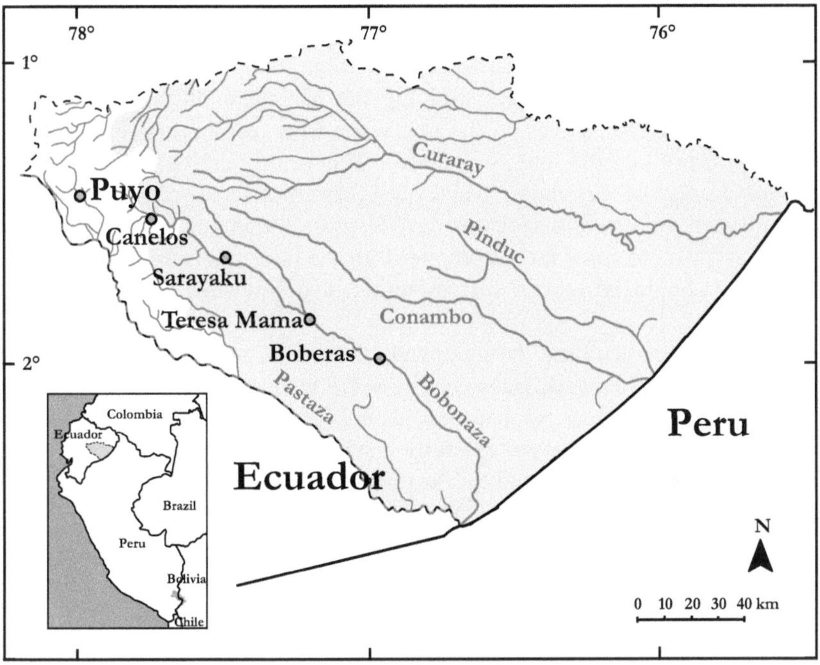

FIGURE 7.3 Location of the settlements, mentioned in this chapter, in the province of Pastaza in Ecuadorian Amazonia. Grey colour indicates territories of indigenous peoples, and white colour indicates land that mostly has been colonised by non-indigenous settlers.

thus it greatly facilitates the process of harvesting and transporting the leaves. Still another alternative is to use the leaves of the *lisan* plant (*Carludovica palmata*), the same species that is used to make Panama hats. This plant grows abundantly, almost as a weed, in agricultural gardens on the alluvial plain along the river, but these leaves do not last as long as palm leaves, often rotting within just five years.

The *wayuri*, however, has a patchy distribution and very specific habitat requirements. Along the Bobonaza River (Figure 7.3), it grows in only four distinct patches, and right in the middle of two of these patches, people settled long ago to create villages, something that is hardly a coincidence. Even within these four areas of occurrence, the *wayuri* grows only in old-growth forest on the crests and upper flanks of ridges, on soils with a thick root mat and a thick layer of fallen leaves. In addition, as an indication of the specific requirements needed regarding soil characteristics, including texture, drainage, and chemical components (edaphic conditions), *wayuri* seems to be associated with a number of particular species of ferns.

LEAF SUPPLY RUNNING OUT

José had become excited when he heard that his son got employed as a field assistant in a research project studying the harvest of the *wayuri* leaves. 'I always leave four leaves on the tree when I harvest *wayuri* (Figure 7.4), but others don't care!' José exclaimed. 'The youngsters cut all the leaves and leave only the leaf shoot. When I tell them they must spare four leaves, they just tell me I should not be stingy. Then I tell them that it is not for my sake that I want to spare the *wayuri* trees for the future: it is for their sake, so that there should be *wayuri* left for them also when they become adults like me'.

'I think about the future,' he continued, 'if there are any mature *wayuri* seeds on the trees that I harvest, I always throw the seeds around, to places where there are no *wayuri* trees, so that there should grow more *wayuri* back.' 'Once when I had harvested *wayuri*, carefully leaving four leaves on each tree, just afterwards somebody came and cut the rest of the leaves on the same trees! Can you imagine?'

FIGURE 7.4 Harvesting *wayuri*.

José lived in one of the five hamlets making up the community called Asociación Boberas, on the Bobonaza River. The nearly 1000 inhabitants of the community exerted considerable pressure on the scarce resource base of *wayuri* palms that grew in only a limited area south of the settlement. Much of what once had been suitable *wayuri* habitat near the Bobonaza River had been replaced by cultivated fields and secondary forests, and continuous old-growth forest does not begin to dominate until one walks a couple of kilometres away from the river. When José was young, there had been plenty of *wayuri* here, but now there was just one adult tree here and there, few of which bore fruit, and saplings were almost absent. Dense stands were found only at 3–4 km distance from the river, and these stands were subject to intense harvest. Unfortunately, the patch of *wayuri* did not extend much beyond this area. Further away from the river, the soil conditions changed, and the *wayuri* was absent. If the harvest front continued moving further and further away from the river, soon there would be no unharvested *wayuri* stands left.

Some 15 km up the Bobonaza River, at a place called Chiun, there was another patch with *wayuri*, and sometimes the people from the Asociación Boberas went there to harvest leaves and transport them back by canoe. Here, the primary forest reached all the way down to the river, but the *wayuri* stands were nevertheless sparse within some 800 m from the river. A count of the remaining stumps of cut petioles revealed that most leaf harvest took place between 800 and 1200 m from the river. Beyond that, all signs of leaf harvest disappeared, and some of the *wayuri* trees bore enormous racemes loaded with fruit, such as never could be seen in places subject to harvest.

Some people in Asociación Boberas were painfully aware that the *wayuri* was vanishing. When the officials of a government program came to offer houses with tin roofs (Figure 7.5), they saw that this might help save the remaining stands. Tin roofs also have the great advantage that nailing a few tin sheets takes just a fraction of the time it takes to put tens of thousands of individual palm leaves in place, one by one. Moreover, the roofs of the houses provided by the government are made of a new type of tin that is covered by a thin layer of an aluminium–zinc alloy, and therefore they are allegedly more durable. 'I will check now', said a woman in her 40s who was a native to the community and worked as a school teacher, 'how long these roofs last.' 'If they last as long as the *wayuri* leaf roofs, I think I will forget about *wayuri*.'

FROM FREE-FOR-ALL TO PRIVATE BUSINESS

When his father died, Agustín became at the age of 52 one of the oldest men in Chubakucha. He gives the impression of being a sort of patriarch: he has eight children, and some of them are already married, so he seems to be constantly surrounded by his children, grandchildren, and sons- and daughters-in-law. Agustín has a monthly salary as a teacher, and he is thus economically somewhat better off than most others in the community, owning a 25-hp two-stroke

FIGURE 7.5 A thatched-roof house flanked by tin roof houses provided by the Ecuadorian government.

Yamaha outboard motor in addition to a simple *peque–peque* motor. Right now, it is vacation time at the school, and he is taking the opportunity to build a new kitchen house. Although he recognises that there are certain advantages offered by tin roofs, he does not hesitate to build his kitchen using *wayuri* leaves. 'This is the way our fathers taught us to build houses, and we want to teach our children the same. This is indigenous culture', he says.

Building a medium-sized house of about 70 m^2 takes between 20,000 and 50,000 leaves. The harvest itself does not take much time, but the weaving of the roof does (Figure 7.6), not to mention the most labour-intensive of all the work involved – the transport of the leaves from the forest. Often, houses are built by arranging work parties, inviting friends and relatives to help, and providing nothing but a party with food and drink in exchange. Agustín, however, is instead paying a daily wage to a group of men who are helping him to harvest the leaves and to build the house. They bring the leaves by canoe from a place a few bends down the river where Agustín has his *pistu*, which is a small hut and a garden with manioc, plantains, and other crops.

Chubakucha is one of three communities, adjacent to each other, that are a few hours' ride by motor boat upriver from Asociación Boberas. A decade or so ago, here there was just one single community, called Teresa Mama, but then there were two community fissions, such that a separate community developed on each side of the river (Chubakucha to the south, and Teresa Mama to

FIGURE 7.6 Putting the leaves in place on a new house.

the north), each one with its own president and community board. And then, just a few hundred metres upriver, yet another very small community, Ishpingu, claimed its independence from the others. All in all, however, these communities had a population of just some 200 inhabitants, much less than that of the Asociación Boberas. Here, there was also a much larger area of *wayuri* habitat, especially south of the Bobonaza River, where the wayuri stands extended at least tens of kilometres, all the way up to the limit of the Kichwa people's lands and beyond, into the lands of their neighbours, the Achuar.

Unlike in Asocación Boberas, most *wayuri* stands in the Teresa Mama area nowadays have an owner. This has not always been so, as previously the *wayuri* was an open-access resource that anybody could harvest anywhere.

Even people from the community of Sarayaku, 40 km upriver, who do not have any natural stands of *wayuri* nearby, used to come freely to harvest leaves near Teresa Mama. The fact that they made this long journey, which, before outboard motors became common in the early 2000s, took two days downriver and three to four days in the upriver direction, tells something about their appreciation of *wayuri* leaves (Figure 7.7).

In 2002, however, the people in Chubakucha decided that they would charge a fee from the people of Sarayaku who came to harvest *wayuri*, and the money would go to a community fund. Defying this decision, one community member,

FIGURE 7.7 Unloading *wayuri* leaves in Sarayaku after the long journey on the river.

who had his *pistu* near one of the best stands of *wayuri*, soon after began to sell leaves on his own. He prohibited anyone to harvest from his *pistu* without permission, and he charged directly the people coming from Sarayaku to harvest leaves. At first, other community members did not accept this, but finally, after long discussions, the community members agreed to change the hitherto established system of property rights: all stands of *wayuri* now became the property of the family having its *pistu* nearby. Only the stands near the Chubakucha village itself would remain as common property where any community member could harvest leaves as they liked.

Soon after this agreement was made, the people in Teresa Mama proper decided to follow suit. Teresa Mama had always had less *wayuri* stands than Chubakucha; also, a few years earlier, a storm had felled vast areas of forest, destroying much of the *wayuri* stands. People in Teresa Mama, however, adopted somewhat different rules than in Chubakucha. Also here, the *wayuri* stands became divided into family properties, but the prohibition to harvest leaves from somebody else's property concerned only cases when the leaves were harvested in order to sell them. For one's own subsistence needs, any community member could still harvest *wayuri* leaves from anywhere. When one family later broke away from Teresa Mama to form Ishpingu a short distance upriver, many accused them of doing this only to illegitimately take possession of *wayuri* stands. This view was firmly rejected by the founder, Ishpingu, who claimed that this had always been the *pistu* of his family and that anybody from Teresa Mama who wanted to harvest *wayuri* for his own needs still was welcome to do so, just like before.

Indeed, the people of Ishpingu made good business selling *wayuri* leaves to the people from Sarayaku. The demand increased in an unprecedented manner in 2010 because of the government-sponsored housing program mentioned in this chapter. Although many people in Sarayaku already had built board houses with tin roofs beside their traditional thatched-roof houses, the response of the community to the government's offer to provide free tin roof houses was mixed. Many thought that this would weaken the indigenous culture and sense of autonomy, and also make the community less attractive for tourism, which in the end would hamper economic development and lead to ever-increasing dependence on hand-outs from the government. Negotiations with the Ministry of Housing led to a modification of the project, such that instead of providing tin sheets for roofs, the Ministry provided funds to buy *wayuri* leaves. Teresa Mama thus came to supply leaves to Sarayaku for some 25 houses within just a few months, instead of the two or three houses annually as had been normal during the previous years.

Each of the 25 houses required 100 bundles of *wayuri* leaves. These leaf bundles, locally called *wangus*, are made as large as a man can carry through the forest (Figure 7.8), and each *wangu* normally contains some 360–400 leaves. The total harvest for the housing program was thus approximately 1 million leaves, affecting on the order of 100,000 trees. The Ishpingu community supplied leaves for no less than 10 of the houses. As 100 bundles per house were needed and the price per bundle of leaves was $5, this should in theory have yielded an income of $5000. The president of the community complained, however, that only $1300 were actually received; as more money than expected had gone to transport and various unexpected expenses, when the time came to pay for the leaves, the project funds were already running out. But this was nevertheless quite a substantial income in a place far away from any markets, and where it is often difficult to find any salaried work opportunities at all.

This sudden and massive additional *wayuri* harvest left very visible traces in the forests. Whereas the harvest front in Ishpingu prior to the housing project was at about 1.5 km from the river, now all *wayuri* trees up to almost 3 km from the river had been harvested, too. The stands of *wayuri* were still dense, but although some leaves already had re-sprouted after harvest, the trees looked very different from how they look at unharvested sites, where they typically bear some 20 large leaves. Although this case was perhaps the most extreme

FIGURE 7.8 Arriving to the village carrying *wayuri* leaves.

example, similar expansion of the harvest front had occurred everywhere around the Teresa Mama communities, as everybody had taken the chance to earn some money by selling *wayuri* leaves.

CONCERN FOR THE FUTURE

In the three communities of the Teresa Mama area, everybody seemed to be very conscious that it was important to leave at least three to four leaves on the palm when harvesting leaves, although some argued that this was not enough and that at least five leaves should be spared in order not to hamper continued leaf production. But people were concerned that, nevertheless, the *wayuri* was becoming increasingly scarce, and they put the blame on people they said were harvesting the leaves in a careless manner, either leaving too few leaves on the stem or harvesting leaves from trees so high that they must be bent down with one hand in order to reach the leaves with the machete, something that often, they said, results in breaking and killing the tree.

In the surrounding forests, however, although one could see the traces of *wayuri* leaf harvest everywhere, there were very rarely any traces of the kind of careless harvest that so many people complained about. The real problem was not that people had been breaking the rules; rather, the problem seemed to be that the rules were insufficient to ensure sustainability. Indeed, *wayuri* leaves may grow back quite fast after the first harvest. Adán - a man in his 40s who grew up in an almost peri-urban indigenous community that had lost much of its land to colonist farmers, then married a woman from Teresa Mama and now thought it was lovely to live a 'free' life in a small community surrounded by large forests - claimed that if you just spare four leaves and the leaf shoot, the leaves will re-sprout so rapidly that in three months, the tree will look as if it 'had never been harvested'. Others disagreed, saying that it actually took seven months or a year. In any case, it is not a matter of eons.

However, although the tree may look fine again soon after having been harvested, its viability may be reduced, which may finally kill it, something that the local people were just partially aware of. As one elderly woman put it, 'Some people say that after you have harvested a tree five times, then it dies'. Although the traditional rules restricted the intensity of each harvest event, they did not put restrictions on the frequency of the harvest. When a tree just barely had recovered from the previous leaf harvest, it could be harvested again, and this hampered its growth and reproduction. Whereas the abundance of adult *wayuri* trees near the communities was clearly reduced, the impact was even more marked if one counted the flower- or fruit-bearing racemes, or the number of young individuals. The density of racemes, as well as of seedlings and saplings (i.e. *wayuri* trees that had not yet developed an above-ground stem), was about one order of magnitude lower in intensively harvested areas than in areas unaffected by leaf harvest (Sirén et al., 2013).

In addition to leaf harvest, hunting may potentially be a contributing cause of this scarcity, too. The local people know that several birds, including toucans (*Ramphastos* spp.) and various cracids (including *Mitu salvini* and *Penelope jacquacu*), eat *wayuri* seeds and then 'plant' them elsewhere when defecating, and this probably both contributes to the dispersal of seeds and speeds up their germination. Unfortunately, however, such birds are appreciated as prey for hunters and therefore are not very common near human settlements.

When the people of the Teresa Mama area decided they would charge money from the Sarayaku people for harvesting *wayuri* leaves, many in Sarayaku became upset. Their forefathers had harvested *wayuri* in that area since time immemorial, and although Teresa Mama now had grown from a handful of huts that were inhabited just part of the year into a permanent settlement with almost 30 families, many thought it was unfair that they now took possession of what previously had belonged to everybody. The then-president of Sarayaku, however, argued that it would be good that the Teresa Mama people would take charge of managing the *wayuri* stands and use them as a basis for economic

FIGURE 7.9 *Wayuri* planted from seeds, after clearing the underbrush in the primary forest near the community of Sarayaku. *Photo: Eriberto Gualinga*

self-sufficiency. And this way it came to be, although not everybody in Sarayaku was really happy about it.

Some people had, however, already begun to plant *wayuri* on a small scale in Sarayaku in the 1970s, and the results seemed to be encouraging. Thus, when the *wayuri* leaves were no longer available for free in the forest, the community responded by initiating a project, funded by a foreign nongovernmental organisation, for planting *wayuri* in its own vicinity, outside the natural range of the species. For one part, paid staff plants *wayuri* seeds in a determined area (Figure 7.9); and for another, the project provides seeds to interested community members who then plant them, thus establishing private plantations of *wayuri*. The mortality of seedlings and saplings has been pretty high, however, especially when the palms have been planted on unsuitable soils. But when they are planted on soils that resemble those where the *wayuri* naturally grows, survival rates are much better. Although the *wayuri* cannot survive if the forest canopy is cleared, it seems to benefit from some thinning, allowing for more sunlight to seep down to the understory.

About 35,000 leaves for a medium-sized house require harvesting some 2000–3000 trees, meaning that a lot of *wayuri* must be planted if it is to contribute a substantial amount of raw material for the houses in the community. But will there really be a demand for *wayuri* leaves in the future? Aren't thatched roofs something of the past, anyway? Won't they disappear as the people become better off economically, and as industrially produced roof materials become better and cheaper? For sure, such economic and technological development will cause changes in the demand for *wayuri* leaves. In the Kichwa community of Canelos, still another 35 km upriver from Sarayaku at the headwaters of the Bobonaza River, houses of concrete blocks and tin roofs are today just as universally present as thatched roof houses were still in the 1990s, before the community had a road connection to the town of Puyo. However, here you can also see some newly constructed thatched roof houses beside the houses of concrete and tin, so the demand for leaves for thatch does not seem to disappear completely.

PALM LEAVES FOR CITY ROOFS

Similarly to Ecuador, in Peru there are housing projects that intend to replace thatch roofs with tin roofing. In the Nanay River headwaters, for example, a regional government-promoted project has donated tin sheets for roofing rural dwellings. The name of this project is nothing less than *Techo digno* (Roof with Dignity), revealing quite an ideological divide between some governments and those who, like some of the people living along the Bobonaza River, on the contrary think that it is a sign of dignity to live in houses made of local materials such as palm leaves. The forests flanking the Nanay River are actually famous for accommodating many areas where the small understory palm *irapay* (*Lepidocaryum tenue*) grows abundantly (Figure 7.10), and palm leaves constitute one of the most important forest products for the rural communities of the basin (Oñate Calvín, 2012; Pyhälä et al., 2006). Here, the harvest of palm leaves is not so much for local consumption but mainly in order to sell the leaves in Iquitos, a nearby city with almost half a million inhabitants. Thus, the harvest of *irapay* palms is likely to remain high despite the roofs provided by the regional government, and the common sight of rafts and riverboats transporting leaves on the Nanay is going to remain, too.

The particular importance of the Nanay forests as a source of palm leaves for Iquitos can be seen in the distribution of the urban palm leaf trade: the main markets of this product are in Morona Cocha and Bellavista de Nanay, the two port neighbourhoods that specifically serve river transport from Nanay (Baluarte and Vásquez, 2000). The growth of the city means a constant demand, as not only are the houses in most of the marginal neighbourhoods of Iquitos made with thatch roofing (Figure 7.11), but also this material is common in the city centre where, in addition to many private houses, numerous bars and restaurants have thatched roofs because of both the rustic look of this material and the comfortable microclimate it helps to create. Good *irapay* roofing lasts up to eight years, after which the leaves have to be replaced. In most cases, however, the durability of the thatched roofs is less than that.

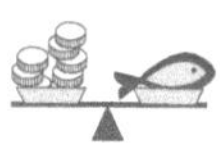

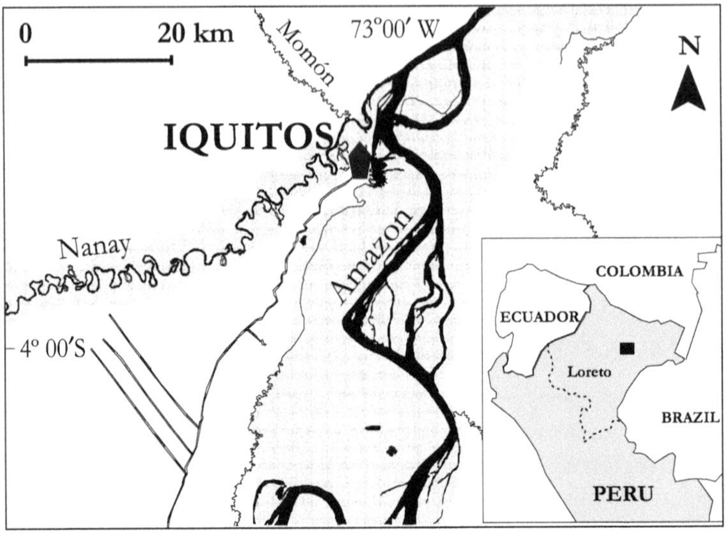

FIGURE 7.10 Iquitos is the main market of *irapay* leaves in Peruvian Amazonia. The most important harvest areas are found in the Nanay River basin.

A good day of work in a typical *irapayal* can yield 2000 leaves (Álvarez Alonso et al., 2007), which then are bound together to form *crisneja*, a sort of one-sided feather-like fan (Figure 7.12) using 2.5-m poles made of the walking palm (*Socratea exorrhiza*). In one day, it is possible to extract leaves for up to 20–40 *crisnejas*, depending on how many leaves are used for each. To get leaves for 100 *crisnejas* thus takes more than two days of hard work, after which at least another two full working days are needed to prepare the fans. Tons of this material are sold in Iquitos on a daily basis, and 100 *crisnejas* are valued at some 100 nuevos soles (equal to about 30 euros) that are normally paid to the extractor by a wholesale buyer, who then can sell them further to the final consumers maybe at a price of 150 nuevos soles (45 euros) (Álvarez Alonso et al., 2007).

From the harvester's viewpoint, this means that collecting the leaves for one *crisneja*, preparing it, and transporting the product to Iquitos are worth less than 30 cents of a euro. If collecting the leaves and preparing 100 *crisnejas* took a total of four days (a very good pace of work) and the transport just one day, the collector selling the 100 *crisnejas* would have made 30 euros in five days, a daily wage of 6 euros, without counting the costs of transport (Álvarez Alonso et al., 2007). In spite of the low profitability, however, *irapay* keeps on appearing at Iquitos' ports and markets, day after day and year after year.

The urban demand has implied that *irapay* stands found within a walking distance of approximately 1 h from the Nanay River are heavily exploited (Oñate Calvín, 2012). One of the places where *irapay* is extracted is the Allpahuayo Mishana National Reserve, where the community members of the six *ribereño* settlements that are located inside the reserve on the banks of Nanay River use the leaves for roofing and for extra income. Some of these communities

FIGURE 7.11 (A) *Irapay* palm stand in the Allpahuayo Mishana National Reserve in Peruvian Amazonia. *(Photo: Petra Mikkolainen.)*. (B) Typical *irapay* roofing competes with tin roofs in the outskirts of Iquitos. (C) An *irapay*-roofed hardware store selling construction materials, including *irapay* leaves.

FIGURE 7.12 *Irapay* fans, locally called *crisnejas*, are sold in the outskirts of Iquitos. The roofs have to be renewed after less than 10 years, keeping the demand constant.

were founded more than 80–90 years ago, after the bust of the rubber boom left many rural people out of employment; yet others are of more recent origin, from the 1980s, when the Peruvian government promoted agriculture-based colonisation. Because *irapay* grows mostly on weathered clayish or sandy soils, most of the good stands happen to be on this same side of the river. The same reason also explains why, these lands do not produce well in agriculture and, consequently, there is an important role for forest products, particularly palm leaves and roundwood for construction, in the community economies (Pyhälä et al., 2006).

PROMOTING SUSTAINABLE HARVEST

In recent decades, extractors of *irapay* have collected leaves in the area using a number of distinct methods; some, mostly local community members, have practised the traditional habit of not cutting all the leaves of an *irapay* palm but rather leaving a few so that the plant does not die. On the other hand, when no property rights were enforced, people from Iquitos or from other communities along the river often used to cut all of the leaves and kill the palm. This was common practice in some of the recent settlements of the reserve, too. As a consequence, it became common wisdom in the area that *irapayales*, as the stands are called, had become scarce in the vicinity of practically all of the communities (Álvarez Alonso et al., 2007).

As this can be explained easily by excessive harvest and destructive harvest methods, it has been relatively easy to promote the general adoption of the methods used by the more careful extractors. Moreover, when the communities became part of the reserve that was established in 1999 (Salo and Pyhälä, 2007), it became clear that if the law were to be followed, then the use of these resources would have to be formalised (i.e. only formally planned uses of resources would be allowed). This aim was quite ambitious because the informality of such activities had been the rule. Nevertheless, from the early 2000s, two projects in particular, the Nanay Project and the BIODAMAZ Project, started to promote adaptive management of resources, including *irapay*, in the communities. Basically, the idea was that the communities were encouraged to agree on internal rules for *irapay* harvest (Raygada Guerra et al., 2007).

The first recommendation was that the traditional way of cutting only a part of the leaves was to be made the general rule. The reserve administration and the projects recommended that the communities agree to cut leaves in such a way that at least five of them were always left, and in case there were more than 10 leaves, one-third of them were left intact. Furthermore, it was recommended that only the most mature leaves should be cut and the most mature palms should be left in peace altogether as seed bearers. As to how much to extract, the administration recommended that communities establish quotas for each family based on the state of the resource. When and where to cut *irapay* are two related questions, and the reserve administration recommended that the communities zone their *irapay* stands for a planned rotation so that the plants had time to recover, and this could also be reinforced by replanting activities.

Finally, the communities were also recommended to make sure that the internal rules were respected and that any outsiders who approached community members with the aim of extracting *irapay* were told to seek authorisation from the community and, if granted access, to respect the community rules. Making all of these recommendations more than what they are – recommendations – is of course easier said than done, but the communities took the initiative, and currently most of them have internal rules in place. In practice, the communities are responsible for the control and monitoring of compliance, and there are great differences from time to time and place to place on how well the community agreements are being followed (Oñate Calvín, 2012). In any case, the harvest of *irapay* is now formally legal for the first time in the history of Peruvian Amazonia. Its sustainability is another question, which requires follow-up in the years to come.

Chapter 8

Collect Locally, Eat Globally – The Journey of the Brazil Nut

RAINFOREST GIANT LEFT ALONE

A massive orange steel bridge extends 723 m over the sediment-loaded brown waters of the Madre de Dios River just one block from the main square of Puerto Maldonado, a burgeoning Peruvian jungle town. After the bridge, the Interoceanic Highway – as the road is called – runs north towards the Brazilian border (Figure 8.1). It forms part of a gigantic endeavour recently finished connecting the ports on the Peruvian Pacific coast with the Brazilian ports and metropolises on the Atlantic side of the continent. Coursing through slightly undulating lowland landscape, the road is constantly bordered by small villages and farms alternating with plantain, maize, and manioc, but just a little further north, a mix of secondary forest, farmland, and pasture starts to dominate along the edges of the road. A large part of the land from which forest has been cleared is pasture, but only here and there skinny cattle chew the pasture of exotic grass species sown on the poor soils kept open by periodic burning.

Although natural vegetation here has disappeared from the immediate vicinity of the highway, there is something in the haphazard agricultural landscape that directly reminds one of the lost forest. Massive straight trunks stand majestically as dark columns that support immense round crowns; of all the rainforest trees, it seems, only the Brazil nut trees (*Bertholletia excelsa*) have endured the change forced by the colonisation frontier. These forest giants, which can outlive complete cycles of human civilisation, have avoided the axe and the chainsaw, but many of them have now been left utterly alone in an almost open landscape. Many have suffered so many fires that they stand dead, yet others still hold out as silent witnesses of the destruction of the biotic community of which they were a part.

Further north, however, the landscape soon starts to change, and whatever the intentions of the human colonisation and despite the advancing agricultural frontier, looking right and left from the Interoceanic Highway, it can be seen that the rainforest is still there. It is just around the corner, standing green and indifferent, just as if it was only waiting to recover from the instantaneous loss caused by the human intruders. And as the highway continues, the lush primary vegetation starts to lurk closer to its edges. This is where José has his farm and

Diagnosing Wild Species Harvest. http://dx.doi.org/10.1016/B978-0-12-397204-0.00008-5

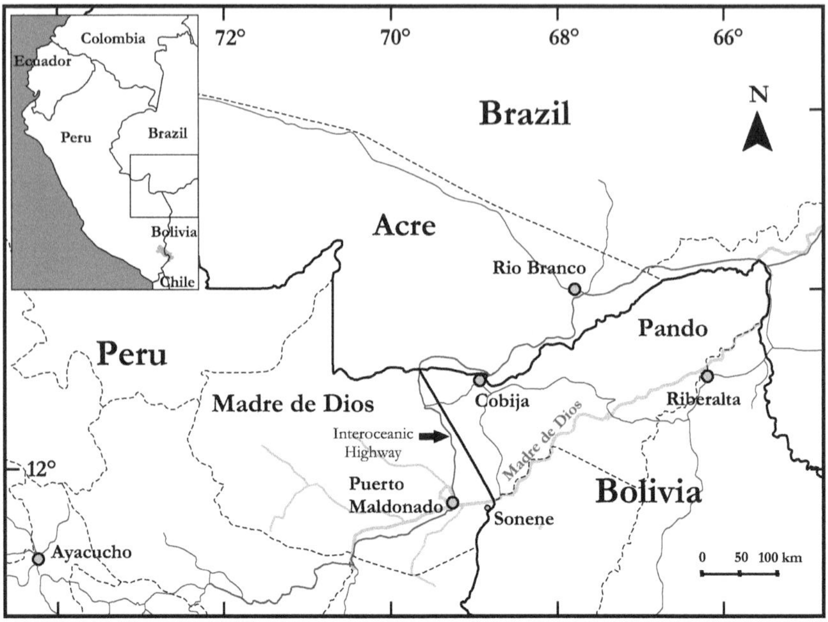

FIGURE 8.1 Brazil nuts form an important source of income in the neighbouring region of Madre de Dios in Peru, State of Acre in Brazil, and Department of Pando in Bolivia.

his Brazil nut concession. A small dirt road diverges from the highway about 5 km north from the village of La Novia, and here, in contrast to most of the journey from Puerto Maldonado, the primary forest reaches the road's edge on both sides, with huge Brazil nut trees growing here and there on its verges (Figure 8.2).

The Brazil nut tree is an impressive detail not just of the natural landscape but also of the human landscape of this region. Its extremely hard-covered fruit, which weigh up to 2 kg, contain large edible seeds, and during one season a single tree can produce more than 200 fruit, each holding between 10 and 25 seeds (called 'nuts'). In fact, botanically, they are not real nuts, but they are tasty and nutritious, and they are under constant demand in the international edible nuts market. Brazil nuts are widely sold in supermarkets across the world, and they are common components in nut mixtures.

The global demand, in turn, has given rise to extractive economies across the range where Brazil nut trees are naturally abundant. The species is found practically all over Amazonia, but the areas of particular abundance roughly cover central Brazilian Amazonia to the south and to the east, northernmost Bolivia, and the south-eastern corner of Peru. In all of these regions, the value of Brazil nuts has been considered so high compared to the value of timber from the same species – an excellent product as well – that it is illegal to fell them. Due to these same reasons, the Brazil nut is considered a prime example of a nontimber

FIGURE 8.2 (A) Large round-crowned Brazil nut trees line the Interoceanic Highway in Madre de Dios, Peru. (B) In the south-eastern parts of the Brazilian State of Acre, numerous Brazil nut trees stand alone, surrounded by extensive cattle pastures.

forest product from Amazonia with both global economic importance and local value for income generation and conservation. Some environmentalist's writings put it simply like this: 'Eat Brazil nuts and save the Amazonian rainforest'.

The individual seeds, or "nuts" are called *castañas* in Peru and Bolivia and *castanhas* in Brazil making a reference to the taxonomically unrelated old world sweet chestnuts (*Castanea sativa*). On the other hand, the entire fruit, which contains several "nuts" is, in the Spanish-speaking countries called *coco*, but in Brazil they are known as *ouriços*. This word means 'hedgehog', which is also what the hedgehog-looking fruits of the sweet chestnut are called in Portuguese. The form of the Brazil nut fruits, however, is very different. They do not have spines, and their form and size are often compared to those of a cannonball. They are very hard, and when a woody fruit speeds dozens of metres in free fall from the canopy, it can easily crush the skull of an unwary collector, something that, according to many, has actually happened in the past. The eventual blast given by a falling Brazil nut fruit would gain a fatal force due to the tree's sheer size: it is one of the largest tree species growing in the rainforests of Amazonia, easily reaching heights of more than 40 or even 50 m. Its massive trunk – sometimes 2 m in diameter at the base – rises branchless for dozens of metres, finally bursting into a huge round crown that emerges over the rest of the forest canopy.

In addition to being one of the largest living creatures found in the rainforest, the Brazil nut tree is also among its champions in terms of longevity: there are reports of a tree that is estimated to be up to 1600 years old, according to its circumference of over 16 m (Clay, 1997) (meaning a diameter of 5 m). Brazil nut trees are also considered to be a keystone species in the forests where they occur. This means that their presence is important for many other species living in the same forest. For example, their reproductive biology is complex, and it involves a number of other organisms that take care of such functions as pollination and seed dispersal.

The flowers of the Brazil nut trees are robust and fleshy, which makes their pollination limited to large-bodied euglossine bees (Euglossini, also called orchid bees) and some bats. Brazil nut trees growing in degraded forests will likely not receive these vital visitors, more so when the trees are located far apart from each other which troubles cross-pollination (meaning that pollen is exchanged between different individual plants) (Maués, 2002). Another group of animals that is important for Brazil nut trees are the agoutis (*Dasyprocta* spp.). These large rodents are among the few that are able to open the hard Brazil nut fruits, and they also frequently engage in this activity (Peres and Baider, 1997). Some of the seeds they hide for future feeding are forgotten by the agoutis and then can give rise to new Brazil nut saplings.

JOYS AND FEARS OF *CASTAÑEROS*

Don José's farm is inside the forest, about a 10-min walk from the highway. The path is in good shape, and here and there its wetter parts have been improved with Brazil nut husks. The family lives on the farm only during the harvest season, whereas they spend the rest of the year in La Novia or in Puerto Maldonado, where there is work and where the children can go to school. Don José was not born in Madre de Dios, but he has lived here for 35 years. Originally, he is from Ayacucho in the central mountains of Peru, and when he came to this place, the highway was not there but only a rough dirt road that was impassable during the rains. When the government started to promote agriculture and cattle ranching in the 1980s, José did not want to jump onto that wagon: 'I had to fight for my forest', he tells outside his house. And the forest stayed in the end.

The land here is legally entitled as the property of José and his wife, Ricardina, and although it is their farm, in fact most of the 550-ha area extending as a 0.5-km-wide and 2.2-km-long stripe along the highway is primary rainforest. Thus, there is no way to see where the farm ends and the family's adjacent 1000-ha Brazil nut concession begins. José cultivates rice and fruits, but his main income is from Brazil nuts. Now it has been raining a lot, and there are still fruit falling from his trees. The easiest way to collect them would be to let most of the fruit fall and then pile them up all at once. That would also diminish the risk of getting a hit from a falling fruit. But not everybody respects the limits of the concession, and that is why José has to make sure he is the first to find all of the fruit.

The collecting period lasts from January to late March or early April, when all of the nuts should be out of the forest if their quality is to be guaranteed. The network of paths linking the scattered trees to one another has to be cleared in November and December every year. The *estradas*, as the collecting trails are locally called, are actually an indication of how José's concession (the area over which he has exclusive rights) came into being in the early 2000s. The forest was not a clean sheet of paper, but the collecting rights were granted to people who had been harvesting in specific locations for years, often decades, and thus had also established their trail networks and their *estradas* (Pedro Flores, personal communication, April 2013).

Many of the Brazil nut trees in José's concession are gigantic. In total, they number about 700, and a trail reaches each one of them. Young trees are also marked with metallic identification plates, and they are added to the *estradas* when they start producing fruit. Don José does not think there is any lack of young Brazil nut trees in his forest but instead claims there seem to be new ones that can start bearing fruit any year now. Ecologists doing field research in other Brazil nut-collecting areas, however, have shown that removing almost all nuts from the forest affects the regeneration of the species. If this continues, the abundance of Brazil nut trees in such areas will definitively decrease, very slowly, because of the longevity of the trees, but steadily if the proportion of seeds that are removed from the ecosystem due to harvest remains too high.

About a 40-minute walk from José's farm, dim green light filters to the forest floor through the distant canopy. A dark brown layer of dead leaves covers the ground, and hanging lianas and vines are accompanied by saplings, small palms, and ferns that thrive in between the trees of different sizes that rise like pillars. Two young men sit on the ground breaking *cocos* with machetes (Figure 8.3). One of them is Don José's nephew, and the other is one of his friends. Both are from the region of Ayacucho, but only José's nephew is from the region's rainforest area. His friend instead is from the Andean highlands and is only slowly getting used to the lowland heat, rain, humidity, and insects. Both men have come to Madre de Dios for the first time, drawn by the possibility to work in the Brazil nut harvest. They are experienced users of the machete, but even so, when opening the fruits with firm chops, the impact of the blade at times hits the shells that protect the valuable seeds.

The fruit piled on the path by the men are from two nearby trees. Normally, the pile is from only one tree, but now the trees grow close to each other and the job is easier this way. The pile is covered with large leaves to protect the fruit from the rain. After opening the *cocos*, the seeds are stuffed into sacks which the company that is buying the seeds has provided. Two dogs accompanying the men have curled up among the sacks. When full, they have to be carried to José's farm. The sacks, called *barricas*, weigh around 70 kg, and carrying them on one's back is hard work even for a strong man. Sometimes, a motorcycle can be used to ease the job, but now the paths are so wet that this is impossible. The two men have been working here since early February and will continue until late March. They are not sure what will happen then. Maybe they will stay and seek work as loggers or gold miners, or maybe they will go back home.

But the work is not just that simple when it comes to how Brazil nuts are commercialised. 'The buyer pays in advance so that we are able to start working. We need money to get things done before we can sell a single Brazil nut. The *estradas* have to be cleared, and everything has to be ready for the harvest when the *cocos* start falling', Don José describes, illustrating the Brazil nut

FIGURE 8.3 (A) Brazil nut fruit are collected around the trees scattered in the forest. Often, the pile is from one tree only. (B) Opening the cannonball-like covers of Brazil nut fruit by machete chops is hard work, and the shelled seeds inside should not be broken.

collector's fragile position *vis-à-vis* the companies that buy the harvest: money is needed, but the advance payment easily ties the producer to the buyer, who then sets the prices (Kalliola and Flores, 2011).

Brazil nuts form an important source of income, and although their prices have recently been on the rise (Gómez and Torres, 2009), they are volatile and not considered high enough for viable harvesting by everybody. Thus, many collectors see an opportunity in organic certification and fair trade. Achieving

Box 8.1 Consumers' Perspectives Far Away

One of the authors of this book noticed Brazil nut bags for sale in an ordinary supermarket in Finland.

Wow, this product comes straight from the Amazonian rainforest, great to have it here! So a bag of this product was bought and happily opened. First taste: mildew. Darn, maybe it was one individual only. Second: The same experience repeated. Mixed feelings of sympathy for the cause and sadness for the problematic content, the rest of the bag got simply discarded.

It was just one bag, but the unfortunate incident illustrates a real problem. Quality requirements, price formation, and consumer conscience are all intertwined. Brazil nuts are collected in the midst of humid tropical forests and often transported in plastic sacks. Despite being carefully sun-dried in the field (Figure 8.5), some nuts may remain moist, or a shower can fall on the bags that are being transported on a pickup truck or in a boat. Things happen. Then, the humid seed is a perfect substrate for moulds that, in addition to damaging the taste, also can produce highly toxic and carcinogenic substances called aflatoxins.

These kinds of problems are taken seriously by different actors in the production chain as the entire chain from the forest to the consumer should be able to prevent the growth of moulds by keeping the harvest dry all the way through. As Brazil nuts do not usually undergo any industrial processing before their consumption, the entire production chain just has to be perfect. This demand resembles the requirement of an uninterrupted cold chain in the supply of heat-sensitive food or chemicals. This is also reflected in the price: on one hand, long chains involving many intermediary actors added to any fair-trade or other premiums paid elevate the price, while, on the other, consumers on the end of the chain expect consistent quality.

Both Brazil nut exporters and importers are well aware of these issues. Amazonian harvesters and buyer companies as well as NGOs working in the field pay much attention to the quality of the commodity chain, and most importing countries require sanitary certificates to guarantee good quality; and in cases of a change in quality, consumers also easily notice anomalies in taste. Market reputation is crucial for the survival of any product, and extracted wild species-based products are no exception (Figure 8.4).

certificates that open the doors to these added-value markets is a considerable effort, and while certification means more work, the benefits are not always so easy to see (Nunes et al., 2012), at least within the immediate time horizon (Box 8.1).

'The lives of many collectors remain like they were. Now they are just older', as Don José puts it. Moreover, their children either split the concessions or, more commonly, move to urban areas looking for better paid work and an education. In Madre de Dios, the collectors are rapidly ageing, their average age being over 55 years (Elías, 2008), and some of them feel that not much has changed with the certificates. 'The organic product premium paid to us is 50 cents of a

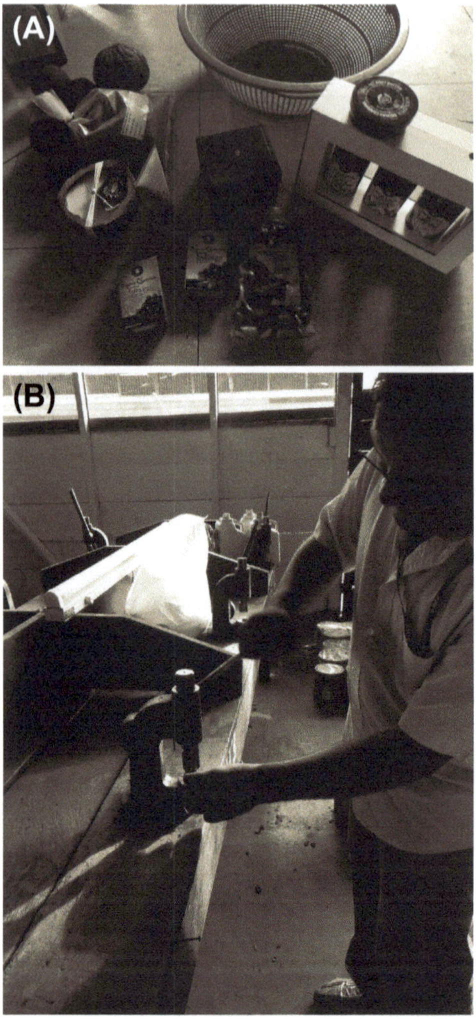

FIGURE 8.4 Brazil nuts are processed to be sold in many different forms, of which the peeled whole seed is the most common one. (A) Other products include cooking oils, cosmetic creams, and chocolates, to list a few *Photo: Natalia Aravena.* (B) The most basic form of processing is the shelling of Brazil nut seeds. This is traditionally carried out mechanically using a simple vise that breaks the shell of the seed.

nuevo sol (0.15 euro) per kilogram. That is minimal. It is half of what is paid to the producers that belong to other unions in spite of the buyer being the same company', he claims. The Brazil nut producers know that their work can help to save the forests from being heavily logged or converted altogether into farmland or pasture. They also know that the world wants them to maintain their forests. 'The world owes us a great deal', Don José says.

The discomfort arises from the suspicion that the hard work and sacrifice become apparent only further on in the commodity chain. Similarly, others – but not the collectors themselves – may be praised as the saviours of the rainforest and exemplary promoters of sustainability. José is aware of projects and sources of funding that seek to promote Brazil nut production as an alternative to land conversion and forest degradation, and he is curious about how these projects linked to the so-called REDD (Reduction of Emissions from Deforestation and Forest Degradation) scheme work. He is also interested in other alternatives as his place is located close to the highway, allowing him to develop tourism, lodging, or even fish farming. But for the time being, he does not have much choice except to try to collect his nuts in order to be able to survive until the next harvest season.

Another Brazil nut concessionaire, Doña Magda, lives in the small roadside village of Mavila. Her children moved away years ago, and now she lives alone; her husband Manuel had passed away just a few months earlier. Just like José, Magda is a colonist but not a newcomer. After arriving in Madre de Dios as a small girl with her family from the capital city of Lima in the early 1950s, her story is that of many other settlers in the area. She has learned how to live off of the forest. 'My father did not want us to study', Magda says in her little tin-roofed wooden cabin on the bank of the Manuripe River. 'He thought we girls were more useful working than at school. When I asked him about going to school, he gave me a machete and told me that it was my pen and the forest was my notebook', she laughs, but there is a trace of blame in her voice.

The times have changed, and Magda's son has been to school. But he uses a machete, too. He lives in Mavila, on the other side of the highway, and during the harvest season he takes part in the collection and transport of Brazil nuts in his mother's 2800-ha concession comprising a good stand of some 800 trees. This is way over the minimum number of trees, 320, estimated to be enough to sustain a collector's family, even in remote areas where other income is scarce (Kalliola and Flores, 2011). The concession land is state owned, and the concessions have been granted upon request, mostly to those collectors who had pre-existing customary rights to the areas, just as José as well as Magda and her husband did. And now Doña Magda's son works to keep the concession producing. The fruits fall during rainy weather, and once, piling up after a heavy shower, he got hit by a *coco* straight in the face. Luckily enough, it was a rebound from a branch that absorbed a part of the fruit's kinetic energy; but even so, the blow broke his nose, and he still sometimes suffers from strange headaches. When Magda herself was 12 years old, a coco hit her in her hip. 'But I was a fat child, so I did not get badly hurt', she laughs.

Now the harvest season is in its peak, and the village is infested by small blackflies (Simuliidae) biting without mercy. Doña Magda's son, however, is not in the forest. It is impossible to enter his mother's concession; heavy rains mark the beginning of the harvest, and this year they have fallen abundantly. The dirt road to Magda's stand has turned into an impassable pool of

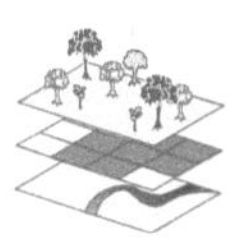

FIGURE 8.5 (A) A tool made of a wooden stick split in three or four parts from one end is locally called *payana*, and it can be used to collect the fruits from the ground to prevent snakebites. (B) The Brazil nut seeds from the opened fruit are still covered by hard shells. They can be dried using a platform with a removable roof *Photo: Natalia Aravena.*

mud, and to make things worse, flooding streams have cut the road altogether in a number of places. The only alternative route, a road used by loggers, is in bad shape as well because of the traffic by heavy forest tractors. If this was not enough, the loggers insist on fees to be paid for the use of the road – which is very bold considering that, commonly, the timber they extract is actually stolen from the Brazil nut concessions and outside of the authorised logging areas.

FROM A FRONTIER TO A TRANSIT ZONE

Madre de Dios was one of the last frontiers that were integrated into the Peruvian economy. Blocked by the Andes and with much better connections by river to the neighbouring countries of Brazil and Bolivia than to Peru, the region is geographically isolated. Apart from barely passable trails winding through the rugged terrain of the Andean forelands from the direction of Cusco and Puno, rivers formed the only transport network for the region until a road finally reached Puerto Maldonado in 1962 (Morcillo, 1982). Further on from Puerto Maldonado, much before there was a road or even a dream of a highway, a dirt trail led to the north, and despite the hardships caused by the climate, it had subtly started to scratch a scar in the face of the forest. First, it had allowed small settlements to be established in the forest away from the river, but because of the heavy rains falling during the wet season between December and May, it was at best a path difficult to keep open and, at worst, a nightmare for anyone desperate enough to try to use it during the rainy season.

The settlements were thus extremely isolated. Many of them first served the local patrons organising the extraction of wild products, such as rubber and timber from places like Iberia, a village lost in the rainforest and only connected to 'civilisation' by this fragile path, and later by a rudimentary airstrip. When Doña Magda came to Madre de Dios in the 1950s, wild rubber tapped from the *Hevea brasiliensis* tree still employed many, including members of her family. Powerful local patrons ruled the tapping economy based on an age-old system of debt dependency. Magda tells how back in those times, she never saw money because everything was paid with rubber or other forest products that were then transported on the backs of mules through the muddy paths connecting the rubber estates to Puerto Maldonado. This is how her father worked and also her husband-to-be, Manuel.

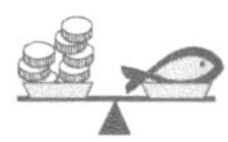

But the prices of rubber were becoming too low to sustain the extraction, and hand in hand with the decline of the rubber economy, rural dwellers started to look elsewhere for alternative sources of income. One of these sources was found in Brazil nuts, whose stands grew further south towards Puerto Maldonado and which had already been collected for export markets since the 1930s, even when there was no established national market. Madre de Dios was still an isolated place, better connected to Bolivia and Brazil than to the Peruvian centres of economic power, and most of the Brazil nuts collected in Madre de Dios were transported to Manaus, a long arduous journey that the nuts travelled – unpeeled not only to maintain their quality but also because there was no processing industry in Peru. This product grew increasingly important, and its production surpassed that of wild rubber by the 1950s (Duchelle et al., 2010).

Brazil nuts were increasingly transported to the neighbouring regions of Acre in Brazil and Pando in Bolivia, where the product had a consolidated role in the formal economy. Later, processing plants started to emerge also in Peru, with the direction of the trade shifting to Lima. Mules had first kept the muddy trail open north from Puerto Maldonado but, little by little, the first tractors

appeared and started to transform the path north of Puerto Maldonado into a kind of a road. New colonists arriving to the region begun to convert the landscape with increasing determination. As a result, the forest retreated further from the road's imposing path. By the late 1980s, with the Peruvian government actively promoting colonisation, the process further accelerated. However, Brazil nuts were always collected by the people, including some of the newcomers.

Today, Brazil nuts are collected in approximately 1000 concessions extending on both sides of the road and covering almost 10,000 square kilometres, roughly 40 percent of the 25,000 square kilometres of forest in the Madre de Dios region that has been estimated to have commercially significant densities of Brazil nut trees. In the visions of the Peruvian government and many nongovernmental organisations, the concessions, which can be granted for 40 years, do not just generate income for the rural families but also serve to halt deforestation. In the rainforests of Peru, most people combine several economic activities for subsistence and income, depending on the location and seasonality of the environment. Nonetheless, the Brazil nut remains today as the main source of income for thousands of people in the region. The recent paving of the Interoceanic Highway has further facilitated the transport of products to Puerto Maldonado, and in general it has given a boost to the local economy, which has experienced a growing demand for services such as restaurants and roadhouses serving not only the travellers but also the wage labourers employed by logging companies and working in the Brazil nut stands during the annual harvest.

INDIGENOUS BRAZIL NUTS FROM PERU

Not all of the Brazil nut collectors, however, are roadside concessionaires. Martín Huaypuna is the head of a Peruvian NGO called AFIMAD (Asociación Forestal Indígena Madre de Dios) that assists many of the region's indigenous communities in many matters predominantly related to land titles and economic activities. One of the main fields of interest that keeps Don Martín busy is the extraction and trade of Brazil nuts from community lands, and now it is time to visit the indigenous community of Sonene on the banks of the Heath River on the Bolivian border. He wants to see the season's harvest and explore the general mood in the community. Martín works hard to promote the idea that the Brazil nut is not only a natural product that can be certified as 'organic' but also a link that can catalyse local and international collaboration as well as enhance the organisation of indigenous communities.

The small port of Candamo in Puerto Maldonado teems with canoes carrying fruit, timber, Brazil nuts, and other products from the communities along the Madre de Dios and Tambopata Rivers and their tributaries. Dozens of cases of beer are also piled on the riverbank. A group of men unload timber from a canoe, carrying the boards on their shoulders to a three-wheeled motorcycle taxi. The majority of the boards are large pieces of *tornillo* (*Cedrelinga catanaeformis*), but there are some smaller ones, among them a few boards of big-leaf mahogany (*Swietenia macrophylla*). One of the men, an extremely skinny Arturo, carries heavy boards alone. After finishing the job, he sits down and opens a bottle of Inca Kola.

He points to the beer cases and says that they are going to a mining camp. 'They have everything in those camps', he continues, 'fridges, generators, stores, laundries, bars, girls, and – everything'. A myriad of informal gold-mining camps proliferate along the rivers of Madre de Dios. Arturo says that a miner can easily spend 1500 nuevos soles in a party, equivalent to some 400 euros. It is no wonder that masses of young men rush into the camps, disregarding the dangers and hardships of the informal work. According to Arturo, a recent shipment of beer was sent to a mining camp for an anniversary celebration of the camp, during which 500 cases were emptied by the miners. The three main motors driving the region's economy are present in the port: timber, gold, and Brazil nuts. Many indigenous communities are involved in all three activities, but conflicts often arise between communities and outsiders, mostly loggers. Mining and Brazil nut collection, in turn, mostly take place in different areas – luckily for the latter, because the big money that motivates the mining often dominates in any conflicts.

It takes about 4 h from Puerto Maldonado to reach the mouth of the Heath River and then about one more hour to get to Sonene. A couple of hours down the river from the port, near the village of Palma Real, Martín stops to have a chat with a group of men travelling in a large canoe packed with 45 sacks of Brazil nuts. The community's Brazil nut production is affiliated with AFIMAD, and Martín is interested in what they say about the harvested volumes and the prices paid in Puerto Maldonado. The situation had looked good in Puerto Maldonado: the price the companies paid per kilo for peeled Brazil nuts was at a comfortable level of 21 nuevos soles. But Martín says that it would not be a great surprise if, in a couple of months, it was down to less than 10. However, the price that is important for the indigenous collectors is the price *por barrica* – what is paid per sack of unpeeled Brazil nuts. The Ese'eja collectors of Sonene and Palma Real have commonly had no choice but to sell their unpeeled Brazil nuts to intermediary merchants who have paid low prices. Now, with a stronger organisation, they increasingly sell their sacks directly in Puerto Maldonado, which gives them more room for bargaining on the price.

But even so, the prices are always likely to fluctuate. 'This is a constant cause of worry for our indigenous brothers. When the prices plummet, we are sad because Brazil nut is the primary economic activity for many, and we know that the prices are manipulated without us having ways to change the situation', Martín says. The boat is soon pushed slowly upriver by its small *peque-peque* motor; the waterline approaches the top of the boat's line. Before heading to the Bolivian border, Martín still asks Rubén, the man managing the boat, to make a short stop in Palma Real, where a group of men unload Brazil nuts from another canoe. The 75-kg sacks have to be moved up the 5-m riverbank, and then an oxcart is used to get them to Palma Real. All of this effort is necessary just to deposit the Brazil nuts here before they are transported further to Puerto Maldonado (Figure 8.6).

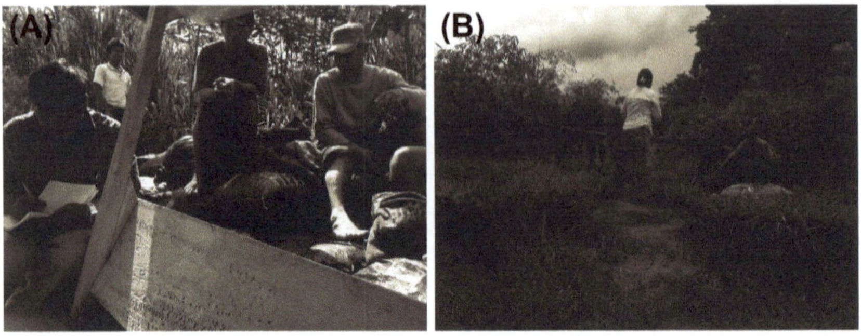

FIGURE 8.6 (A) Unshelled Brazil nuts are transported in sacks to Puerto Maldonado by river from many indigenous communities *Photo: Natalia Aravena.* (B) Brazil nut sacks are transported by oxcart for storage in the community of Palma Real on the banks of the Madre de Dios River. From there, the accumulated sacks will be transported by river further to Puerto Maldonado.

The journey continues down the Madre de Dios River to the border and then to the south to the Heath River, a river named after a North American explorer, Edwin Heath (1839-1932), who travelled in the area during the early years of the rubber boom in search of new transportation routes, areas and resources to exploit. The river marks the limit between the Bolivian and the Peruvian territories. For the inhabitants of the area, this limit used to have no meaning. But now the community of Sonene is on the Peruvian bank of the river that the indigenous inhabitants, the Ese'ejas, call with the same name. Ese'ejas are one of today's indigenous groups of the Madre de Dios region, and they have inhabited the area since before the arrival of the Europeans. They have gone through drastic processes of demographic decrease and loss of cultural identity (Desmarchelier et al., 1996), and they currently number fewer than 700 in a total of three communities (INEI, 2009). Moreover, the border between Peru and Bolivia cuts their ancestral lands in two.

Any boat arriving in Sonene is received by a bunch of smiling Ese'eja children. Men also approach the visitors, while women smilingly contemplate the scene from a distance with their babies in their arms. Visitors are accommodated in the school building, and although classes have not yet started, the walls of the classroom are full of children's drawings from the past semester, showing plants and animals with their names in Spanish and in Ese'eja. The oldest community member, a grandfather named David, is 79 but with his boyish smile looks younger than that. His Brazil nut stand is close to the village, and the trail that leads there runs through the primary forest with some clear signs of resource harvest.

Don David stops here and there along the path to show trees that serve various purposes; there are some that produce latex, such as the *Hevea* rubber trees and cow trees (*Couma macrocarpa*), and others that serve for construction or medicine, such as the cinchona tree (*Cinchona* sp.). But Brazil nuts are no doubt the most important nontimber forest product here. It takes time and effort to collect the fruit, open them, and then carry the sacks to Sonene, but now David has already piled the

FIGURE 8.7 (A–C) Grandfather David collecting Brazil nuts in Sonene *Photos: Natalia Aravena.* (For (B) please see the colour plate at the back of the book.)

fruit, and now he only has to open them using a machete. Three to five firm chops and a fruit is opened, and not a single seed is damaged; countless opened fruit have trained his hand. In little less than half an hour, the grandfather has filled a sack of some 35 kg (Figure 8.7). He prepares slings using the bark of *misa* (*Couratari* sp.) and lifts the sack on his back. Luckily, it has to be carried only for half a kilometre, after which Don David hides the sack in the woods along a smaller path so that an assistant can later transport it to the village on a motorcycle.

Grandfather David's Brazil nut trees are near the village, while some others have to walk hours to reach their stands. This is not a coincidence but a community decision. Those community members who are older or for some other reason in need of easier access are given this privilege. Moreover, the community levies a toll of one-tenth of the total harvest of Brazil nuts so that it can be distributed among those community members who have no productive trees assigned (Pedro Flores, personal communication, April 2013). In many indigenous communities such as Sonene, Brazil nuts are not collected on a concession basis but with a permit granted by the state. Although the trees have been allocated through community decisions, similarly to what happens in the concessions, growing families often mean that the same trees have to be shared with ever more family members. Added to the fact that the indigenous communities often have fewer trees per collector than is the case in the concessions (Pedro Flores, personal communication, April 2013), the further division of the stands among new members obviously means decreasing per capita earnings.

Grandfather David now lives in the village with his wife only, which makes the work harder because no other family members are permanently around to help during the harvest season. The couple nevertheless gives an impression of serenity and even happiness. They have also held to many Ese'eja traditions that otherwise would have been rapidly lost. One of these is the rich repertoire of songs in Ese'eja, describing the surrounding forest and its creatures – as well as the human life. The traditional lands of the Ese'eja do not recognise the frontier that runs along the Heath River and divides the forest in two. Recent evidence also points to the possibility that there could be Ese'eja groups living in isolation in the frontier area (Camacho Nassar, 2010).

EXTRACTIVE ECONOMIES ACROSS THE BORDER

On the Bolivian side of the border, the Brazil nut industry seems to be better off than in Peru: at least, Bolivia is the world's leading Brazil nut exporter dwarfing even Brazil that has destroyed many of its best stands in the face of agricultural expansion. Brazil's considerable internal markets also absorb a large part of its own production. Compared to Peru, the Bolivian Brazil nut industry has been notably successful and has gone through a more profound process of mechanisation (Pacheco et al., 2009). Brazil nuts form the cornerstone of the region's economy, and if seen at the level of a single species, the Brazil nut is in fact Bolivia's main forest-related export commodity (Soriano et al., 2012). Cobija (in the Department of Pando on the Brazilian border) and Riberalta (in the neighbouring Department of Beni) are the centres for Bolivian Brazil nut processing.

The Bolivian state has recently created the Bolivian Brazil Nut and Derivatives Company (*Empresa Boliviana de Almendra y Derivados*, or EBA), a public enterprise that complements the sector dominated by the private exporting firms and promotes the industry in general. It also buys Brazil nuts in order to help stabilise the prices. The central role of Brazil nuts in the region's economy makes it vulnerable to fluctuations of demand on the world market. Near Cobija, in the small town of Porvenir, Brazil nut collectors are gathered in the house of the COINACAPA cooperative (*Cooperativa Integral Agro-extractivista de Campesinos de Pando*), which was founded in the late 1990s (Figure 8.8A).

Troy is from the community of Holanda on the Manuripi River, where his community has a total of 21,600 ha of land. The 28 families of the community each have a land holding of 500 ha, assigned according to the Agrarian Reform Law. Outside of the Brazil nut harvest season, people are dedicated to the farming of manioc, fruit, and other products for subsistence and local trade. For income generation, however, the community's 7000 ha of Brazil nut forests are priceless as most money comes from Brazil nut sales, and that income has to cover the needs of the whole year. It is March, and the collectors of Holanda are transporting their sacks from the forest to the community. From the area's communities, the Brazil nut sacks are normally transported by river to Riberalta, but this time heavy rains have made the transport difficult, both by road and by river.

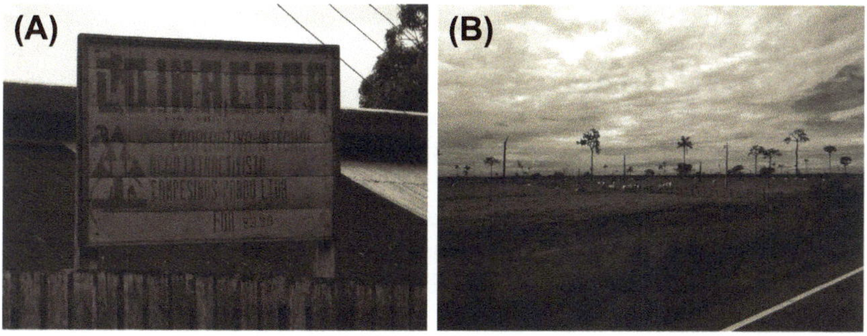

FIGURE 8.8 (A) The agroextractive cooperative COINACAPA in Porvenir, Pando, Bolivia. (B) Roadside pasture and Brazil nut trees in Acre, Brazil.

Troy says that the collector families pay a yearly fee of 250 bolivianos (about 27 euros) for the paperwork to get their permits in order. But that is not the only investment needed. The cooperative is 'enabled' by a buyer company, but every community member has the liberty to choose their 'enabler'. 'Enabling' means in practice a system in which the buyer of the extracted product provides the extractors in advance with what they need to work and expects payment for these goods in harvested Brazil nuts. Many times, the provided goods are overvalued and the extracted product undervalued. Some collectors have taken loans from the bank, but mostly the advance 'enabling' payments consist of food, clothes, and tools, and they are repaid with Brazil nuts. The intermediary buyers still have a lot of power in the area, and many community members prefer selling their Brazil nuts to them. The lower price paid by these merchants is compensated by the fact that the collectors avoid having to arrange transport to Cobija or Riberalta.

One family normally produces between 30 and 150 sacks of Brazil nuts per harvest season, which according to Troy at the moment are valued at 400 bolivianos each (about 45 euros). He says that during a good day of work, two sacks can be filled. Men work with their families or with hired wage labourers, and Troy estimates that if the forest gives more than 200 sacks, there is a need for two hired labourers to prevent the product from perishing due to excessive humidity. In his opinion, it is better to invest money and contract people if possible because that means faster collecting and better prices: when the harvest season draws to its end, the quality of Brazil nuts collected starts to deteriorate and prices plummet. The Bolivian collectors are not the only ones who feed the area's Brazil nut industry; the processing plants in Cobija can also buy Brazil nuts from across the border in Acre, and then trade them further as Bolivian Brazil nuts (de Jong and Ruiz, 2012).

THE LEGACY OF RUBBER TAPPER ACTIVISTS

Compared to Bolivian Pando, the same level of dependency on forest products is not true for Acre, the neighbouring Brazilian state that also shares the border with Madre de Dios. But historically, Acre is maybe even more famous for its forest products, of which wild rubber is undoubtedly the most emblematic one. Recent decades, however, have seen large areas of Acrean forest being lost to cattle ranching and agriculture, which now dominate the economy of the region. This development is most visible along the Brazilian part of the Interoceanic Highway, where the Brazil nut trees are left standing in a practically treeless pastoral landscape (Figure 8.8B).

But, although agriculture and cattle ranching in Acre dominate many statistics, these scenes of gloomy deforestation to some, and flourishing countryside to others, are somewhat misleading – only about 10 percent of the area of Acre has been deforested, which is still well below the total accumulated deforestation of Brazilian Amazonia, which is estimated at 19 percent (Salati et al., 2012). And it was in Acre where Chico Mendes and his colleague rubber-tapping activists managed to bring the extractive forest economies into the world's conscience (Revkin, 2004; see also Chapter 3). The Brazilian state now owns all of the land in the almost million-hectare extractive reserve named after Mendes. Although the rules of the reserve state that all of the individual plots assigned to rural families have to retain 90 percent of their forest cover, the reserve has also had its share of change: its inhabitants' lives are based on a mix of agriculture, cattle ranching, and wild species harvest. Of these, extractivism remains important, particularly with Brazil nuts as the main product having surpassed rubber tapping as the main source of forest-based income long ago (Vadjunec, 2009).

In Acre, as in many other places in Amazonia, there is reason to believe that humans have influenced the Brazil nut populations for a very long time (Shepard and Ramirez, 2011). Although heavy harvest pressure has been shown to affect population structures by disrupting the seed dispersal and consequent sprouting of new trees, there is also evidence showing that some human impacts, such as that caused by shifting cultivation, can be beneficial for the regeneration of the Brazil nut tree populations (Paiva et al., 2011). Brazil nut trees depend on gaps providing light into the forest, and while these gaps can be produced by natural events such as tree falls due to storms, also thinning or selective logging can create such gaps in the forest (Clay, 1997).

Good arguments and sound reasoning point in two directions: humans can positively contribute to Brazil nut populations in relation to their distribution and density, but human activities may also affect the species by bringing it to the brink of becoming threatened (the species is currently classified as vulnerable to extinction by the International Union for Conservation of Nature). The fact that Brazil nuts have never been successfully produced in plantations or in highly human-modified environments renders the species a unique role as a 'social-economic keystone species' with real potential to function as an umbrella under which different actors can gather to seek solutions promoting socio-economic and cultural development as well as forest conservation.

Chapter 9

Changing the Law of the Jungle: Forests and Forestry in Peru

NEW FOREST POLICIES: SAME OLD HABITS?

The townsfolk watched helplessly as protesters burned down the government agencies' offices, harassed workers from nongovernmental organisations, persecuted reporters, and turned into ash hundreds of cubic metres of mahogany and other valuable timber confiscated by the authorities, worth hundreds of thousands of dollars. It was 2002 in Puerto Maldonado in southeastern Peru, and the region's forest sector made it clear that they had their own law. The downtown streets, lined by tin-roofed one-story houses, were filled with forest workers, truck drivers, carpenters, timber merchants, river transport labourers, and sawmill employees. They were accompanied by students and other citizens, and all seemed to be worried about the future of the timber-based economy that dominated in this isolated region.

The hot and humid air was thick with black smoke soaring up from burning tires as the protesters marched under their banners. For many observers, the new forest law that was finally entering into force in Peru was striving to change the age-old logic of the timber business in Peruvian Amazonia. The spirit of the new law seemed to contradict the very essence of the extractive economy that has reigned in Amazonia since times immemorial. Newspaper headlines all over Amazonia declared how the law's promoters did not know about the Amazonian reality, how Amazonian issues could not be solved at an office in Lima, and how any reform that would really make a difference should be planned with the participation of the Amazonians (Salo et al., 2013) – and not just any Amazonians. The different regions such as Loreto, Ucayali, and Madre de Dios are so different from each other that, in the view of many people, the same rules could not be applicable in all of these regions (Figure 9.1). Yet others thought that it was time for the old system to be taken down. These people also claimed that the protest was orchestrated by timber industrialists who were jealous of their lucrative business and manipulated the masses (Salo et al., 2013).

The history of Peruvian forestry goes back centuries. The lush and humid valleys on the eastern flanks of the Andean mountain chain and their adjacent lowlands have been a source of precious timber, such as mahogany (*Swietenia* spp.) and cedar (*Cedrela* spp.), at least since the times of the Spanish colony but likely much longer than that (Cleary, 2001; Lathrap, 1973). In the course of time, with Peru obtaining its independence from the Spanish crown, the increasing need for timber continued

Diagnosing Wild Species Harvest. http://dx.doi.org/10.1016/B978-0-12-397204-0.00009-7

FIGURE 9.1 Forestry and timber processing are important economic activities in Iquitos. (A) The port area of Masusa is an important entrance point of both legal and illegal timber to the city's numerous sawmills. (B) Many of the sawmills are found near the mouth of the Itaya River in the Belén District of Iquitos.

to expand the logging frontier into new areas, mostly to the east of the Andean valleys. Gradually, the eye of the loggers was turned increasingly to the Amazonian lowlands, where the evergreen, monotonous carpet of rainforest is only broken by the serpents of countless meandering rivers. Nonetheless, it was not until the collapse of the rubber boom in the 1910s that the precious Peruvian timber became the focus of an export economy.

The first Peruvian laws directly addressing forests were issued in the early 1900s. Although they provided regulations for timber extraction in natural forests, their main scope was in and around urban areas, with an emphasis on the protection of tree cover and on reforestation (Dourojeanni, 2009). Until the 1940s, the forest governance virtually meant reforestation of the coastal and mountain regions of Peru. Since the 1940s, however, the pressure to intensify the colonisation of Peruvian Amazonia grew and, along with the increasing interest in harnessing the eastern rainforest areas to benefit the nation, the focus of forest management began to move into Amazonia. An important landmark for forest management was the creation of the Ministry of Agriculture in 1943, an institution that ever since has held the responsibility for administering the Peruvian forest resources.

However, the emphasis of the nation, by and large, was still on agriculture and cattle ranching, while forests were an issue of secondary importance. This changed drastically in 1963, when the first Peruvian Forest Law was decreed (Dourojeanni, 2009). It defined many mechanisms that, in one way or another, are still applied in the Peruvian forest regime. The law recognised national parks as an important land use designation within the country's forests and, furthermore, enacted logging contracts and forest concessions as the principal means of utilisation of the forest resources.

In the 1970s, Peru recognised the rights of indigenous communities to the natural resources that are found in the territories in their possession. Although this was more about theory than practice in most cases, the 1975 Forest Law created the concept of Communal Reserves that further addressed the rights of local forest-dwelling populations. In general, the Peruvian 1975 Forest Law had a clear tendency towards a social focus in the use of forest resources. The law also stipulated that forest resources belong to the public domain. This principle was later reinforced in the new political constitution of Peru in 1979, in which all natural resources were included inalienably into the national wealth. Unfortunately, the enforcement of the new law was no more successful than that of the previous law from 1963. This illustrated a serious problem that affected – and still does affect – most countries with major tropical forests: forest legislation, progressive as it is, cannot be effectively enforced. As an additional problem, the logging contracts that the Peruvian law directed to the so-called small extractors did not require any management planning whatsoever.

From the rest of the 1970s through the 1990s, the Forest Law from 1975 was in effect; however, as it was progressively modified, even the law's initial good intentions were little by little diluted, as logging was atomised and highly mobile with no long-term concern about resource regeneration. In 2000, another new Forest Law was decreed, making many actors in the Amazonian forest industry suspicious about the reform. Although the new law was intended to promote private incentive and long-term commitment, it was the private sector that reared up in the protests that spread through Peruvian Amazonia. Those powerful actors, who had been used to the relative freedom – if not to say anarchy – prevailing in the sector, saw the reform as a threat to their privileged positions. The most concrete change brought by the new law was the elimination of the 'under 1000 ha contracts' that had become the perfect means of evading the spirit of the old forest legislation, which in theory was promoting sustainable use. These contracts, small both in temporal and spatial scales, were almost impossible to control and monitor in practice, opening the vast lowland forests to indiscriminate exploitation.

There was no doubt that the prevailing system was in need of reform. In the old system, short-term logging contracts authorised the felling of trees in areas smaller than 1000 ha during periods of just 1–2 years, after which the loggers went on to other areas where valuable timber could be found. Many times, they worked through a system called *habilitación* (literally meaning "enabling") in which a person that has capital invests it in logging through a local patron or a merchant, equipping a team of loggers on a basis of advance payment. As is the way of Amazonia, the loggers are commonly enlisted by the local headmen, who in turn are accountable to other middlemen with direct contacts to the urban timber merchants.

This is how the investor tries to distribute the risk of not getting returns as planned through the whole chain of actors (Sears and Pinedo-Vasquez, 2011), extending from the humble *machetero* opening skid trails with a jungle knife and an axe to the final buyer of the timber in the city. This is also how the resource, its primary collectors in the forest, and the urban traders were connected to each other by a chain of intermediaries. The arrangement, based on a circle of debt, ties forest workers and timber industry intricately together, but it also works to separate the ends of the production chain, thus also setting the forest workers and the timber industry against one another. On the one hand, the industry depends on a workforce pool that can guarantee a steady flow of raw material from the forest; on the other hand, the forest workers depend on the profitability of the forest industry. If companies or sawmills are closed, jobs are also lost. In this way, the harvest of wild species can become highly political (Salo et al., 2013). One of the things that caused such an upset in the Peruvian forest sector of the early 2000s was that the small contracts, upon which the functioning of the system was largely based from a legal viewpoint, were to be replaced by larger 40-year concessions.

Timber extraction in Amazonia is often very selective. Because of the low densities of trees in the case of most of the commercial timber species, large areas of forest are needed if profitable logging is to be carried out in a sustainable manner. One of the goals of the new forest legislation was to make long-term large-scale forest concessions the cornerstone of Peruvian timber production. More recent history has shown, however, that the Peruvian forest reform has not managed to entirely change the old logic of the sector (Sears and Pinedo-Vasquez, 2011). Old habits based on informality, mobility, and high-grading (the selective removal of the most valuable timber) still persist over a large part of the region, and it seems that many of the concessions do not work as they are intended to work; in fact, it seems that timber increasingly comes from community lands (Salo, unpublished data). However, the forest sector is no doubt going through a structural change, and there are people who increasingly question the existence of a clear-cut trade-off between low-impact forest practices and profitable timber businesses.

LOGGING TEAM IN THE WOODS

After just a quick chop to the neck with a bare hand, the chicken is dead. Then the man, wearing a black balaclava in the tropical heat, dunks the unlucky bird into a battered metal pot full of boiling water and starts to rip off the feathers. The Peruvian logger in his 40s is called *El Colombiano* by his fellow workers. He is an amazingly fast cook, and it does not take him more than 20 minutes to have the chicken plucked, cut into pieces, and cooked with some spaghetti, salt, and spices. The loggers have their improvised meal with a dose of lively banter, and their good mood is clearly for a reason: in just a few more days, they will be visiting home and families after three months in the forest (Figure 9.2).

FIGURE 9.2 Fresh meat is high on the list of favourite raw materials for a logging team's meal. Sometimes there are chickens to be used, but it is also common for the loggers to use a 'subsidy from nature' – game meat provided by professional hunters. The logging team's brunch is made at the Tacsha Curaray River, this time from chicken *Photo: Natalia Aravena.*

The chicken also tastes particularly good now because the men have not been able to diversify their daily diets with fresh game meat lately. Normally, a logging camp would be home not only to the logging team but also to at least one professional hunter, who should be able to guarantee a constant flow of meat from the forest into the camp kitchen. This is how the almost free-of-charge forest wildlife can often help to reduce the forest workers' dietary costs for their employer. But now, the timber company works under new standards and has prohibited such activities from its employees. The authority that the company has over the forest area, however, does not extend beyond its own people and other possible outsiders. The subsistence hunters that occasionally enter the concession area from the nearby rural communities of the area do have the legal right to hunt.

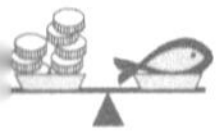

Only a little over a 100 km (as the crow flies) separate the logging team from the delights of the city life in the regional capital of Iquitos, with almost half a million inhabitants (Figure 9.3). In practice, this distance means a full day's river travel for a person moving by motorboat and a few more days for timber floated as rafts – and this is a place with easy access by the region's standards. To get where the company operates, you first take a motorboat in the downriver direction from the Port of Producers in Iquitos. Then, after an hour or so, you get off the boat on a small landing with a few houses on the Amazon riverbank, from which a small road runs 5 km through an isthmus that separates the main Amazon River channel from a huge meander of its large tributary, the Napo. There, in the mouth of a river called Mazán, is a small commercial town with the same name. Finally, one takes another boat up the Napo River, which is busy with boats of different sizes, including large dredgers that tirelessly process river sediment in search of gold. Many, if not all, of these crafts work informally, often using prospecting permits to conceal actual gold extraction, and they constantly play cat and mouse with the government authorities.

Every once in a while, rafts of logs float down the river. Forestry is one of the most important economic activities in the Loreto region in northeast Peru (Figure 9.3). All timber that can be seen floating down the Napo comes from natural forests. More precisely, it comes either from state-owned production forests or from community lands – at least if the official logging registers are to be believed. Informality, however, is not only common in mining but also in forestry. Technically, informality can be either deliberate fraud, simple negligence, or even ignorance; sometimes, paperwork of perfectly legal activities means an extra cost that does not really pay off when control is lacking. But many people also extract timber outside of the formal framework, just avoiding occasional control or bribing their way out of trouble. This kind of unfair competition is a problem for those actors who try to follow the rules. It is difficult to quantify illegality, but illegal timber is generally cheaper than legal timber. Informality causes substantial losses in the form of lost government revenues and other economic inputs to the formal economy (Gutierrez-Velez and MacDicken, 2008).

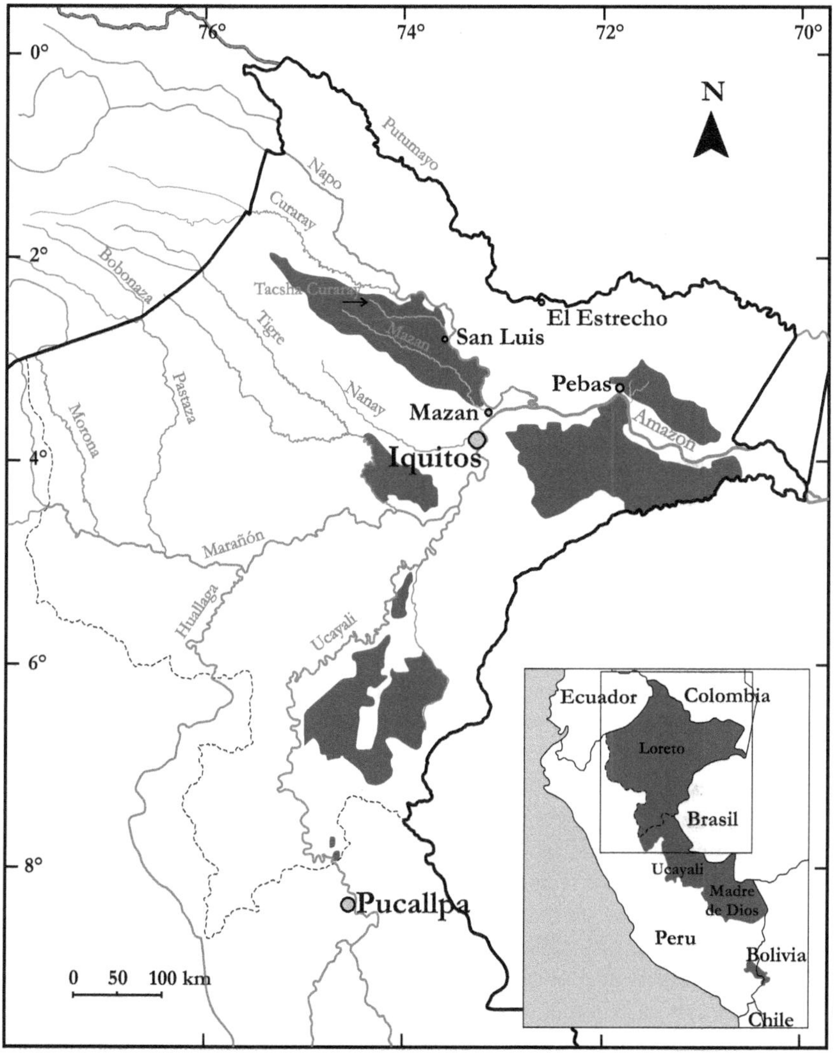

FIGURE 9.3 The region of Loreto is the largest of the Peruvian administrative regions, surpassing in size the area of Germany. Forest concession blocks (grey areas on the map) offered for tender cover over 40,000 square kilometres and are accessible almost only through the region's extensive river network. The Ucayali and Madre de Dios regions are two others of the three most important regions for forestry in Peruvian Amazonia.

It would be too dramatic to say that the Napo forms a dividing line between two forest regimes. Rather, there is a difference of grade. From the Napo to the northeast, there is the Putumayo River and the Colombian border. There, the state-owned forests have not been allocated for timber production in concessions; instead, all formal contracts are made in agreement with the local communities.

These permits allow extraction of timber, such as *Dipteryx* hardwood for parquet, in entitled community lands, often by loggers who enter and pay the communities for the timber harvested. Many times, the amount of money paid is very small, or the payment is only in the form of material provisions. The communities, furthermore, have to cope with a complex social setting that characterises the troubled border zone that not only forms a theatre for the Colombian guerrilla and important illegal coca plantations but also attracts illegal loggers operating on both sides of the Putumayo River. The river itself then offers a highway for illegal timber, gold, drugs, and arms. To get to Iquitos from the province capital El Estrecho, it takes either a one-hour small-plane flight (in which timber hardly fits as cargo) or two to three weeks and more than 2000 km of river-travel, first down the Putumayo to Colombia and Brazil and then up the Amazon back to Peru.

If this is the reality of the Putumayo area, the forests to the southwest from the Napo seem to be just around the corner from Iquitos (Figure 9.3). The Peruvian forest authorities have evaluated the watersheds of the Mazán, Curaray, and Tacsha Curaray Rivers as suitable for commercial timber production and designated them as forest concession areas. The concept of concessions is simple in principle: they are areas that the forest owner, most commonly the state, leases to private actors such as logging companies or individual entrepreneurs for a determined period of time, granting these concessionaires the exclusive right to use the area's timber and often also other resources. This privilege comes with an accompanied responsibility to follow legally established management practices and to pay determined concession fees to the owner of the area. Many other tropical countries, including neighbouring Bolivia and Brazil, have put their hopes in forest concessions as a market-based mechanism through which to organise the access to state-owned forests.

The Peruvian reform of the early 2000s first involved the reclassification of the country's forests (Salo and Toivonen, 2009). This meant that some forests were labelled as permanent production forests while others, because of their characteristics or location, were deemed unsuitable for timber production. Within the permanent production forests, the authorities further selected areas where they drew a grid that delimited areas to be offered as forest concessions. These concessions were to be allocated in public tenderings, in which companies or people interested in a particular forest area presented documents that include both an economic offer and a technical tender describing a working plan and the capacity of the applicant to implement this plan. These proposals were then evaluated to determine the applicant with the highest overall score – the winner (Salo et al., 2011).

Thus, the tenderings were not mere auctions where the highest economic bid wins. Rather, they put emphasis on other relevant issues too, such as the experience and technical capacity of the applicants (Salo et al. 2011). The concession areas range from 5000 to 50,000 ha, and in Loreto, the tendering allocated 2.8 million hectares of forest to individual entrepreneurs and companies, with the annual concession fees offered ranging from 40 cents of a dollar to more than $1 per hectare – in theory meaning direct public revenue of over $1 million per year.

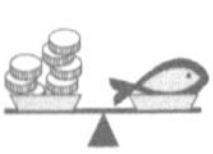

This, however, is not the case because the concession holders are entitled to different kinds of discounts, many cannot pay at all, and yet others have lost their concessions. The main idea of the concessions was not only to raise money through concession fees but also to provide an incentive to the whole sector, of which benefits could then be seen in the regional and even the national economies at large.

As a way to promote the formalisation of informal actors, the concessions tendering system allowed groups of individuals to participate if they were committed to establishing a formal organisation in the case they were granted a concession. In a few words, the idea was to make long-term logging contracts available for both very small actors and medium-sized forest companies. Both of these groups are very important in Peruvian Amazonia, where timber companies typically buy their raw material from the extractors through a network of middlemen. This is another way in which the Tacsha Curaray forest concession is a little different. Here, the forest company takes care of the whole timber chain from standing trees to export shipments. Most of the other concessions in the area, however, are small (only the minimum of 5000 ha) and the concession holders are small and local actors. Many of them won their concessions mainly based on the advantage that they had gained in the tendering by showing that they were experienced timber extractors that either had resided or had previous logging contracts in the area (Salo et al., 2011).

This was an explicit policy of the Peruvian forest authorities when the concessions were allocated. The idea was to promote the formalisation of small informal actors and improve their direct access to and control of forest resources. In practice, this would then empower the small extractors in relation to the middlemen and the forest industry at large. Many doubted, however, that the small and local actors would have the means to assume this role because they lacked capital and technical capacity that would be essential if the concessions were to be run according to the rigour and spirit of the law. According to these views, substantial capacity building would have been needed to make the system work for such small actors (Salo et al., 2013).

PUTTING THEORY IN PRACTICE

The village of San Luis is located in the mouth of the Tacsha Curaray River, and it forms a stepping-stone to the forest concessions further up the river. San Luis is basically a single long street along which the village is built. A heavy tropical shower has just broken and a completely wet man waits on the rain-swept riverbank landing. The man introduces himself as Nilson; he works for the forest company and has a small motorboat ready to continue the journey up the Tacsha Curaray, a beautiful blackwater river twisting through a lush lowland forest. The rain soon ceases, and isolated rays of sunlight penetrate the thick cloud cover, doing justice to the pink colour of a couple of Amazon river dolphins that surface near a raft of logs slowly floating downriver with two men on guard.

After a seemingly endless series of river bends, the boat reaches a logging camp where a small hut stands on a riverbank clearing. Guillermo takes care of the camp, together with a skinny dog called *Sacolargo* ("Long Sack"). The company has used the camp as a base for the forest tractors to reach the nearby annual logging parcel, but now the clearing has been abandoned by most of the machinery as logging activities have moved further up the river. After another hour or so, Nilson turns the boat into a smaller creek, where a few hundred metres further there is another camp on the riverbank. The forest engineer of the company, recognisable by his white helmet, is supervising a group of 10 loggers preparing rafts in the creek.

The river is the only way out of the concession, so now the previous months' harvest is floated out, taking advantage of the flooding river. The men use slings to attach the logs to one another using large metal staples. After the rafts have been formed, the slings are replaced by strong steel cables. Each raft is made up of some 120 logs of different species, all of them about 4 m long but with variable diameters. The loggers in their red helmets confidently walk on the floating logs, occasionally jumping into the water to swim from one raft to another. Some of the logs are of such dense wood that they hardly float, but the lighter ones support them as part of the raft (Figure 9.4).

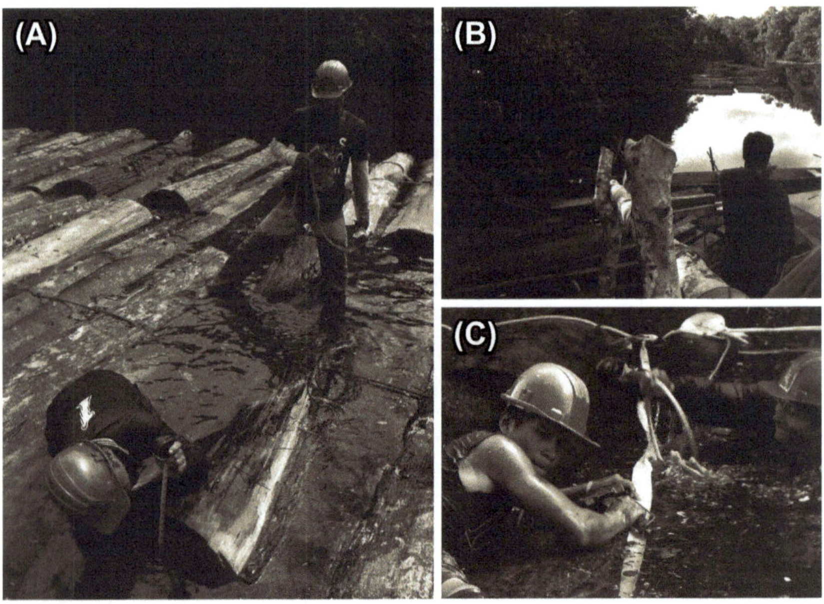

FIGURE 9.4 Preparing the rafts in a small tributary of the Tacsha Curaray River. (A) All logs are measured and labelled by the loggers. (B) The rafts have been formed and the workers have a rest before they finish the job by tying them tightly together using steel cables. (C) The cables are attached to the logs using strong iron staples. When the rafts are ready, the men start floating them downriver *Photos B and C: Natalia Aravena.*

In this part of Amazonia, huge areas of lowland forest are annually covered by seasonal floods that, depending on the location, last roughly from November to June. Water level variations play an important role in the seasonality of the logging activities; some places are only accessible when the waters are up and the smaller tributaries and creeks flood enough for floating the logs through the forest. Similarly, some activities, such as felling the trees and transporting them by land, is mostly carried out during the months with low water levels and less rain, when the soils are more stable and forest tractors are easier to use.

In the region, practically all timber is transported using the area's extensive river network. This is not a choice between cheap river transport and expensive roads but rather an inevitable adaptation to the fact that there are hardly any roads in Loreto; it is estimated that there are less than 2000 km of any kind of road for a region that is larger than Germany. For comparison, Germany has more than 600,000 km of paved roads alone. But in the fluvial network, Loreto's 8200 km of rivers navigable for commercial vessels wins over wealthy Germany's approximately 7300 km of inland waterways. In reality, the advantages of Loreto are multiple if smaller waterways are also included. However, most loggers would still prefer good roads that allow much faster and thus cheaper transport than the slowly flowing lowland rivers.

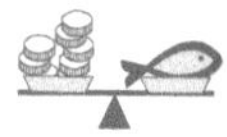

The engineer Samuel boards the boat and the journey continues upriver. The creek is surrounded by increasingly luxuriant lowland rainforest with intact looks. Common squirrel monkeys *(Saimiri sciureus)* rustle the leaves of the riverside trees. A closer look reveals how the branches hanging over the creek have been cut off to make way for boats. But even so, in many places it is difficult to proceed due to the rising water level, which starts to reach the branches of the trees above the level to which they have been cut. Samuel explains how the team was packed into five canoes with *peque–peque* motors a few months earlier and proceeded to open a way along the creek, sometimes with machete chops but often using a chainsaw too. This kind of work also forms part of the transport costs, as the men use days to open the way for future floating of the logs.

Nilson is an experienced motor boat driver so he is able to keep advancing at a regular speed on the heavily meandering river. The landscape is so flat that the river flows slowly. At most river bends, the boat returns very close to the previous meander and meets its own backwash that has passed through the inundated forest between the two bends. Nilson knows all of the shortcuts that connect one river meander to another. Every once in a while, he branches off from the main course of the creek and plunges into the forest. Most of these shortcuts are almost invisible to an outsider. After passing through one, Nilson says that the short deviation from the main course has saved the boat from making 10 bends following the main course, which is maybe a slight exaggeration. What is true is that shortcuts are so narrow that they are not passable for the rafts of logs – a fact that considerably increases the transport effort in time and costs.

Here and there along the course of the creek, there are fresh signs informing about the limits of the concession and the annual felling parcels; others

FIGURE 9.5 A logging camp is home to the team for long periods of time. (A) Mosquito nets hang under the plastic roofs of the sleeping shelters. (B) There are also other small-scale forms of wild species harvest visible in the camp, such as palm leaves for roofs and an orchid for ornament.

declare the prohibition of hunting in the area. After about an hour's journey, the boat reaches another camp. Just as the previous ones, also this camp has been constructed on a clearing on a high riverbank (Figure 9.5). In this place, however, the *tierra firme* (noninundated forest land) forms an island completely surrounded by water; on one side is the creek and on the other is the floodplain forest, isolating the campsite from dry land. Among the eight men left in the camp are the tractor driver Darío and the cook Víctor, who is preparing supper for the men. The campsite clearing has enough space for a rustic hut and two simple shelters. An open tin-roofed shelter forms the kitchen, a small tool shed accommodates a communications radio, and a row of mosquito nets is hanged over pallets made of rough boards, all covered with black plastic. The camp is clean and well organised – something you would not necessarily expect from all logging camps used for several months.

Samuel unfolds a map on the kitchen table (Figure 9.6). It indicates the limits of the annual felling area – the extent of forest within which logging takes place during one year of operation. Although the concession period is 40 years in Peru, the calculated rotation time is only half of that; thus, the area of the annual parcel roughly represents a twentieth of the whole concession. In this case, this means a little over 2000 ha. The idea is that timber should be harvested once from the annual logging parcel, after which the company plans not to come back until 20 years have passed, during which the stocks should have recovered. There is not much scientific evidence to back the 20-year rotation time, however, and different timber species regenerate at different paces.

Small symbols on the map indicate the location of economically important trees found in the annual felling parcel. Among the marked trees are the ones that will be felled, as well as the ones that should be saved as seed bearers. The map is full of small triangles that stand for the different species of *cumala* (*Virola* spp.), while the next most abundant symbols seem to be the small stars that indicate *tornillo* (*Cedrelinga catenaeformis*). Samuel says that, all in all, the felling of approximately 40 species has been authorised by the forest

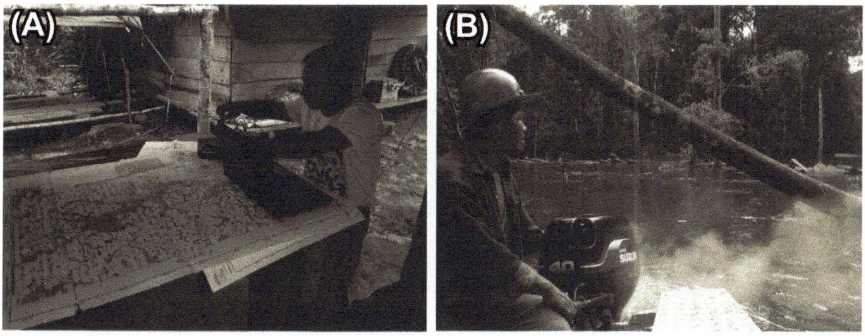

FIGURE 9.6 (A) Map of the annual felling area shows the locations of the trees to be felled and those to be left as seed bearers, as well as the creeks, floodplains, and significant topographic features. (B) Boards of hardwood are left on a flooded log yard by the creek; the wood is so dense that the logs do not float away. (For (B) please see the colour plate at the back of the book.)

authorities, but for the moment only about 10 are targeted by the loggers. The inventory method is simple. The engineer walks along 100-m wide transects with a measuring tape and a compass, indicating trees of interest on the map. He is assisted by a *matero*, a local tree identification specialist, who is the person actually finding the trees. During these walks, the limits of inundated forest and *tierra firme* are also carefully mapped in order to facilitate the planning of forest roads. The increasing availability of cheap global positioning systems has not radically changed this way of working. But it will, sooner or later.

Just a few kilometres up the creek is one of the roads used by the forest tractor, habitually called *la máquina* (the machine) by the loggers. Another forest clearing there has been used to stock the timber before it is floated downriver, and the place can be reached through the creek. On their way to the clearing, Nilson and Samuel lift an empty fuel barrel into the boat before it floats away. Now, most of the logs are gone and the whole clearing is completely flooded, with only the densest hardwood timber still there. The logs have been sawed into large boards that now form small piles here and there, lying partly underwater. The company uses a large chainsaw attached into a mount to transform the heaviest logs into boards that are easier to transport.

TROUBLED TIMBER IN THE BLACK MARKETS

The transformation of logs into boards using a chainsaw has been prohibited for years because of the waste of timber it supposedly implies. However, this practice has been and still is widespread throughout all Amazonia. Every day, below the impenetrable canopies of the Peruvian lowland rainforests, innumerable logging teams are on the job at any given moment of time. Equipped with chainsaws, winches, canoes, and outboard motors, the loggers make their way deep into the remote corners of the rainforest, habitually entering protected

FIGURE 9.7 Illegally logged mahogany waiting for further river transport in the Pacaya Samiria National Reserve, Peruvian Amazonia.

areas and indigenous territories extracting mahogany (*Swietenia macrophylla*) (Figure 9.7), cedar (*Cedrela odorata*), *shihuahuaco* (*Dipteryx micrantha*), *tornillo* (*Cedrelinga catanaeformis*), and whatever species the merchants want to purchase. These trees grow scattered across the landscape, in increasingly distant locations with arduous access. There is no specific harvest time for these loggers; they do not rest and there is always something you can do in the forest, either cutting or transporting the logs or boards.

The informal logging teams often follow a common procedure. The trees to be felled are identified in the field. Then, while a team of two fellers cut them using a chainsaw, two more labourers use their time opening trails along the forest floor, so that the timber sawn into boards can be transported to the river. Many times, the loggers have to carry the boards on their shoulders. In such a case, the logging teams operate at relatively short distances from rivers or forest roads, yet the boards can only be carried some 500 m through the forest – although in the case of the most valuable timber, such as mahogany, the distances can be much longer. These methods are used by logging teams that have little or no heavy machinery at their disposal. Sometimes, the work is carried out by community members in entitled community lands, but it is also common that the loggers are recruited elsewhere, commonly from urban areas where underemployment among young men is widespread.

This kind of small-scale logging is very selective, concentrating only on the most valuable species. It causes relatively little damage to the structure of the forest compared to more mechanised logging. Legality, of course, is not always a synonym for low impact or sustainability, and similarly, everything that is illegal does not necessarily cause high environmental impact. For example, processing logs into boards directly in the forest using chainsaws can sometimes save the forest from collateral damage because boards are easier to transport than logs. Once the timber transported by the small-scale teams reaches the riverbank, it can be floated downriver as rafts. In the case where timber is extracted from unauthorised places, the control posts established by the forest authorities are often passed by at night, using the darkness to conceal the troubled merchandise. Then, at the first sawmill on the riverbank, a middleman can step in, purchasing the whole load at a relatively low price. The Peruvian timber industry is still getting raw material this way, and the middlemen are an important part of the commodity chain.

Informally and illegally obtained and processed timber must increasingly be 'laundered' and given a legal origin before it enters the formal system and further trade, particularly when it is going to be exported. The complicity of concession holders, forest authorities, and timber companies is often needed to conceal the complicated origin of timber (Hiedanpää et al., in preparation). The fraud works through the use of falsified permits and transport documents, or by using real documents of legal origin in a fraudulent way. A flourishing black market exists for these documents (Sears and Pinedo-Vasquez, 2011). For example, it is not all that rare that a concession and its authorised logging volumes are used in timber laundering. The trick is to show the volumes to be logged in a concession area, while not actually extracting any timber oneself but instead selling the documents to be used elsewhere; selling documents is much easier and often less risky than engaging in the business of serious logging. Because there is always timber that needs a legal label, the supply and demand easily find each other. When the timber and the documents meet, it is difficult to prove that the timber in reality comes from an unauthorised location. Checking the originally authorised locations and verifying the existence of stumps is costly, and thus this kind of game goes on.

Some of the most tragic outcomes of illegal logging operations in Peru have fallen on the indigenous population inhabiting the vast wilderness areas in the remote rainforests of Ucayali, Madre de Dios, and Loreto regions. A number of these indigenous groups avoid contact with the outside world in order to survive (Huertas Castillo, 2004), but logging teams frequently stumble upon them. There are dreadful histories of how some of these encounters have turned into violent clashes, leaving dead on both sides. Even when there is no direct violence involved, infectious disease spread by the loggers can easily decimate such isolated indigenous populations.

THE QUEST FOR BETTER FORESTRY

The flooded clearing in the Tacsha Curaray concession is the terminal point of a forest road. A muddy trail leads up from the water and continues into the forest. The ground is not completely flat and small hills and depressions force the road to take a twisting line. Here, for the first time, one can see some of the physical impacts of logging (Figure 9.8). Repeated passing of the forest tractor

FIGURE 9.8 (A) The forest tractor leaves a wide trail into the forest floor. (B) Sometimes, the trunks of trees are hollow. This decreases the total volume of timber obtained from a tree compared to what has been estimated before felling. Hollow trunks can significantly reduce the value of timber obtained from such trees.

has opened the ground, transforming it into a pool of mud. Roots of the trees standing on the sides of the track have been damaged. The burning rays of the tropical sun reach the forest floor covered by dry litter, and on both sides of the road skid trails lead to the stumps of felled trees.

The fact that there are stumps indicates at least one thing: the felling parcel has been used for logging operations. Picking one tree here and another there seems not to affect the landscape too dramatically. However, selective logging can also cause considerable changes in forest structure by increasing gaps and opening the canopy. Collateral damage adds to this, and as a consequence logging causes changes in microclimate, soil characteristics, and biotic communities. Analyses based on satellite imagery have found widespread changes caused by such logging throughout the Amazon Basin (Asner et al., 2005), and all this activity cumulatively affects the region's carbon stocks (e.g. Asner et al., 2010).

Back in the creek, the company workers are finalising the rafts. All logs are measured and marked with plastic tags containing an identification number so that the journey of the timber can be tracked from the forest to the end user. The rafts are then bound to riverside trees until they are ready to be floated down the creek and then along the Tacsha Curaray River to San Luis. From there, they continue down the Napo to reach the Amazon, from where a barge pushed by a tugboat transports the timber to Iquitos. The journey takes a few days, after which the loggers will still participate in a capacity-building workshop organised by the company. Then, the loggers will be on vacations for a couple of weeks, after which they return to the concession – this time to work with the dense hardwood timber left behind.

If Loreto is famous for its troubled forestry, there are also logging companies that have become increasingly interested in changing this image. For example, one way to promote the change of attitudes towards how the companies work is to seek forest certification. The Forest Stewardship Council, for example, has been recently evaluating forest concessions in the region in order to certify their practices. The Tacsha Curaray concession holder aims to show, in more general terms, that there are alternative ways to produce timber. The company representative claims that being legal – something that one would expect everyone has as a standard – is actually more profitable for the company than the business as usual based on circumventing the rules. If this is true, it means that the costs of planning and following the plans are lower than the costs of covering the unauthorised forestry practices through concealment and bribery. The Tacsha Curaray concession is in the beginning of its anticipated 40-year lifespan, and it remains to be seen whether it can be a model for new forestry in Peru.

Chapter 10

Biodiversity and Business: An Experience with Medicinal Plants

AT THE END OF THE ROAD

Villaflora is a village at the end of the road. It is an indigenous Kichwa community located right at the colonisation frontier that once advanced rapidly as the Ecuadorian government promoted the construction of penetration roads into the Amazonian territory. Along the road, the government then assigned 50-ha plots for free to settlers that came from the highlands and the coastal regions of the country, on the condition that they 'improved' the land by cutting down the forest (Figure 10.1). Here, in the Pastaza province, however, the frontier finally stabilised in the 1990s, when decades of protests by the indigenous people finally led to the national government granting them communal land titles. Some 30 families live in Villaflora, most of them in simple tin-roofed houses with an elevated floor and walls of boards. Many also have a separate kitchen house with a dirt floor and a roof made of the leaves of the Panama hat plant (*Carludovica palmata*), locally called *lisan*. The pride of the people in having a good thatched-roof house for receiving visitors, so common in the Bobonaza River valley (see Chapter 7), seems to be absent here – as is much of the self-confidence of the indigenous culture. This community is in a process of a language shift, with many young people preferring to speak Spanish among themselves rather than their own native Kichwa language.

Just the other week, Villaflora got connected to the national electricity network. Taking advantage of this supply of cheap energy, somebody now has a radio on even though it is daytime, letting the visitor to the village enjoy the South Korean megahit 'Gangnam Style' in this once so remote place on Earth. The primary rainforest is, however, just nearby, and a few hours walk from the village it feels like you are in another world. Here, in a hut at the side of a small river in the middle of the forest, Simón, a 50-year-old man with a wide smile, sits surrounded by his wife and several of his children, sons- and daughters-in-law, and grandchildren, with one of the youngest dressed in a Winnie-the-Pooh suit (Figure 10.2). The hut looks as if it is not going to remain there for too long. Several of the poles keeping up the thatched roof have broken, and the pouring rain leaks in here and there such that it is a challenge to find a place to sit where one does not get wet. This is Simón's *tambu*, a place where he sometimes

Diagnosing Wild Species Harvest. http://dx.doi.org/10.1016/B978-0-12-397204-0.00010-3

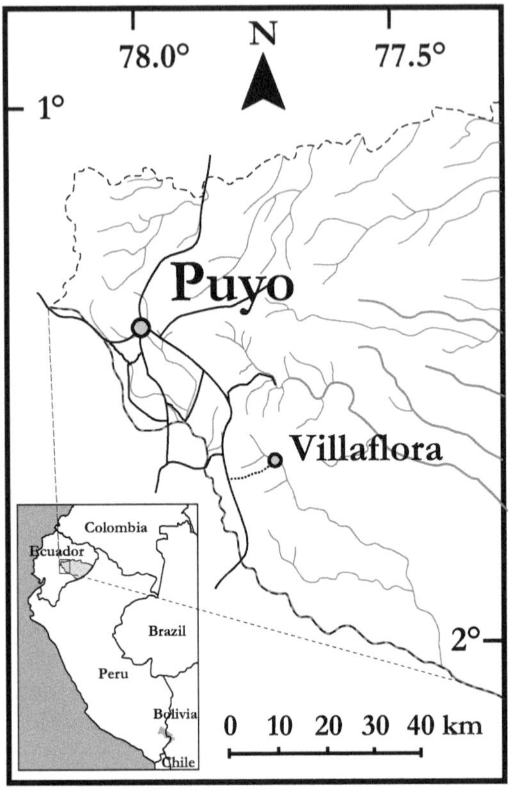

FIGURE 10.1 The location of the village of Villaflora at the end of the road network in the western part of the Pastaza province in Ecuadorian Amazonia.

comes from his home in Villaflora to hunt and fish, and now he has come here to build a new hut. There is quite a bit of game around, he says. For example, he spotted a howler monkey (*Alouatta seniculus*) on the way here, although he was not able to shoot it. Woolly monkeys (*Lagothrix* sp.), however, which are more susceptible to overhunting, are absent.

Near Simón's hut at the side of the trail, there is a tall tree with a few distinctive scars in its bark (Figure 10.3A). This is the *chuchuwasa* tree (*Maytenus* sp.), whose bark is appreciated by the locals as a cure for common colds and various other illnesses, as well as for boosting the body's own resistance against disease. It takes just a few minutes to cut a piece of bark of a few square decimetres size from the tree (Figure 10.3B). The bark can be stored for a very long time by preserving it in liquour. Recently, however, Simón has also started to exploit this resource as a source of cash income. On the request of a man he calls 'Doctor Didier', he sold 20kg of *chuchuwasa* bark for three dollars per kilogram. And then he sold much more bark to other merchants, although these people paid a little bit less. Altogether, for only two or three days of work, he earned the equivalent of at least a couple of weeks of wage work.

FIGURE 10.2 Simón surrounded by some of his grandchildren.

FIGURE 10.3A A chuchuwasa tree bearing scars as vestiges of past harvest of bark for household consumption.

FIGURE 10.3B Simón's son-in-law Fredy cutting *chuchuwasa* bark from a standing tree.

'Yes, I cut the tree down,' says Simón candidly. 'There is no other way to get so much bark from the *chuchuwasa* tree. If you want to sell *chuchuwasa* bark, you must cut down the tree,' he continues. Just as frankly, he comments on the sustainability aspect of this practice: 'They will run out, because there are not very many of them,' he says. People in this region have varying attitudes towards the conservation of natural resources. Some tend to feel threatened when the sustainability of their practices of natural resource use is brought into focus. Others are very aware that the ways of using many kinds of natural resources need to be changed if their children and grandchildren are not to be left with impoverished lands. Simón belongs to the latter sort. 'Caring for the trees such that they don't run out would be good. I would like to work with that,' he says.

Simón had not felled the *chuchuwasa* tree here near his *tambu* but had instead chosen a tree much closer to the village, in the plot of land he has there. Inspired by the way the government previously granted private land holdings to settlers, the Villaflora community has internally assigned, out of the legally granted communal lands, one 50-ha plot called a *finca* to each family. 'It is very far to carry the bark from the *tambu* here,' Simón explains as the reason why he chose to cut the tree in his *finca* instead. All of the *fincas* are located near the village and, thus, close to the road, and only the respective owners have the right to extract resources there. On the other hand, the more distant forests are held in common and resource extraction is open to all members of the community. 'Here, everybody may take what they want. This is our reserve,' Simón's son-in-law Fredy explains while walking through the forest, revealing a somewhat different interpretation of the term 'reserve' than what is common among the conservationist community.

A DREAMER AND BUSINESSMAN

'I have told them that I don't want them to cut the trees, but I fear that they do it anyway,' Didier had said the same morning while driving his four-wheel-drive car along a road southwards from the town of Puyo. A few years earlier, the same road had been just another bumpy dirt road, and a few decades back it was just a vision in the minds of the national elite dreaming about the integration of Ecuadorian Amazonia into the rest of the nation. Now, it is a perfectly level paved road, and the forest has long ago been replaced by pasture and fields of sugar cane and other cash crops – although in some places, a secondary forest has grown back.

Didier is not really a doctor, but he does not seem to be very disturbed by this title of honour given to him by the local people. Actually, he dreamed of becoming a doctor when he was a child in his native France, and he also dreamed of visiting some of the most exotic places on Earth. This latter dream, at least, has come true as Didier first went to India in the 1970s; after two and a half years there, he went on to Peru in 1979. Peru had captured his imagination ever since he had read the Tintin comic album *Prisoners of Sun* as a child. He thought that a journey to the country of his dreams would satisfy his need for adventure, after which he could return to France and reintegrate himself into society. But destiny had something else waiting for him. In Peru, Didier was employed by a tourism company as an administrator of a lodge deep in the Amazonian rainforest, and he stayed. A couple of years later, the same company gave him the task to plan a thematic tour around the country, focusing on the topic of traditional medicine. Thanks to this task, Didier got the opportunity to travel in Peru for six months, all the time learning about traditional medicine.

During this period, he spent some time among the Shipibo-Conibo people in Amazonia, and he became the apprentice of a local shaman – a relationship that continued long after he completed the job for the tourism company. One thing led to another, and soon he was contracted to work with the AMETRA- 2001 project, which focused on the recovery of traditional medicine among the indigenous peoples of the department of Madre de Dios in the southeastern part of Peruvian Amazonia (Alexiades and Lacaze, 1996). As his reputation as an authority on traditional medicine grew, he was hired in 2000 by an international nongovernmental organisation to work for one year on a similar project with an indigenous federation, the *Federación Shuar*, in Ecuadorian Amazonia. However, after his contract ended, he continued providing advisory services for several additional years, during which he also spent some time in the United States receiving training in the processing of plant material into medicinal and cosmetic products. Didier liked Ecuador and decided to stay. He bought a piece of land and built a house, and somewhere along the way he also married Rosa, who is a distant relative of Simón. Today, in addition to doing health-related consultancies for oil

companies and local health authorities, among others, he and Rosa also run their own business producing natural medicines and cosmetics.

For being a businessman, Didier is remarkably uninterested in talking about business. He is more interested in the seemingly paradoxical situation of the indigenous people in Amazonia, who live in the midst of such a cornucopia of medicinal plants of which they possess a rich knowledge, yet they very often remain in poverty and have a poor health status in comparison to the national averages. Moreover, in Didier's view, they all too easily turn to western medicine and transport the sick to hospitals in town instead of making use of the plant medicines available in their own surroundings. Although Didier has worked in numerous projects recovering traditional medicine and has observed still other similar projects from a distance, he thinks that there have been little results compared to the efforts and resources invested. All of this concerns him very much.

Starting the company was a way to get experience in actually making a business out of medicinal plants. Didier's goal is to share these experiences with people from the indigenous communities, arranging courses and workshops, hoping that one day there will be people interested in taking active roles in the business. For the moment, most of the people involved are providers of raw materials, either to Didier himself, other merchants who visit the communities every now and then, or shop owners who receive raw materials in their stores in the town of Puyo. The company is not a big business. Its monthly sales are just about $3500, but after costs are paid this is still enough to permanently employ three local people in addition to Didier, who only sometimes takes out any salary for himself. One employee is in charge of the garden, where much of the raw materials are cultivated or harvested. Another employee works in the laboratory, where the materials are processed and bottled or packed, and a third runs a small shop selling the final products in the center of Puyo.

THE GARDEN AND THE FOREST

The company's products are based on a total of some 30 plant species; more than 20 of the plants, mostly herbs, are cultivated in a 1-ha garden. Most of these plants belong to species that are commonly cultivated by the indigenous peoples in the region, but that does not necessarily mean that they are native to Amazonia. For example, the plant commonly called the air plant or life plant (*Kalanchoe pinnata*) and locally known as *pakipanga* is very frequently used by the indigenous people to treat wounds and injuries, but it is actually native to Madagascar. Another example is ginger (*Zingiber officinalis*), which is amazingly effective against flu symptoms and originates from Asia. Two species used by the company are brought from the Andean highlands, one is imported from France, and only five species are actually harvested from the wild. Out of these five species, three are pioneer species that are common in secondary forests: the cat's claw vine (*Uncaria tomentosa*); the dragon's blood tree (*Croton lechleri*), known as *sangre de drago* in Ecuador

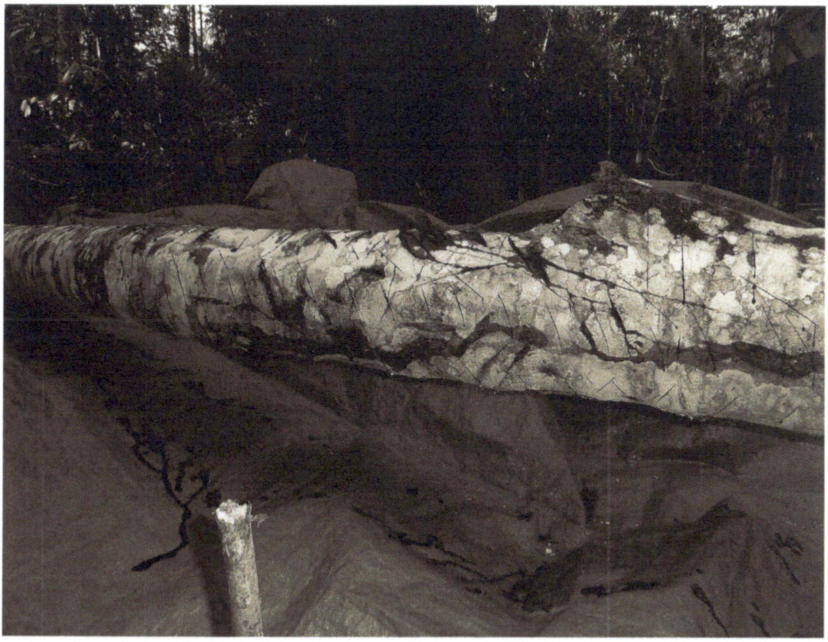

FIGURE 10.4 *Sangre de drago* is often tapped from standing trees when the latex is needed for personal or household use. When the latex is extracted for commercial purposes, it is done by felling the trees, as in this photo.

and *sangre de grado* in Peru (Figure 10.4); and sarsaparilla (*Smilax* sp.). Only two species are harvested from the primary forest: a palm locally called *ungurahua* (*Oenocarpus bataua*) and the *chuchuwasa* tree. Didier is not directly involved in the harvest of these species; instead, he buys the material partly processed from local indigenous communities near the town of Puyo. And, he is concerned about the sustainability of the extraction.

The *ungurahua* is a tall palm that bears oily fruits that are tasty and nutritious, and they are a popular snack among the local people. Sometimes, the locals also extract the oil in order to use it as a moisturising hair care product. For this purpose, *ungurahua* oil-based products are also sold on the national and international market. This tree grows slowly and is uncommon in secondary forests. Although many people have one or a few trees planted near their houses, this is not enough to satisfy even the household consumption, much less to manufacture products for commercialisation, so inevitably most of the extraction takes place in the primary forest. *Ungurahua* can reach 25m in height. Although it is often perfectly possible to harvest the fruits from even the highest palms by climbing a nearby tree, it is much faster to just cut the tree down. This practice is common, Didier says, and raw oil and shampoos based on such unsustainable extraction of *ungurahua* are even exported as certified organic products. All Didier himself can do to prevent unsustainable harvest of *ungurahua*, though, is to ask his providers not to cut the trees down, but he is not sure that they follow his advice. Some organisations, Didier tells, have recently tried to promote the use of climbing irons for ascending the stem of the ungurahua tree itself, which should be more sustainable than cutting the trees down, as well as much faster and easier than climbing a nearby tree using just hands and feet.

The other species harvested from primary forest is the *chuchuwasa*, the tree that Simón had felled in order to sell the bark to Didier. Although Didier asks his providers not to cut these trees down, he realises that cutting the tree is, by far, the most practical way to harvest the bark in any quantities beyond mere household consumption. Didier hopes to be able to introduce some sort of sustainable management practices to replace cutting the trees. However, he fears that even if one does not fell the tree, just removing large amounts of bark will also hurt it badly and eventually kill it. He speculates that it may perhaps be possible to harvest bark by climbing the tree and cutting off branches or even using the leaves of the tree instead of the bark, if it turns out that the same active ingredients present in the bark are also abundant enough in the leaves.

It may be surprising that a company dedicated to natural medicines and cosmetics in Amazonia actually uses mostly cultivated plants, with only two of the plants coming from the primary forest. There are reasons for this, however. For Didier and his company, it was a conscious choice to concentrate on species whose properties are well known and documented – both in order to be able to sustain claims about their effects in marketing and to avoid the risk of getting accused of the theft of intellectual property from the indigenous peoples (so-called biopiracy). Because documentation of a plant's bioactive properties usually starts with recording its use among indigenous or other local people, this implies that the selection of species largely reflects the indigenous peoples' knowledge about them.

Local people are often more familiar with the flora of gardens and secondary forests than with that of primary forests. This is logical because they often spend more time near the former than the latter, even in remote indigenous communities. When some specific medicinal plant is needed, gardens or secondary forests are usually the closest sources for such plants. Thus, the indigenous knowledge about the medicinal properties of plants growing in such familiar habitats therefore easily grows greater than the knowledge about plants in the primary forest. In these communities, men often go to the forest to hunt or fell timber, but these activities often take up a quite modest portion of their time compared to work near homes. Women, in turn, are in charge of the agricultural fields and the home, and go to the primary forest relatively seldom.

It is also often much cheaper to produce medicinal plants in gardens than to harvest them from the forest. Herbs cultivated in the garden occupy little space, become productive at an early age, and then provide a steady supply of raw material at a relatively low cost. For large and slow-growing trees such as *ungurahua* and *chuchuwasa*, the situation is somewhat different, though. There are no large plantations of these trees, and, if they were planted now, it would take decades until they would become productive. Because there is no experience with planting them on a large scale, it is uncertain whether this, in the end, would be profitable at all.

Thus, the only remaining option, at least for now, is to harvest these species from the wild. In spite of Simón's and Didier's concerns about unsustainable harvest, it should also be noted that commercial harvest of *chuchuwasa* is very limited in space. The province of Pastaza, where the town of Puyo and the village of Villaflora are located, covers 29,520 km^2, out of which most of the area is primary forest and far away from any roads. Commercial extraction of *chuchuwasa* is limited to the vicinity of a handful of penetration roads, such as the one that reaches the village of Villaflora. In villages further away from roads, such as Sarayaku (Chapter 4), extraction of various products from wild trees for sale to merchants was an important part of the local economy, at least from the late nineteenth century and up to the late twentieth century; however, it has now basically ceased because it is no longer profitable. Thus, over huge expanses of forest, the stands of *chuchuwasa* as well as other tree species with medicinal or cosmetic properties are unaffected by commercial harvest.

Although the species is not threatened as such, even local depletion of *chuchuwasa* stands would be unlucky for the local people – and, as many other rainforest species, the *chuchuwasa* occurs naturally at low densities. Simón estimates that there are only four adult trees along the 8-km trail from the village to his *tambu*, and Didier believes there may be about two trees per hectare. This is the downside of high biodiversity. There are many useful species, but none of them is very common. As Didier puts it, 'Everybody talks about diversity, but I say that we need to talk about density as well.'

UNUSUAL USE OF A LAUNDRY ROOM

Converting the raw plant material into finished products that are bottled, packed, and labelled is the work of Didier's wife Rosa, who is in charge of the "laboratory," a small room in the couple's house originally designed as the place to have the laundry machine. The benches of tile originally made for folding washed clothes are filled with jars and bottles with liquids of different colours (Figure 10.5).

Some products are commercialised basically unprocessed, such as the *chuchuwasa* bark, sold in cellophane bags containing 100 g. The latex of the *sangre de drago* tree and the oil of the *ungurahua* fruits are just bottled in glass droppers and sold pure. Many herbs are processed through infusion, putting them in boiled water but not boiling them. Many barks and roots are processed through decoction by boiling them in water over a prolonged time and sometimes adding some alcohol (to a concentration of 10%) as a preservative. Alternatively, the process of tincturing involves putting the raw material in a mixture of water and alcohol (60–70%) and shaking it every day for two weeks. Some products are then further processed into lotions, creams, repellents, soaps, shampoos, or 'floral waters' used by local healers in Puyo.

FIGURE 10.5 Didier and Rosa in the laboratory where they process the plant material into products to be commercialised. (Please see the colour plate at the back of the book.)

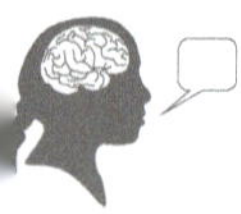

'We do everything ourselves and use fresh products. Therefore, our products are of high quality,' Didier says. He is convinced about the efficiency of medicinal plants. 'The idea of conventional medicine is to use certain chemical compounds with specific properties. The idea of medicinal plants, however, is that they generate processes in your body, helping it to recover by self-healing,' he says. Didier also expresses a positive attitude towards natural science: 'We need more scientific studies. We need to know more about what effects all those medicinal plants really have. So much is based on hearsay, traditions, and accumulated experience over generations, but we need scientific studies to support our claims, in order to be able to provide information to the customers about our products.'

PROFITABLE SUSTAINABILITY?

Sixty years ago, the town of Puyo had just a few hundred inhabitants living in houses with thatched roofs, but today the thatched-roof houses are long gone. Of the houses of wooden boards and tin roofs that dominated the townscape just three decades ago, only a few remain between the concrete houses that grow ever higher each time an old house is levelled to the ground to make a place for modernity. Here is Didier's company shop, housed in a small room with large display windows and walls covered with shelves full of small bottles (Figure 10.6).

FIGURE 10.6 The store in the town of Puyo.

The commerce is not very lively, though. 'The people in Puyo are very interested in natural medicine,' Didier says, 'but they prefer to buy products originating from the highlands or from other countries. Most people living here have come from the highlands; they have a highlander culture. They don't yet appreciate products from Amazonia, they don't know them, and thus are not sure if they can trust that they can be good. There are four or five natural medicine shops in Puyo, but they sell mostly imported products and very few products from Amazonia.'

About half of the sales in the shop, Didier estimates, are to tourists, including national tourists as well as tourists from other countries. Although the shop in Puyo is not very profitable, the company also distributes its products to retailers in seven other towns all over Ecuador and takes orders over the internet to deliver anywhere within Ecuador; it also does some exporting. Didier says that the market is growing, but the main obstacle to expanding, he says, are the legal requirements. Exportation requires a whole lot of paperwork. Sending 20–30 bottles to end consumers in other countries for their own use is feasible, but sending larger quantities to retailers would require having all export papers in order.

Even selling within Ecuador is becoming more complicated, as nowadays every product must have a sanitary registry; products lacking this may even be confiscated. A sanitary registry requires laboratory analyses and much paperwork, such that the total cost reaches some $2000 per product. In addition, the sanitary registry is valid for only 5 years, after which it must be upgraded to a so-called category A registry (the same as for regular drugs sold in pharmacies), which requires clinical studies as evidence of any medicinal properties claimed. For a small business marketing as many as 50 different products, this becomes prohibitively expensive. The company currently has sanitary registries for only 10 of its products; the intention is to acquire it for another 10, but then reduce the total number of products rather than pay for sanitary registries for the rest of the products currently marketed.

'In order to be profitable, we need to increase the volumes,' Didier says. Is there a middle road between the small-scale trade of unprocessed and cheap *chuchuwasa* bark at the Sunday market in Puyo and the large-scale trade leading to rapid resource depletion even in remote places, such as what happened in the past with rosewood (*Aniba* spp.) and quinine-containing *Cinchona* trees (see Chapter 3)? To some extent, there apparently is, as evidenced by the mere existence of a company like the one that Didier is running. Whether this experience will be replicated and whether the indigenous communities will get increasingly engaged in the business beyond being only providers of raw material will only be known with time.

Chapter 11

Açaí: The Forest Farms of the Amazon Estuary

THE *VER-O-PESO* MARKET

A round-bellied shirtless man places baskets made of palm fibre in a row on a cobbled market square. Then he piles more baskets upon the first row, and when the stand is ready he points at the baskets with a flashlight to show their content to passers-by. The baskets, called *paneiros*, are full of small round fruit; in the dim light mixing from the moon and the streetlamps, they first look almost black, but when the flashlight hits them it reveals a bluish-purple hue with a shade of grey. The full moon still hangs over the downtown buildings when the açaí section of the famous Ver-o-Peso market of Belém is at its busiest (Figure 11.1). The açaí palm (*Euterpe oleracea*) produces the dominant item for commerce in this section of the huge marketplace. The merchant middlemen and the rural producers have arrived in their boats and canoes through the darkness of the Guajará Bay using the myriad channels that lead through the archipelago of the Amazon estuary. Many of the boats come from the nearby islands, yet others have travelled much longer. Açaí fruit is available around the year, but the high season is during the dry months from August to January. Then, the boats make this journey every day.

The sides of the açaí market accommodate stands that sell meat, firewood, fish, manioc flour, peach palm fruits (*Bactris gasipaes*, here locally called *pupunha*), plantains, crabs, prawns, and other regional products, but the major part of the square is dominated by dozens of açaí stalls. During the late night and in the early morning hours before dawn, men carry thousands of baskets from numerous boats of different sizes. Their work has been somewhat easier now that the tide is high and the boats are tied up directly to the dock. When the tide is low, the boats cannot approach the dock, and the products have to be carried through a field of squashy mud where bands of black vultures and great egrets watch out for leftovers from the fishermen's vessels.

Nazareno is here before dawn every morning of the year. He owns two restaurants in Belém, his menu depends on an uninterrupted flow of fresh açaí from the surrounding islands, and he wants to buy the fruit personally. Nazareno puts great emphasis on the freshness of his açaí, and this is how he is able to make sure everything goes according to his standard. After the brokers have emptied their *paneiros* into plastic cases, two market workers drag the load in a cart

Diagnosing Wild Species Harvest. http://dx.doi.org/10.1016/B978-0-12-397204-0.00011-5

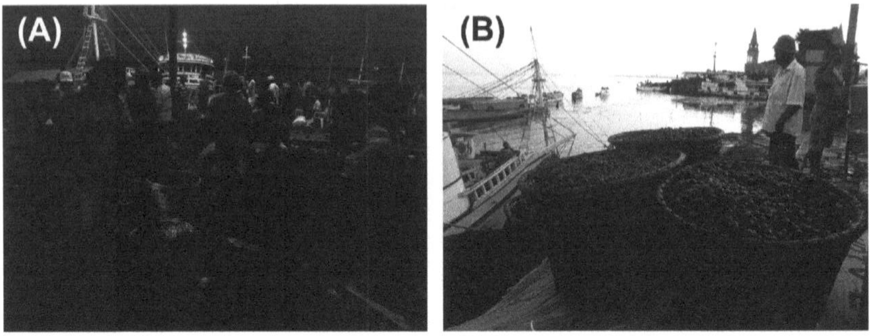

FIGURE 11.1 (A–B) The açaí section of the Ver-o-Peso market in Belém, Brazil, is busy in the early morning, when the harvesters and intermediary merchants arrive in boats and canoes from all over the Amazon estuary. Brokers and buyers swarm among the açaí baskets, and by sunrise the show is almost over *Photo B: Natalia Aravena.* (For (A) please see the colour plate at the back of the book.)

straight to where Nazareno's small truck waits parked. He then drives the product directly to his small processing plant, from where the pulp is further sent to the kitchens of his restaurants.

At the crack of dawn, the show is almost over. Most of the açaí has been sold, and an old man sits crouched on the ground and picks up fruits that have fallen from the vendors' baskets. He has protected his hand with a small white plastic bag, and as he picks the fruits one by one, he drops them carefully into another larger plastic bag. This is the way in which many poor households in Belém get extra income in a way that might be called a real dribble-down effect, as the pickers are mostly people who cannot find other work, many of them either aged or children. Nazareno buys a small plastic cup of black sugary coffee and says that a good day's picking can maybe yield one or even two baskets of açaí, about 40 Brazilian *reais* (BRL; or 15.50 euros) each, which would not be a bad salary compared to the national minimum wage, which is about 20 BRL a day. He remembers a single mother who had all of her children picking açaí this way. There are also other uses for açaí residue: while only about one-fifth of the açaí fruit's volume is pulp, the rest is in its kernel that can be used as fuel.

About 70 percent of the region's açaí production passes through the Belém market in order to supply the growing demand both locally and outside the region. During the peak of the season, thousands of açaí baskets are sold every day in the market; in the early 2000s, the daily volume during the high season was estimated to be between 10,000 and 15,000 baskets (up to 180 tons of açaí), and the volume has likely grown since then. The magnitude of the açaí economy in the estuary is something on the order of 40–200 million euros annually, and it has made a great contribution to the majority of the estuary households (Brondízio et al., 2002; Brondízio, 2011). Açaí is the first natural product since the great rubber boom that has achieved such a scale of economy, and it has been estimated that açaí juice is consumed significantly

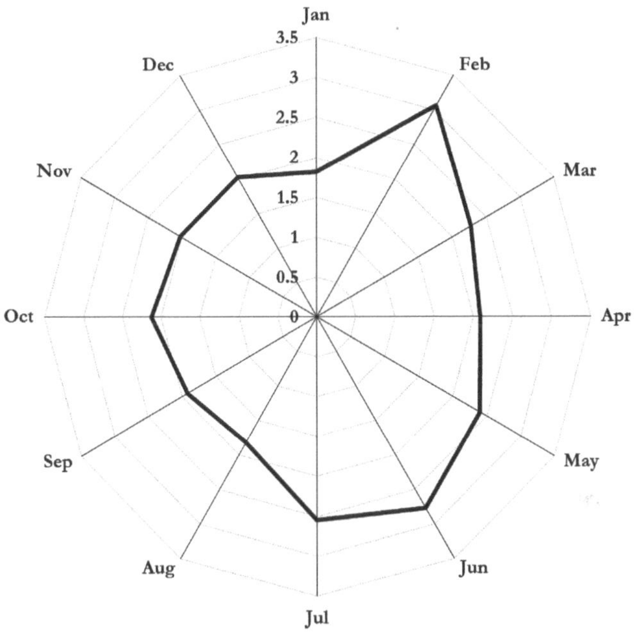

FIGURE 11.2 Monthly average prices paid by one daily buyer in the Belém açaí markets during the year 2011 (BLR per kilogram; data provided to M. Salo by the buyer). In general, the prices paid for açaí fluctuate according to the volumes available – both seasonally and during one single day. According to Brondizio et al. (2002), there are two practically continuous seasons: 'summer' from August to January and 'winter' from March to July. During the main season ('summer'), the prices can plummet. The seasons are somewhat visible in this figure. February as the month between the seasons stands out with the highest average prices in the small data set used to produce the graph.

more than milk in Belém (Brondízio et al., 2002). These volumes could not be processed using the traditional technology of extracting the pulp by hand, and one of the technological innovations that have enabled the açaí boom is the electric machinery used in the production of açaí juice.

Most of the trade is controlled by brokers who form a link between the producers and the middlemen at one end of the chain and the buyers at the other. Middlemen are a critical link in the commodity chain; although many times it can be argued that they diminish the producers' share of the profit, they are also commonly irreplaceable connectors between the backwaters of the estuary and the urban markets (Brondízio et al., 2002). Although improved management and particularly the expansion of açaí production in the regions surrounding the Amazon estuary have extended the supply over the whole year, the high and low seasons still make a big difference, and the price of açaí has strong seasonal fluctuations (Brondízio et al., 2002) (Figure 11.2). Thus, when the harvest is good, the joy of large volumes of açaí may be shadowed by the fact that constant levels of demand mean that increased supply makes the prices plummet.

When the season is low, on the other hand, the lack of açaí and consequent competition elevate prices. These are not the only factors that affect the prices; access to forest and particularly transport costs also have important roles. Only quite recently has the market started to be controlled by larger companies that are able to stock açaí and thus regulate the seasonality of supply (Brondízio et al., 2002). In any case, the local consumer prefers fresh açaí.

Today, Nazareno needs to buy more than two tonnes of açaí, and he cannot acquire all of it in the main market. So, after leaving a dozen large cases of fruit in his processing plant, he heads to another smaller açaí market called Conceição. Here, he says, the açaí is of better quality; it comes from the nearby islets and is thus very fresh. The açaí that comes from further in the archipelago or from the Island of Marajó decreases in its quality during the journey, decreasing more the further it travels. Without freezing or drying, the fruit perishes after three days of storing (Brondízio et al., 2002), but this kind of improved storing technology is mostly used only by larger companies that export açaí from the estuary. Although the good quality makes the product in Conceição more expensive, this time there are no middlemen involved; Nazareno has a deal directly with a producer and not with a broker. Louro lives less than half an hour's journey from the port of Conceicão. Now he has arrived to the market with half a dozen baskets of açaí and a couple of small baskets with live prawns. There are other buyers around, but Louro waits for Nazareno to sell him all his açaí.

Belém is full of açaí vendors; here and there, the streets are lined by red signs with white text: 'Açaí'. Açaí is something that simultaneously takes the city into the countryside and brings the countryside into the city, a phenomenon that has been described under the illustrating title 'Urban Forest and Rural Cities' (Padoch et al., 2008). This is how the fruit that used to be a highly valued staple eminently linked to poor rural dwellers has now entered not just the city restaurants in Belém but also the upper-class circles of Rio de Janeiro and São Paulo, or any other Brazilian city, and, further, the global urban markets as a booming superfood and an exotic supplement to gourmet plates in fine restaurants (Brondízio, 2008). At the same time, the harvest and trade of açaí have started to introduce components of the urban lifestyle into the estuary dwellings, where an increasing number of flat-screen televisions receive a plethora of satellite channels and show movies from DVD players. This development is reinforced by the blurred line between urban and rural as many families have some of their members constantly moving from the country to the city and back, as is typical for today's urban-rural movements in many tropical areas (Padoch et al., 2008).

For Nazareno, the açaí business used to be quite different only some years ago. In the mid-2000s, he sold açaí in his garage, serving people at a couple of tables just like countless other vendors in Belém. But from time to time, he started to wonder whether there was room for expansion; as people in the region have a strong connection to açaí, maybe regional plates could be accompanied with the juice to attract new clients. The idea turned out not to be bad, and soon Nazareno had two restaurants serving a large local clientele.

It is Saturday night, and everybody is out having a drink, dinner, or both. But Nazareno cannot have his free hours quite yet, because he has business to take care of. The following day is Mothers' Day Sunday, and he has to be there again early in the morning in the Ver-o-Peso market to guarantee his load of açaí. Many families will be out for lunch or dinner in his restaurants, and everybody wants to make sure there is enough fresh açaí juice to make the meal complete. Nazareno drives his car to the market and approaches a group of shirtless men playing cards on an improvised table in one dark corner of the market. This is the moment to close a deal about tomorrow morning's açaí baskets. After only a few short hours and another morning breaks, thousands of baskets of açaí will be sold and then many more litres of açaí juice enjoyed in the city; if he struck no deal, it could be that he would be left without his açaí.

AÇAÍ IN THE HISTORY OF THE AMAZON ESTUARY

The Amazon estuary is fed by several large rivers, including not only the Amazon and its tributaries, but also the Tocantins and Pará Rivers, forming an extensive network of interconnected rivers, streams, and channels, and containing the large 50,000 square kilometre Island of Marajó (Figure 11.3). The estuary provides a quite unique environment in that it is affected both by daily tides, due to its proximity to

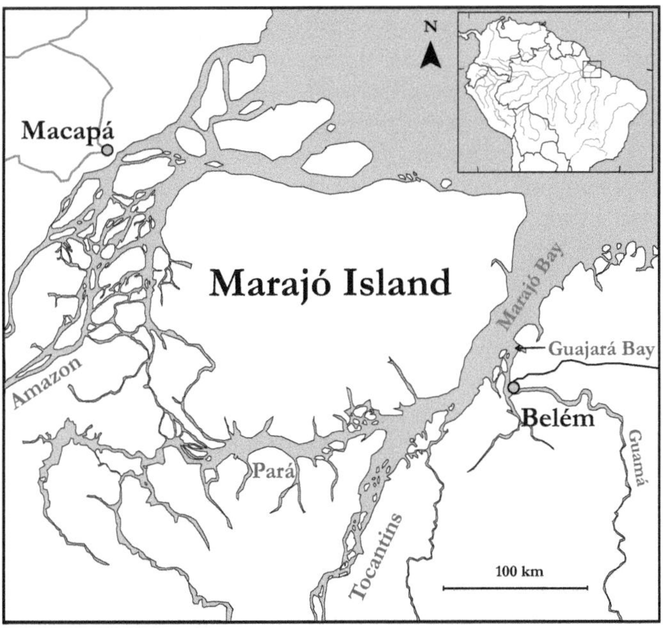

FIGURE 11.3 The city of Belém lies on the banks of the Guajará Bay and the Guamá River on the south-eastern side of the Amazon estuary. In the middle of the estuary, there is the Island of Marajó, which is similar in area to the Netherlands.

the ocean, and by seasonal flooding, due to seasonal variation in rainfall and associated river discharge in the larger Amazon basin. The vegetation is therefore dominated by species that are particularly adapted to these extreme circumstances, and plant biodiversity is relatively low by Amazonian standards. One dominant species is the açaí palm, which is appreciated not only for its fruit but also for its edible palm hearts (i.e. the growing bud at the tip of each stem). Unlike many other palm species, the açaí grows in clumps and each one bears multiple stems, meaning that cutting a stem does not kill the whole individual but instead can stimulate the growth of the other stems on the same clump (Weinstein and Moegenburg, 2004).

Açaí fruits have been an important component of the diet of the people living along the lower Amazon ever since pre-Columbian times (Schaan, 2010). Most of the current rural population of the region is of mixed ethnic origin, including indigenous, Portuguese, and African influences. They form a distinct cultural group often called *caboclos* or *ribeirinhos* that emerged during the eighteenth and nineteenth centuries as the Portuguese royalty called upon missionaries to simultaneously 'civilise' the 'Indians' and protect them from the cruel treatment they had often been subjected to by colonisers and slave hunters (Brondizio and Siqueira, 1997). The missionaries established a language mix of indigenous Tupí and Portuguese as the lingua franca, calling it *lingua geral* or *inhengatu*, and persuaded the indigenous population to take part in cash cropping, ranching, and extractivism under the control of the missions themselves.

In the latter part of the eighteenth century, however, the missionaries were expelled by the Portuguese crown, which instead created a directorate system (*diretórios*) that promoted ethnic-mixing marriages and appointed selected local leaders, who were often of indigenous origin (de Alencar Guzman, 2009; MacLachlan, 1972), to be in charge of organising agriculture and increasing the export of commodities to Europe. The directorates were in place only from 1757 to 1799, but they had long-lasting impacts on the region's society. The *lingua geral* was suppressed and was increasingly replaced by Portuguese, and the vast land areas previously controlled by the missions were transformed into private land holdings. Large farms, encompassing almost all of the floodplain areas as well as some upland forests, were sold or granted to Portuguese settlers for establishing cash-crop plantations or cattle ranches, whereas the indigenous people received only small plots for subsistence agriculture or worked on the plantations of the Portuguese. This was the context in which the *caboclo* culture emerged, subjugated to the large landowners, a condition that was even more reinforced during the rubber boom in the late nineteenth century. Still today, many of the *caboclos* live on land they just rent from absentee landowners.

In the 1960s and 1970s, there was a boom in the extraction of palm hearts due to increased demand in the cities of southern Brazil, where it was considered a luxury delicacy. Harvesting palm hearts from most other palm species inevitably implies killing the tree but, as mentioned above, this is not the case for multistem species such as açaí. Nevertheless, during these decades, many açaí stands became so intensively exploited that there was a marked decline in their abundance over large parts of the estuary. This picture had already started

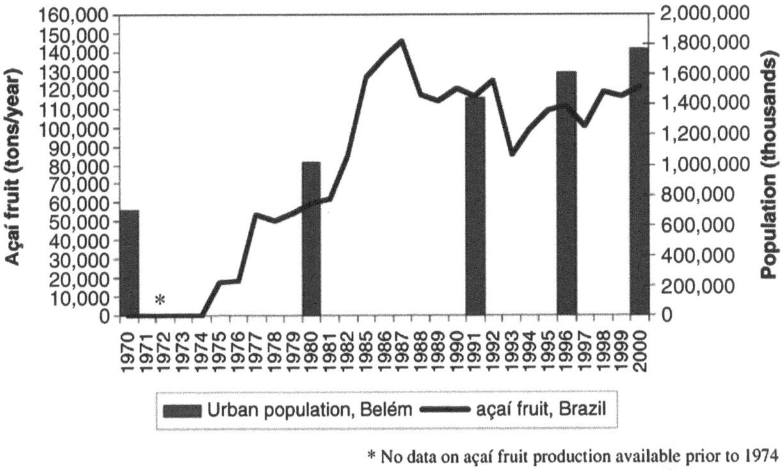

FIGURE 11.4 Brazilian production of açaí fruit (1974–2000) and population growth in Belém and surrounding urban areas (1970–2000). *From Brondizio et al. (2002).*

to change in the 1970s, however, when increasing numbers of rural dwellers migrated to the city of Belém, maintaining their preference for consuming açaí fruit (Figure 11.4).

A PERI-URBAN FOREST FARM

From Belém, Louro's house is just around the corner. His small boat makes the journey in about 20 min through a landscape of low-lying islands bordered by mangrove forest. The contrast is striking when the Belém skyline is left behind and in a few minutes it disappears behind the forested island banks. Louro's house has a landing that accommodates up to several metres of variation in the level of the tide. A small orderly tin-roofed house stands a few dozen metres from the landing, just beyond the reach of the normal high tide. The forest surrounding the house harbours a dense stand of açaí palms but also many other fruit-bearing trees such as cacao (*Theobroma cacao*) and its relative, *cupuaçu* (*Theobroma grandiflorum*). Louro's living room is spacious, and on one side of the room there is a wide flat-screen television. This seems to be a prime example of the style of extractivism that brings material wealth into rural households.

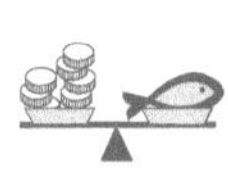

It is not the high season, but Louro's açaí stand bears fruit. His son Flavio uses a palm leaf stalk to make a firm loop that he accommodates around his feet. Then, he just jumps up on the palm trunk and starts to climb it at a considerable speed (Figure 11.5). Once up, Flavio cuts the bunch of fruit with a machete and slides down like a fireman. Then he rips the fruits off the bunch by hand. All this happens in a few minutes, after which Flavio climbs again. He tells about accidents that happen when men fall from the palms. 'Last year, two men died this way', he says. Despite any potential hardship, many seem to want

FIGURE 11.5 (A) The açaí fruit are stripped from the racemes by hand *Photo: Natalia Aravena*. (B) Climbing to an açaí palm is facilitated by the use of slings made of palm fibre.

to work in the açaí economy, and it is undoubtedly an opportunity to generate cash in quantities that are otherwise unthinkable. A good investment can be, for example, a motorboat that provides direct access to the açaí market or a small house in the outskirts of Belém. Such a city dwelling is highly desirable for many because it improves access to many urban services such as, for example, healthcare and education.

Meanwhile, Louro walks down to a small creek to check a prawn trap (*matapi*) made of fibres of the jupatí palm (*Raphia taedigera*). He collects a few prawns and laments the difficulty of getting a labour force to harvest açaí, especially during the peak of the season. He thinks that government subsidies have made people lazy. This may be a politically founded opinion, but according to studies, it is true that many households depend on federal aid channelled through programmes such as the *Bolsa familia*, a programme that aims at reducing poverty and promoting education through direct financial aid (Brondízio, 2011). The açaí economy, though, can also have the opposite effect: many schoolboys miss classes because they are perfect açaí harvesters.

In just a couple of morning hours, Flavio collects a large basket full of açaí. With the moment's prices, he will be able to sell it for up to 100 BLR (around 40 euros). He tells that during the high harvest season, the work is even faster as then you need to climb only three or four palms in order to fill the basket. Sometimes, nearby palms can be harvested by climbing just one of them and using a stick to reach the next stem and draw it closer. In Louro's forest, a typical high-season day yields 60 to 70 baskets. On the other hand, during the peak season the prices are considerably lower.

WILD OR CULTIVATED?

Although açaí is a native plant of the estuary, its production today increasingly involves management activities that can be grouped into three main categories: cleaning, stem cutting, and enrichment, each of which can be further divided into different subcategories (Weinstein and Moegenburg, 2004). Almost all producers weed the understory vegetation in order to reduce competition such that the açaí palms grow faster. Many also cut lianas and vines, and fell other trees in a selective fashion, although they save valuable trees and avoid opening up the canopy too much, as this negatively affects the quality of the fruit. Most producers also cut açaí stems to extract the palm hearts or to remove stems that are too tall to climb or too old to produce much fruit. Finally, many also practice enrichment planting of açaí seeds or seedlings in order to increase the density of the species. Much less commonly, some people cut off inflorescences in order to delay the fruiting to the rainy season, when the price for açaí is higher, and some tie knots in the leaves of young trees, claiming that this makes the adult tree grow shorter, which facilitates collecting the fruits, and makes the palm hearts grow thicker.

Recently, it has become common to plant açaí even outside of the natural habitat of the species, in the uplands that are not subject to floods. Although the growth there is slower than on the floodplains, the expected profit is nevertheless greater than that of alternative land uses such as agriculture producing, for example, manioc. The use of synthetic fertilisers to enhance production has recently been introduced, too (Lewis, 2008). If the palm heart boom in the 1960s and 1970s led to overexploitation and reduced the abundance of açaí

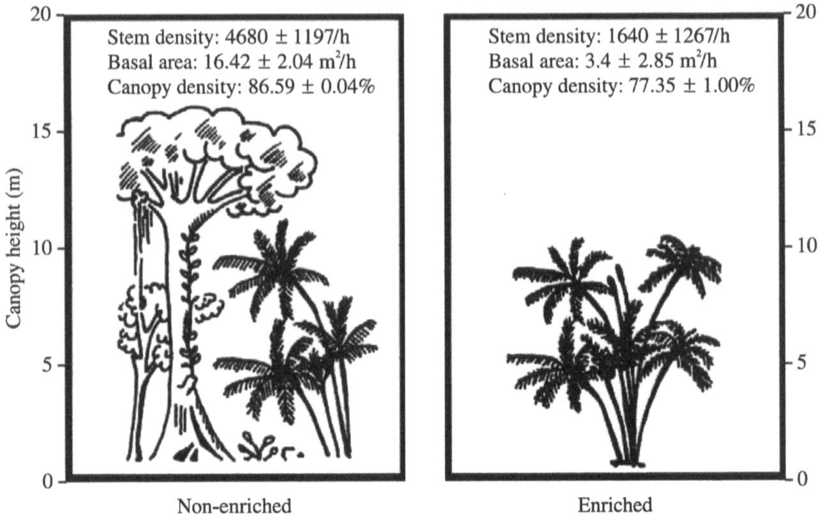

FIGURE 11.6 Typical vegetation profiles in unmanaged and managed açaí forests. *From Weinstein and Moegenburg (2004).*

fruit, the effect of the açaí juice boom has been quite the opposite. The density of adult açaí trees in enriched forests is about six times that in unmanaged natural stands (Weinstein and Moegenburg, 2004).

The obvious downside of the açaí boom is that the increased density has taken place at the cost of other species. The overall number of tree species is greatly reduced in managed açaí stands, as are also the density of stems and the basal area of tree species other than açaí (Brondizio and Siqueira, 1997; Weinstein and Moegenburg, 2004) (Figure 11.6). This process of landscape-level change in Amazon estuary plant communities has been called 'açaízation' (Hiraoka, 1995). However, the second most profitable land use in many parts of the estuary, after açaí management, is hardly to leave the forest untouched but, rather, to use the land for agricultural purposes. The açaí fruit boom has actually, in some places, led to the conversion of sugar cane plantations and areas used for swidden agriculture into managed açaí forests.

THE BOOMING AÇAÍ ECONOMY

The increasing value of the açaí trade has had a profound effect on the estuary lifestyle. One of the implications has been the revaluation of land and access to resources. Louro owns his plot and is proud to tell that his family has a history of four centuries in the estuary. He describes how generation after generation has been engaged in different economic activities in a succession of farming and extraction, and how this legacy has been shifted from father to son. The difference seems to be, however, that the generations before Louro were poor, as he puts it; they did not harvest açaí but rather products from trees like *andiroba* (*Carapa guianensis*) or wild rubber (*Hevea brasiliensis*). Now, Louro has three parcels on the Island of Combú, where he started enrichment planting of açaí in the mid-1970s. Back then, he says, there was not much açaí at all there, not even 'enough for a juice for himself', as he puts it. Now, açaí can be harvested throughout the year and not just during the peak season. Louro says that the palms yield a bunch every 15 days.

Many other families with generations of history in the region are only renting their lands. Many of them have faced the misfortune of being expelled by land owners or claimants when land holdings previously seen as of little use have suddenly become revalued because of açaí (Brondízio, 2011). In Brazil, the peasants' ability to show they have used the land (i.e. converted it to agriculture) has often helped them claim ownership. In this sense, the extractive nature of açaí production works in the absentee landowner's favour and against the sharecropper families, who only enjoy usufruct that they pay for to the land-owner in agricultural or extractive goods. The authorities' willingness and ability to solve disputes of possession or to guarantee exclusive rights of access are also in doubt. One açaí producer refers to this problem by saying that any outsiders who enter and harvest açaí on his lands would need to be expelled by force: 'Stealing my açaí? Nobody does that and if someone did, I know people that can take care of such thieves.' Conflicts between land owners and peasant extractors are frequent; on the one hand, land owners and middlemen can easily undervalue the açaí harvest to increase their profit, and, on the other, extractors can hide a part of the harvest and sell it clandestinely to other middlemen (Brondízio et al., 2002). Everybody is aware of the prices. It is just about who has the right to sell and to whom.

The estuary is constantly fertilised by the sediments brought by the Amazon River. Açaí production competes with other productive systems of the region such as swidden agriculture, selective logging, and cattle ranching. However, the booming açaí economy has not changed the human landscape of the Amazon estuary completely. While Louro's flat-screen television is a clear indication of increased cash income in some households, even the better off cannot escape the fact that the açaí boom has not contributed much to public services; most of the açaí economy is informal, and the municipal authorities lack the means to levy taxes on the wealth that the sector undoubtedly generates. This contributes to a paradox in which demand for public services grows along with the booming açaí economy, but the municipalities that have to respond to this demand languish having no more money than before for accomplishing the task (Brondízio, 2011).

The growing demand for açaí outside of the estuary means that the price hikes for the best quality açaí can increasingly leave the urban poor unable to purchase it, which may thus affect their diet. This has also led to the manipulation of açaí products with additives such as starch and water in a way that increases the volume but decreases their nutrition content (Brondízio, 2011).

The local açaí economy also largely lacks a refined transformation industry; most producers just sell the fruit, and, although in Belém there are processing plants, most of the added value and thus wealth are generated higher up in the commodity chain and even outside of the region. In São Paulo, açaí is not necessarily synonymous with açaí juice, but the products are more elaborated, such as ice cream, yoghurt, sports drinks, and cosmetics, to list a few. Açaí is marketed now as a superfood because of its energetic and health properties, and it is not just an urban fashion anymore; international demand is also growing rapidly (Brondízio, 2008). Thus, açaí definitely has that 'something' that makes it an exceptional case of wild species harvest – but it is still a challenge to devise mechanisms that could be used to more efficiently channel the benefits created by the açaí economy to benefit society as a whole in the region.

Chapter 12

Blank Maps and Desires about Biodiversity Wealth

ROADS INTO THE UNKNOWN

Seen through the small window of a jet airplane, the green carpet of lowland rainforests extends to the horizon in all directions and seems almost intact, only broken by meandering rivers here and there. A few minutes before approaching the airport of the city of Iquitos, however, one starts to see deforested land, yet it only provides a vague idea of the various types of human impacts that lead to forest clearance in the region. In comparison, a flight in a light aircraft over the surroundings of Iquitos is a more eye-opening experience. The details within this type of landscape during such a lower and slower moving flight are intriguing. Around the urban and suburban spheres, forest has been cut for agriculture, pasture, and road building; for chicken farms or leisure resorts; and for many other, maybe less perceivable reasons (Figure 12.1).

In most cases where the forest has been cut, however, the use of the cleared lands does not necessarily look very intensive. Some pastures seem to be almost abandoned as there are only a few cattle, if any, and houses and people abound only near roads and at the river margins. In the places where secondary growth is taking over, all of the trees have the same height, and wherever the human-modified lands have contact with the surrounding multilayered natural forests, the border is crisp. Dense forests seem like an obstacle that does not attract humans; at least evidence of their uses is hard to see when looking from the air.

But no doubt forests are being used there, too; it is only that some forms of wild species harvest are very discreet, even to an experienced eye and on the ground level. Moderate levels of fishing, hunting, and gathering can go almost unnoticed in a natural landscape. Large-bodied mammals may have been hunted out, but this does not change the way the forests look in any obvious way, even when walking therein. Other types of human activities, however, such as the logging of trees cannot easily escape detection. Even selective logging can be identified in the forest landscape using modern high-resolution satellite imagery. The benefit in using such Earth-observing data is also that it provides a synoptic view that allows one to assess landscapes analytically as a whole, which is quite different from just looking at pieces of the landscape at an oblique angle through an aircraft window.

Diagnosing Wild Species Harvest. http://dx.doi.org/10.1016/B978-0-12-397204-0.00012-7

FIGURE 12.1 Different kinds of land use in the surroundings of Iquitos. (A) Forest cover with an intact appearance; (B) a penetrating road surrounded by patches of old farms with secondary growth; (C) forest freshly felled for plantations; and (D) forest cleared for a chicken farm. (For (B) please see the colour plate at the back of the book.)

Satellite images also help one to realise how different the landscape patterns in natural and human-dominated areas are (Figure 12.2). Particularly when natural resources are used in large quantities or infrastructures are built, human impact often becomes strikingly visible. Forest clearances appear like wounds – as drastically altered crisp bordered landscape elements within the natural setting of rather soft-looking forest canopy. Although most of such scars on the face of the Earth directly result from other human activities than wild species harvest, there are multiple ways in which the patterns of clear-cut forests also tell about harvest from the wild.

The chicken farms seen around Iquitos provide an interesting example. Chicken is a popular food, and there are countless chicken restaurants in the city. This mostly urban demand for chicken meat, which is produced using processes of semi-industrial efficacy, does not seem an important case of wild species harvest. But it has its connection; the vast majority of the roasted-chicken plates in the city are prepared using charcoal produced in the forests surrounding the city. Charcoal is generated by burning wood very slowly in a low-oxygen environment, and the more chicken is eaten, the more charcoal is needed and consequently produced when forest is cleared, burning wood from different species of primary or secondary vegetation. In addition, some nearby farmlands may serve to provide feed for the chickens, although most of the fodder used in this region is actually imported.

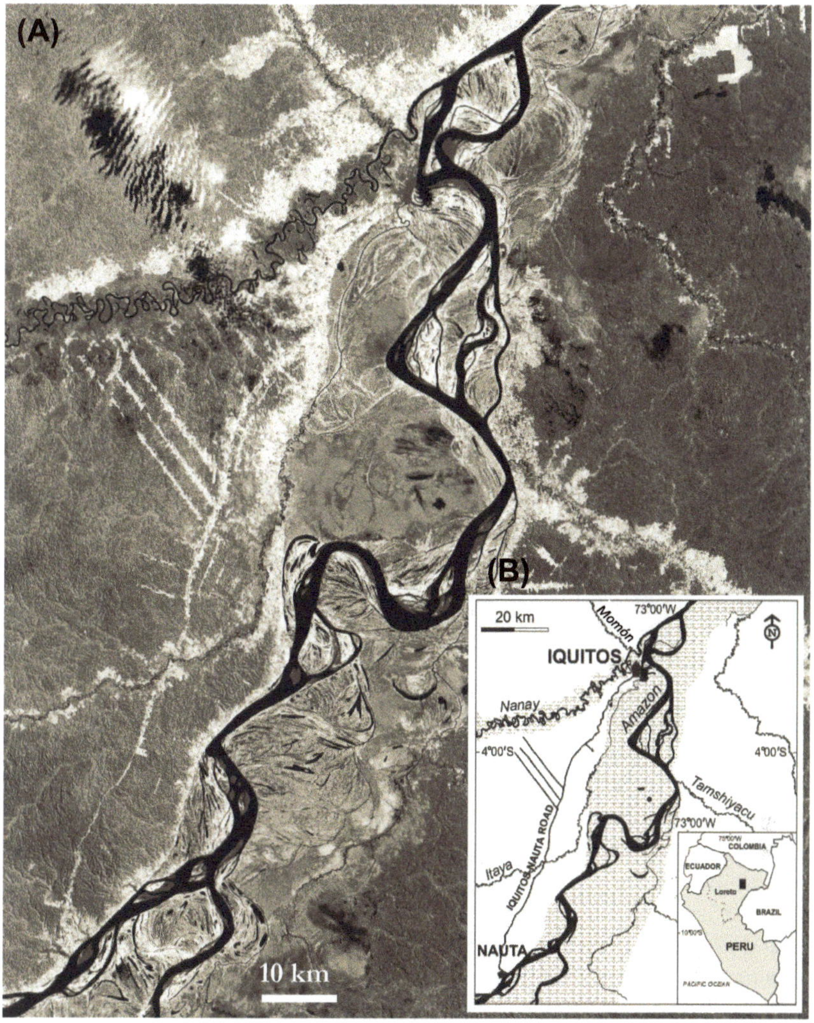

FIGURE 12.2 (A) A satellite image (Landsat TM from 8 August, 1993) showing the regions of the cities Iquitos and Nauta at the time when the two were still not connected by a permanently passable road. The cleared line in the forest is well visible all through, but in practice the road was passable only from Iquitos to a few kilometres south of the Itaya River. The image also shows three parallel penetrating roads running north-west from the main road toward the Nanay River. Inset map (B) shows the locations.

With its almost half a million inhabitants, Iquitos is a huge consumer of the region's natural resources. The city has grown and still functions as an important economic and administrative centre, and its strategic location near the junction of the big rivers of Ucayali, Marañon, and Napo offers good access to large forest areas in the Upper Amazonia. For instance, it is easy to transport products to Iquitos just by floating them and using the river current.

Iquitos is also the furthermost post to which oceangoing vessels can navigate from the Atlantic Ocean. This setting has provided perfect conditions for the export of Amazonian goods to world markets. Nowadays, Iquitos is one of Amazonia's most important cities, and it is also the world's largest city not connected to the national road network. Thus, in addition to fluvial transportation, the growth of Iquitos has been based on a well-functioning air bridge. This situation, however, is not fully satisfactory anymore from the local perspective, and there are constantly evolving plans for terrestrial connections, including a railroad to the city of Yurimaguas located in the foothills of the Andes.

There is one road, however. Towards the end of the twentieth century, the increasing pressure to build roads from Iquitos finally materialised in an endeavour to link the city with the town of Nauta some 100 km to the southwest. Plans for this road existed, among other regional development desires, for decades before they became reality in 2002, when the connection was finished. In addition to connecting two major population centres of the region, improved transport conditions enhanced the interaction between the rural and urban areas within this region by opening new frontiers for settlement, forest extraction, and agriculture. The road did not come into existence at once; its first parts had been built already in the 1970s, but then the construction works ceased for more than a decade. This left the road in such a state that in the late 1990s there was an asphalt highway heading southwest from Iquitos, which after some 20 km turned into a muddy gravel road, and even further on into a mere footpath in the middle of a pristine-looking forest. In Figure 12.2, it is possible to distinguish a narrow cleared corridor from the concurrent road ending toward the town of Nauta.

Each time that the road construction took a step further away from Iquitos, resource extractors and settlers rushed in (Figure 12.3). Wildlife hunters had their best chance right after new forest areas became easily accessible. Then followed the extractors of many other resources. These conditions favoured the harvest of any such forest products that could be transported away by foot or by using motorcycles on the muddy initial road. At some places, there were signs indicating properties on the flanks of the road to-be. As there were not too many guards around, many extractors probably used this window of opportunity without paying attention to the legality of their work. In those places where the quality of the road was better, an increasing number of people began to settle, each time with new types of resource use activities. At the time when the satellite image of Figure 12.2 was taken, a clearly zoned pattern of different development belts existed along the roadside, starting from the urban fringe of Iquitos and ending up with a nearly pristine forest at the periphery (Mäki et al., 2001). Then, after the road was completed, this setting started to change towards another type of regional dynamics, one dominated by two population centres with rural lands in between.

FIGURE 12.3 (A) Where road construction had ended in the early 2000s, a footpath surrounded by some cleared lands continued. (B) The final parts of the road were muddy; and, at places; (C) logs to be transported were laying on the ground. *Photos: Sanna Mäki.*

New roads can make important social, political, and economic contributions, but they also have a peculiar effect on people's minds. The people inhabiting penetration roads are colonists who may not know much about the space around them. Indicative of this, place names are scarce, whilst it is most typical to refer to different locations along the road according to their distances from the road's beginning. Somebody may live 32 km from Iquitos, whilst another family dwells 46 km away. In one way, this also tells how the colonists establish their lives in a 'nowhere' that only little by little starts to become a home. A local sense of place will inevitably develop with time, but nevertheless in the initial phase newcomers may lack place-specific knowledge about the region's natural resources and their use possibilities. Often, this means burning the forest for agriculture out of a perceived lack of better alternatives.

Penetration roads into primary forest can also serve to relocate unemployed people. This reality is illustrated by the settlement called Ex Petroleros near Iquitos, a name referring to a group of people who had earlier worked in the infrastructure-building phase of the oil and gas plants in the Peruvian lowlands. When the production phase of these resources started, the labour used in the construction became redundant for the companies; when these labourers were left jobless, they sought new opportunities, and the new road was there to help. Many of these new colonists were of Andean or coastal origin, probably with limited traditional knowledge about how to live by using the wild species of the diverse Amazonian nature.

Policies supporting road construction, with eyes closed to the consequences, may prevail under the illusion of endless resource (land) availability. Amazonia is endlessly big, or so it seems. When there are always new pristine areas where the same recipe can be repeated, losses of resources due to short-term use do not seem like a problem. Over time, however, untouched forest will become a scarcity. Regions facing these problems are already common in many parts of Amazonia, but in the region of Iquitos, the transition from the extraction-based forest use that is characteristic of remote locations to more intensive agricultural use is still underway, as the example of the Iquitos–Nauta road serves to illustrate.

An eye-catching feature in the satellite image of Figure 12.2 is the set of three parallel roads from the main road in the direction of the Nanay River in the northwest. No doubt, those roads have been designed by using a ruler on a planning map, and later they were implemented in the field with the help of some targeted project money. There is reason to believe that the planning map was probably pretty blank, with hardly any information about land suitability. Just looking at the satellite image, it is clear how deforestation follows these penetration roads as very narrow belts and almost disappears where dark spots in the forest occur. The story behind this setting provides an important lesson about the need to use solid facts in land-use planning and not underestimate the significance of environmental constraints.

In large parts of lowland Amazonia, although by no means everywhere, the soils supporting tropical rainforests are poor in nutrients. The situation is at its extreme in places where the soil is formed of pure white quartz sand. In these areas, the particularly harsh growth conditions give rise to highly specialised ecosystems with reduced biomass and many endemic species (Fine et al. 2011). Such spots are portrayed in Figure 12.2 as dark patches west from Iquitos. It is common knowledge in the region that the soils in these areas are unsuitable for agricultural production. However, the road designers hardly realised that their plans cut through many such patches of land. The building of these penetration roads opened a rapidly expanding deforestation front that very soon, however, got restricted into a narrow belt in the roads' vicinity. The reason for any clearing taking place at all was that, at the time, it was supported by subsidies whose payments required colonists to show that their work in the field had started. So there was an incentive for the colonists to burn some forest, but there was no good reason to stay and continue farming: it would not be successful anyway. That rich tropical rainforests got destroyed just so that somebody should be able to get a cheap loan may sound absurd. Yet from the perspective of the individuals involved, given the incentives they faced, it made sense.

Another indication of the region's variable soil properties and their significance for land-use decisions can be found near the Nanay River in the northern part of the same satellite image (Figure 12.2). Down the river, on the right-hand (southern) side of the river channel, hardly any cleared lands occur. Instead, on the left-hand side, a belt of white deforested lands consisting of small agricultural plots and fallows occurs. The local people, when asked, indicated that the reason for this difference is obvious. Everybody knows that the soils on the right-hand side of the river are rather impossible for agriculture, whereas the other side offers much better soils. A woman in the village of Mishana made this point very clear, stating that the soils on the right side of the river lack 'vitamins'. This notion unmistakably reveals how residents living in a region are aware of many such ecological realities that the planners working behind their desks may ignore.

BIODIVERSITY'S PROMISE

In a regional expert meeting discussing the use possibilities of biodiversity resources in Peruvian Amazonia, a researcher from the Research Institute of Peruvian Amazonia (Instituto de Investigaciones de la Amazonía Peruana, or IIAP) listed promising wild species from the region: 'We have *aguaje, castaña, pijuayo, camu camu, copoazú, guaraná, capirona, caucho, shiringa, cedro, chambira, uña de gato, paiche*…'. This list includes moriche palm fruits (*Mauritia flexuosa*), Brazil nuts (*Bertholletia excelsa*), peach palm fruit (*Bactris gasipaes*), camu camu fruits (*Myrciaria dubia*), cupuaçu (*Theobroma grandiflorum*) and guaraná fruits (*Paullinia cupana*), wild rubber from the *Castilloa* and *Hevea* trees, Spanish cedar (*Cedrelinga odorata*), chambira palm fibres (*Astrocaryum chambira*), cat's claw vines (*Uncaria tomentosa*), and arapaima fish (*Arapaima gigas*). Altogether, over 30 high-potential species were identified, all well-known regionally and with abundant studies on their production potential. They are also known to have genuine market demand, mainly regionally and nationally but some also internationally.

The message from this meeting, as from many other similar meetings, was that biodiversity is not only important in itself but also a potential source of wealth that can bring prosperity to the region; one just has to find the right ways to harvest, produce, process, and market this treasure. Moreover, growing demand and markets should then give rise to new exports and new industries, with added value and associated benefits. Everybody in the audience was highly interested in these possibilities and realised that the list of genuinely promising species was just provisional. A good reference for this notion is the main marketplace of Iquitos, the market of the Belén district that is famous for its amazing variety of wild species on sale (Vasquez and Gentry, 1989), including foodstuffs, spices, natural drugs, and ornaments (Figure 12.4). Many of these items are extracted directly from the wild by experienced harvesters. No doubt, some of these products could have market potential also outside of the region. Evidence supporting this prospect is provided by the high number of Amazonian products that have broken into the international market (see Chapter 3).

The backgrounds of the participants in this meeting constituted an interesting mix. There were researchers dedicated to the study of biological diversity, specialists of specific species or product types hoping to share or update their knowledge, politicians interested in new development possibilities in the region, businesspeople who understood that this meeting was about future perspectives

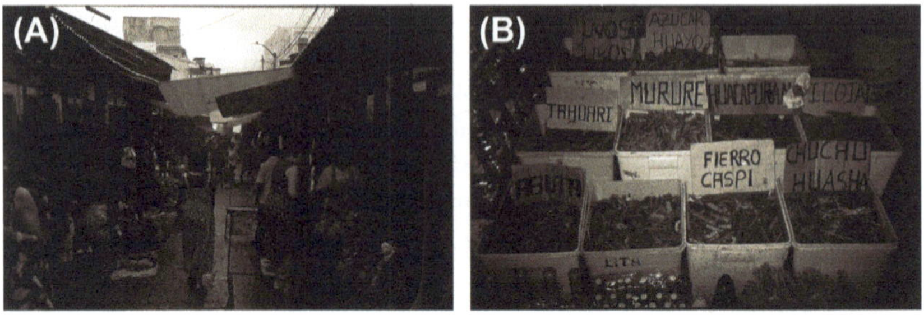

FIGURE 12.4 (A) Biodiversity wealth in the Belén marketplace in Iquitos. (B) Medicinal plants are sold for traditional healing in the alleys of the market. *Photos: Ilari E. Sääksjärvi.*

FIGURE 12.5 The years after the Rio de Janeiro Earth Summit of 1992 saw how biodiversity resources increasingly emerged as development assets in Peru and elsewhere in the tropics. The IIAP has offered space for numerous discussions and events dealing with the topic. *Photo: IIAP–BIODAMAZ.*

and potentially productive investments, harvesters with genuine know-how about the work in the field, conservationists who believed that economically successful extraction of nontimber forest products (NTFPs) could contribute to combatting deforestation, promoters of tourism who hoped that the fame of biodiversity resources in the region could attract more visitors. There were also students and interested laypeople of different kinds.

This and related seminars about the potential of Amazonian biodiversity wealth featured an atmosphere of hope that there was and still is in the region (Figure 12.5). By highlighting the potential at hand, it is easy to uplift the spirit of optimism, believing in extractive forest uses rather than in those alternatives that lead to deforestation. As time has passed, however, it has become increasingly clear that the challenge is to show that this is not wishful thinking but something real and feasible. Economies based on wild species harvest just have to be more sustainable than they have been thus far, avoiding the paths that lead to a collapse that follows after prosperous times with flourishing extraction, trade, and consumption. Everyone in the region is aware of the history of booms and busts that have so drastically shaped Amazonian development ever since the times of the *Cinchona* bark and wild rubber (Chapter 3), and all agree that wild species harvest should now be built on a much more stable foundation. This may require, for instance, working with several different cash-generating products simultaneously rather than with just a few.

With these thoughts in the back of their minds, many experts and those involved in projects from Peru and elsewhere have sought to push forward on multispecies extractive economies in Amazonia. Despite success in some cases, experiences have also led some actors into disappointment and confusion, and sometimes even to withdrawal of their activities altogether. On one hand, there is undeniable potential: only a few places on Earth can compete with Amazonia in terms of the wealth and variety of the potential biodiversity products. But, on the other hand, fulfilling industrial-level demand and production of any forest-related goods may be full of caveats. Resource availability may collapse after a few years of successful harvest if management is overlooked. Land ownership and usufruct regulations may cause restrictions, or transportation costs may become too high. It may also be that the availability of the species in question is too restricted or its quality variations are intolerable for industrial buyers. When any of these problems emerge, there may be no profit to be made. As a consequence, it is often very difficult to withstand the competition caused by cultivated goods or, in some cases, even artificial products replacing previously valued wild species. Extractive activities thus often remain local, or they may even vanish rather than expand.

If Amazonia is to be developed based on economic expansion and growth, the contribution of wild species can only be based on profitable and sustainable harvest. Almost everybody agrees about this, but views are different when it comes to the implications of this statement. Unambiguously, there must be a third way or a balance between strict protection and forest destruction. Where forest is to be maintained outside of protected areas, the extraction of products of natural origin often must show its ability to outweigh the value of alternative forms of land use. In theory, this is, if not a win–win strategy, at least a promise of something for everyone: conservation for those interested in the future of the biodiversity, and extractive products for those willing to secure socio-economic development. This means that there must be sufficient incentives to help people and companies commit to levels of use that can be sustained over time. And all of this needs to take place while controlling the impacts of harvest and management, and taking into account the variety of resources used. Easier said than done, it seems.

Advice could be sought from the local residents, people who live in the region, use its local resources, and maintain some of the experience and wisdom generated through the millennia of human presence in the rainforest. Scientists working with Amazonian indigenous peoples have tried to understand how they make a living in the primary rainforest. Each of the region's hundreds of ethnic groups knows the uses of tens or even hundreds of different species for purposes like food, construction, tools, handicrafts, or drugs. For them, the diverse rainforest is what a supermarket is for urban dwellers – a place to find anything needed. This knowledge is also unique from the perspective of an outsider. Many contemporary medicines have their origin in wild species, and often hints of their medicinal use potentials have actually originated from traditional cultures (see Chapter 10). Ethnobotanists and drug seekers have walked in the forest with traditional healers, trying to learn what they harvest, how they prepare their medicines, and how they are finally used.

Scientists and companies involved in such work are sometimes accused of unauthorised appropriation of genetic material and associated local knowledge. Few such cases, however, have actually been discovered. The message nevertheless is clear: the wealth of species with already existing use value is undeniable. These kinds of prospects have led researchers to suggest that in some places, the value of NTFPs in economic terms could be superior to that of cutting timber from these forests (Peters et al., 1989). If this was true, such products not only would yield higher net revenues to harvesters but also could be harvested with considerably less damage to the forest. Unfortunately, thus far there is little evidence to back such optimism in practice (see e.g. Pyhälä et al., 2006). If it is true, however, why do we not see biodiversity being transformed into wealth?

IIAP, in whose auditorium the aforementioned meeting was held to discuss Peruvian biodiversity resources, should be a good place to look for answers. It is a public institute with the function to support the development of Peruvian Amazonia through scientific research and the dissemination of its results to society. The personnel of the institute have worked hard to enhance the use of wild species and to promote their management. Other Amazonian countries carry out similar work via their respective research institutions, such as the Brazilian National Institute of Research of Amazonia (Instituto Nacional de Pesquisas da Amazônia, or INPA) and the Colombian Amazonian Institute for Scientific Research (Instituto Amazónico de Investigaciones Científicas, or SINCHI). Institutions like these have produced numerous publications about Amazonian species of interest and their optimised exploitation (Figure 12.6).

Many of their studies convincingly reveal the high potential of Amazonia's biodiversity resources. However, the commodity chains, including processing

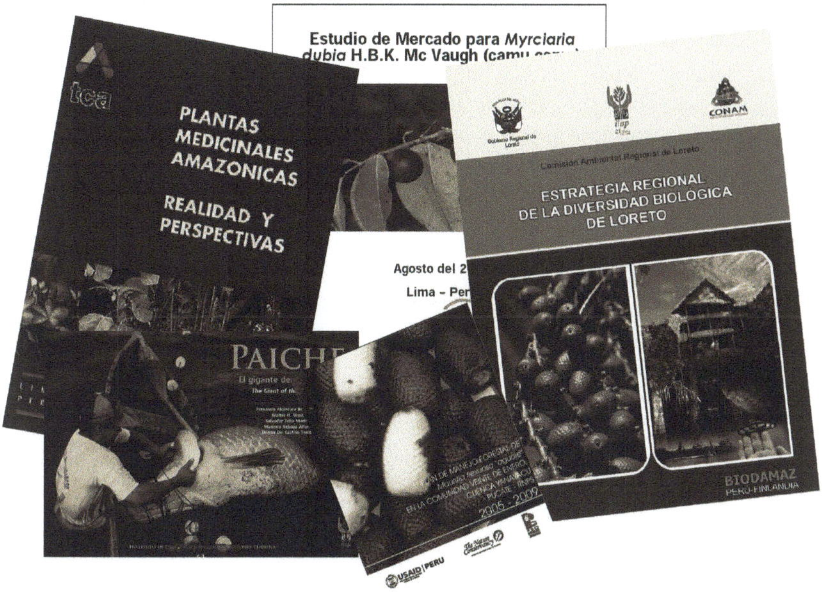

FIGURE 12.6 A steadily growing body of publications about Amazonian biodiversity resources adds to the accumulation of knowledge that can support better use and conservation. (For colour version of this figure, the reader is referred to the online version of this book.)

and marketing, are often fragile, and for success stories to emerge all of the pieces must fit together: ecological settings, market demand, extraction, processing, quality assurance, transportation, distribution, labour, and capital, just to list a few of the most important variables. When the market starts to 'pull', one has to produce large quantities of high quality and with punctuality. Where species are many, each with specific characteristics and low population densities, management and logistics costs easily become high, making the list of obstacles increasingly long. One of the few examples of wild species harvest in Amazonia that truly contributes to conservation while also being economically profitable is the aquarium fish trade (Box 12.1 and Figure 12.7).

A frank comparison with the boreal forests of the Nordic countries may illustrate the challenging nature of wild species harvest in the tropics. Species-poor, relatively low-growing, and simple-structured boreal forests have supported the growth of large forest industries which have significantly contributed to the economic development of these northern regions. Big wood-processing plants have been built with close ties to nearby forest resources as well as to the markets of their products (Figure 12.8). This engagement has been successful in many respects, yet many criticise that optimised wood production has sacrificed other important forest values, including biodiversity. Nevertheless, when increased wood production was set as a national goal, governments started, by the late nineteenth century, to invest in scientific

Box 12.1 Aquarium Fish Trade

Most of today's aquarium fishes are bred in industrial facilities in South-east Asia. However, several important locations in tropical countries are still sources of collected natural aquarium fish, and large-scale trade of natural freshwater aquarium fishes is an important source of income in the African Great Lakes, Amazonia, and several Asian countries. Marine fishes and invertebrates are also collected in the Pacific and Indian Ocean regions and, to a lesser extent, in the Caribbean.

The collection of freshwater fishes is sometimes part of consumption fishing. However, in most cases, the collection is a highly demanding skill and involves specialised fishermen and communities. In Amazonia, more than 30,000 families are estimated to be involved in the collection of ornamental fishes. The number of fish species subject to the aquarium trade is around 500 species, of which about 100 are common in the world aquarium fish trade.

Most of the fishes in Amazonia are caught by hand nets or various forms or traps. Many fishing techniques involve night fishing using flashlights. In the flooded forests, some fishermen use natural plant extracts or commercial piscicides like rotenone to stun the fish and collect them from the surface. Many of the skills and practices to catch the fish are carefully guarded family secrets that pass from fathers to sons.

The artisanal aquarium fish trade in Amazonia often involves everybody in the villages. The role of women and girls is often to guard and feed the fish in storage tanks, and to wait for the boats that collect the fish from the remote tributaries. The fish are sometimes kept in ponds to reach the optimal size, as the trade is highly seasonal. The collecting boats bring the fish to national hubs like Iquitos in Peru or Manaus in Brazil, where the wholesalers typically store the fish for up to three months. The fish are packed in plastic bags and polystyrene foam boxes and are sent as flight cargo to the wholesalers of the customer countries. The modern packing technique allows up to 48 h of transportation time. Often, the fish are kept in quarantine for up to two weeks by wholesalers and aquarium shops before being sold to customers.

The aquarium fish trade has raised some concern about the effects of fishing on the natural fish population. There are a few cases, like the red-tailed black shark (*Epalzeorhynchos bicolor*) in Thailand and some varieties of the discus fish (*Symphysodon* spp.) in Amazonia, that have declined due to ornamental fishery.

Generally, however, ornamental fishery is considered one of the prime economic activities in rural Amazonia with a positive impact on conservation. The fishing communities are fierce defenders of aquatic ecosystems against commercial overfishing, deforestation of the floodplains, and excessive use of piscicides. One of the best examples of positive conservation impact has been the Piaba project that is operating in the Rio Negro floodplains in Brazil. The fishing communities have been granted usufruct rights to their traditional aquarium-fishing areas. The local communities are able to control the fishing in flooded forests and help to control the illegal logging and charcoal burning that elsewhere have destroyed many floodplain forests in river systems that do not support ornamental fishery.

By Jukka Salo

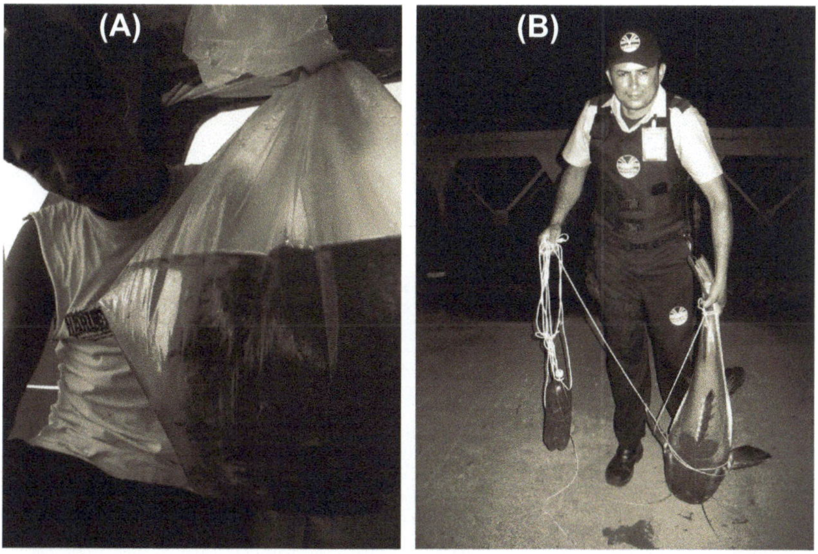

FIGURE 12.7 (A) Aquarium fish are an important source of income for many *ribereño* households in Peruvian Amazonia. (B) A red-tailed catfish (*Phractocephalus hemioliopterus*) caught for food in the port of Óbidos in the Brazilian State of Pará; despite its size as an adult, the species is also a common aquarium fish.

research to reinforce the formulation of management procedures that would better guarantee selected benefit flows. National forest research institutions were created in quite the same way as the Amazonian countries have created their respective research institutions.

In addition to supporting the prosperous wood-processing industry, the boreal forests also provide a number of other extractive forest products. This is largely because forest management is focused on native tree species, and it has furthermore tried to mimic some of the natural processes occurring in the region, including the effects of natural forest fire. In consequence, even though these forests are intensively managed, they are seminatural, having a species composition that is somewhat impoverished but nevertheless fairly similar to that of natural forests. Wild berries such as lingonberries (*Vaccinium vitis-idaea*) and bilberries (*Vaccinium myrtillus*) as well as edible mushrooms are harvested in great quantities by locals and by seasonally contracted professional pickers (see Box 1.1). These NTFPs are much used both at the household level and in food industries. Moreover, elk (*Alces alces*), grouse (e.g. *Lyrurus tetrix, Tetrao urogallus, Lagopus lagopus*), and many other game animals are hunted as an outdoors activity while also providing meat both for local consumption and to be sold at a high price in specialised shops and restaurants. These examples show that even if the northern forests are species-poor

FIGURE 12.8 Paper mill complex in Jämsänkoski, central Finland. When wood resources are of the same kind and plentiful, raw materials can supply the establishment of such big industries. This condition is met in simply structured seminatural boreal forests but less so in species-rich forests in tropical areas. *Photo: Aarno Isomäki.*

and much less productive compared to the tropical rainforests, they too offer plenty of resources for humans. Actually, it is hard to find any solid evidence to show how the wealth of species and their potential uses would be superior in Amazonia, as compared to the boreal forests, at least in quantitative terms.

When experts from the north started to assist tropical countries in order to strengthen their forest sectors in the late twentieth century, reference was often taken from the north. Obviously, this was a mistake, as tropical forests are different and much harder to comprehend. They are messy looking and multilayered, with hundreds of species of trees, shrubs, lianas, and herbs. Furthermore, seen from the northern perspective, many tropical trees turned out to be too big, too hard, or otherwise too difficult to process when taken to sawmills. Chain and circular saws did not always resist tropical hardwood. Paper mills required even-quality raw materials in large quantities, not just logs of all kinds of trees with dissimilar technical properties. In short, species-rich forests were not compatible with the specialised requirements developed to fit the 'easy' northern settings. Moreover, the transportation of wood from the forest is extremely difficult when the trees are huge and the most wanted species have sparse distributions. All of these problems forced companies to consider tree plantations as an alternative to logging in natural forests. In the 1970s, many projects in different parts of Amazonia began to replace diverse tropical forests with monocultures that resembled temperate or boreal forests rather than the natural environments of the region. However, both the ecological and economic results from these investments were discouraging.

In the late 1980s, researchers from the University of Turku had an opportunity to collaborate with the Faculty of Forest Sciences of the Iquitos-based National University of Peruvian Amazonia (Universidad Nacional de la Amazonia Peruana, or UNAP). One day, there was a visit to the university sawmill with machinery donated by a former Finnish development project. Some of the local timber species were sawn using the machinery, but in general the mill was much better suited for the processing of managed fast-growing trees from the university's experimental plantations. During the visit, the dean of the faculty opened the northern visitors' eyes to some further problems. According to his view, the international forest science of that time suffered from a northern bias related to both management and technology.

The most important textbooks at that time were published with boreal or temperate environments as their focus. Learning about the growth of coniferous trees, as well as their measurement, management, and wood processing, was of little use for the students of the UNAP faculty. Moreover, many of the economic models used in forest management did not adjust well to the conditions met in Iquitos. The challenge faced by the Amazonian forest engineers in every possible context was the diversity of everything. This implied considerable effort in the search for suitable individuals of a wanted species as well as significant transportation costs and difficulties in maintaining the quality of the timber harvested in the hot and humid forest.

The dean continued by explaining the difficulties of learning about the technical properties of the local forest trees and about the ways to exploit them in practice. Knowing about the technical characteristics of the different species is but the first step, however. Other challenges include finding good uses for the different species and developing methods for their harvest, transport, and processing. These questions are cumbersome even if there was demand for each identified product. Finally, there is always the danger that the populations of the most desirable tree species are easily overexploited if attention is placed more on harvest than on management that guarantees that (1) the populations of the harvested species remain viable and (2) ecosystem disturbance is kept at acceptable levels. These issues have been addressed, and knowledge has increased, both locally and internationally, involving work by such organisations as the International Tropical Timber Organization (ITTO). But there is still much to be done.

DOES IMPROVED KNOWLEDGE MAKE A DIFFERENCE?

As knowledge on the characteristics of the diversity of species grows, this opens up more options to use forest resources. However, it may also mean that there are increasing levels of impact on the natural environments of the harvested species; a longer list of target species is far from a guarantee of sustainable harvest.

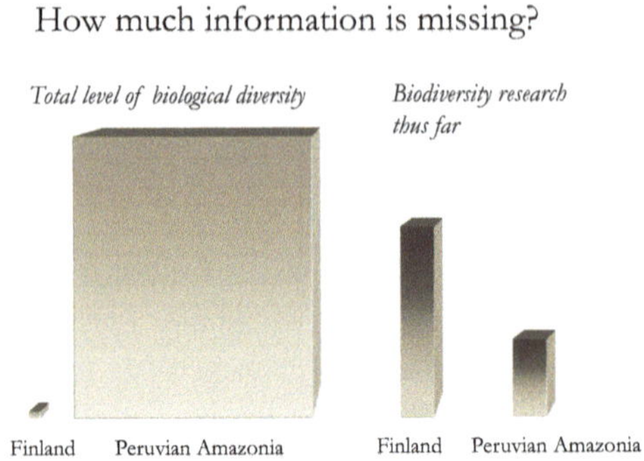

FIGURE 12.9 Example of awareness-raising material used to highlight the big differences in the levels of biodiversity (left) between Finland and Peruvian Amazonia as well as the amounts of scientific effort made in each one of them (right). The main message is that the more challenging system has been studied less.

Clearly, the use of tropical forests requires even more know-how, patience, and scientific facts than the considerable amount of such inputs needed in the simpler northern forest ecosystems. This is illustrated in Figure 12.9, which was originally designed to be used among scientists and practitioners working in the Amazonian lowlands for awareness-raising purposes. The point to be made was to highlight the relative lack of scientific efforts in the Amazonian lowlands. The ecological systems in this region are so complex that any resource use and management efforts easily lead into problems. This setting was discussed when the dean of the Faculty of Forest Sciences at UNAP later visited Finland. A particularly interesting question for the dean was the heated debate among forest companies and conservationists on the resilience of the Finnish forests in the face of intensive management. If there were such important deficits in understanding concerning the sustainability of the simple Finnish forests, how could the Amazonian reality be better understood?

Recent years have shown positive signs, however. Institutions such as IIAP, INPA, and SINCHI are highlighting the need to link science, planning, and users into one dialogue. Scientific progress and a growing appreciation of evidence-supported planning should increasingly help overcome the problems that are caused by the often inevitable but suboptimal ways of acting before thinking that occur in forest use. In some cases, it is also important to make uncertainty visual, so that insufficient information basis does not get overlooked in planning (Figure 12.10). Although it is unrealistic to expect that all potential impacts of decisions could be considered prior to decision making, there are indications of approaches that at least aim to find ways to use biodiversity resources sustainably. These kinds of approaches have been embedded in the Convention on Biological Diversity (see Chapter 2), emphasising the value of biodiversity and the adaptive management of biodiversity resources as parts of functioning ecosystems. As an outcome from the CBD, countries have

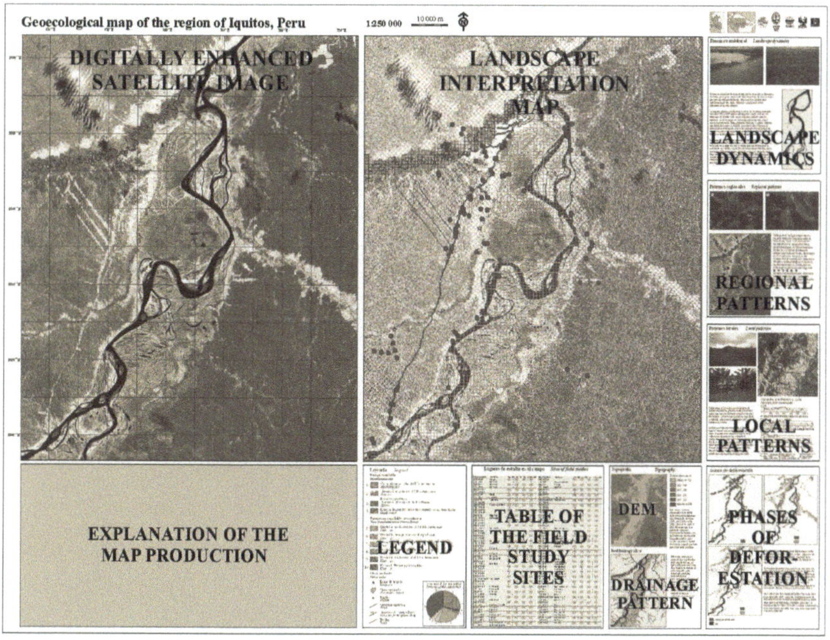

FIGURE 12.10 Visualising uncertainty: a geo-ecological map of the region of Iquitos representing aspects of its environmental heterogeneity. The left side shows the same satellite image as in Figure 12.2; and on the right, some environmental units are tentatively interpreted and shown upon it. The margins show additional supportive information about the region. *Mäki and Kalliola (2000).*

formulated strategies and action plans for the conservation and sustainable use of their national biodiversity resources. In Peru, regional assessments along these lines have been made separately for the different regions of Amazonia, with IIAP being the institution in charge.

The result has been not only a regional biodiversity strategy and action plan for Peruvian Amazonia but also, maybe even more importantly, the enhancement of institutional skills and capacities. People taking part in the activities involved in this process are usually committed and inspired, and there is an atmosphere of truly positive expectations. 'We are killing the chicken that lays golden eggs' and 'Peru is like the beggar who sits on a golden bench' were among the common metaphors used by the participants in some of these dialogues. In the end, if the accumulating new knowledge is to make any difference, there must be ways to transmit knowledge through information that is directly useful as a basis for decision making. But how are the researchers, professionals, civil servants, and other stakeholders going to bridge the gap from their roundtable discussions to the places where decisions concerning the use of resources are made?

Part III

Seven Thematic Perspectives

Learning deeper lessons from real-world cases of wild species harvest is much more efficient if there are ways in which stories can be read or empirical cases assessed in a systematic and analytical manner. This part of the book offers seven complementary thematic perspectives that aim at theoretically structuring the analysis of wild species harvest.

We prefer to call the chapters found in this part of the book 'thematic perspectives' without using the names of any particular scientific disciplines. This choice in terminology seeks to emphasise that the issues dealt with in each perspective finally require truly interdisciplinary understanding; theories and methodologies must be included from many different fields instead of one or a few only. It is not only that the scientific thinking from natural and social sciences has to get together, but also that the needed methods require a well-equipped interdisciplinary toolbox.

The seven thematic perspectives are not fully independent, but rather they are mutually interdependent, tied to one another. For example, resource dynamics is inherently involved in any economic calculations regarding wild species harvest, all management operations require knowledge, and space and legacies are present in almost everything. This is not a problem, however, as the aim is not to classify the details of wild species harvest strictly into one or another of these categories. Instead, we invite our readers to consider wild species harvest from all of these different viewing angles, and through this approach to appreciate the variety of issues and settings that are involved.

Preview to the Chapters of Part 3

The chapters in this part of the book offer theoretical backing for the diagnosis of wild species harvest. We invite readers, while reading the following chapters, also to browse back through the book and find examples of how the real-world cases relate to their theoretical assessment. This helps to see how simplified mental representations of reality can be used to highlight some aspects of the real world, while disregarding others. Theoretical thinking also helps us to make the world understandable in spite of its enormous complexity.

In science, conceptual models are commonly used to help to structure our thinking whilst causal relationships are often expressed as mathematical formulas and as graphs. Often the core message of the models can be understood by just visually observing the graphs. The mathematical formulas can, nevertheless, also be useful in order to achieve a deeper understanding, and, in particular, if one wants to examine the interplay between different variables or further develop a model for additional analyses. We have included some examples of such models in the following chapters. It has been our ambition to explain each model, in the first place, verbally and graphically with enough clarity to make them accessible and useful, even to those readers who are not very accustomed to using the formal language of mathematics. However, we also provide the mathematical foundations of the models, having in mind that some readers would otherwise be disappointed.

Chapter 13 discusses how resource dynamics are behind the provision from nature and how the harvester's transactions with nature and harvest interventions have effects on the resource system. **Chapter 14** deals with the harvester's decisions when weighing the costs of harvest against the expected benefits obtained. **Chapter 15** discusses management, i.e., the purposeful actions taken by harvesters themselves or others, in order to improve the harvest system. **Chapter 16** explores the role of governance in creating incentive structures affecting the harvesters' decisions. **Chapter 17** lays out the ways in which knowledge, information, and communication are necessary for harvest systems to function. **Chapter 18** analyses the role of spatiality in wild species harvest. And, finally, **Chapter 19** adds to the six other thematic perspectives the considerations of time, how past events and developments affect the present, and how today's decisions shape the future.

Chapter 13

Resource Dynamics behind the Provision from Nature

TRANSACTIONS WITH NATURE

From the human viewpoint, ecosystem services offer a constant provision of resources to be harvested from nature. Any further analysis added to this simple premise reveals how a multitude of factors influence this stream of benefits. First, the state of those ecosystem functions that are vital for resource growth and regeneration determines how the provision from nature develops in time and space. These functions can involve both biotic and abiotic processes that lead to and support reproduction, growth, or dispersal, among other things essential for resource regeneration (e.g. Blüthgen and Klein, 2011; Jordano et al., 2011; Schleuning et al., 2012;). Second, because harvest always implies not only that the desired biological matter is subtracted from nature but also that there is an associated human intervention, harvest in itself tends to interfere with the above processes and associated ecosystem functions. The effects of harvest interventions can be direct or indirect, and they can be anything from an almost unnoticeable momentary change in the biomass of an individual organism to a drastic transformation of a complete ecosystem. Eventually, it is the balance of the above transactions with nature that determines how sustainable harvest activities can become from an ecological viewpoint (Figure 13.1).

People generally have an intuitive understanding of at least some of the fundamental conditions that set the limits of wild species harvest, even though this understanding is not always shown as concrete actions and concerns. The most obvious of these is that if the off-take of any living organism or its parts from nature is to be sustained in the long term, it must not exceed the capacity of these organisms to recover, maintain their populations, and regenerate. A common metaphor for this kind of thinking is the statement that sustainability requires living off the interest rather than the capital (e.g. Norse et al., 2012). This comparison is often used even though living resources certainly do not behave in ways fully comparable with financial assets. Although one can take out the interest from a bank account and leave the capital intact, the harvest of wild species always implies some degree of ecological impact; when humans remove biological material from an ecosystem, this inevitably affects not only the abundance of the target species, but it also has indirect

Diagnosing Wild Species Harvest. http://dx.doi.org/10.1016/B978-0-12-397204-0.00013-9

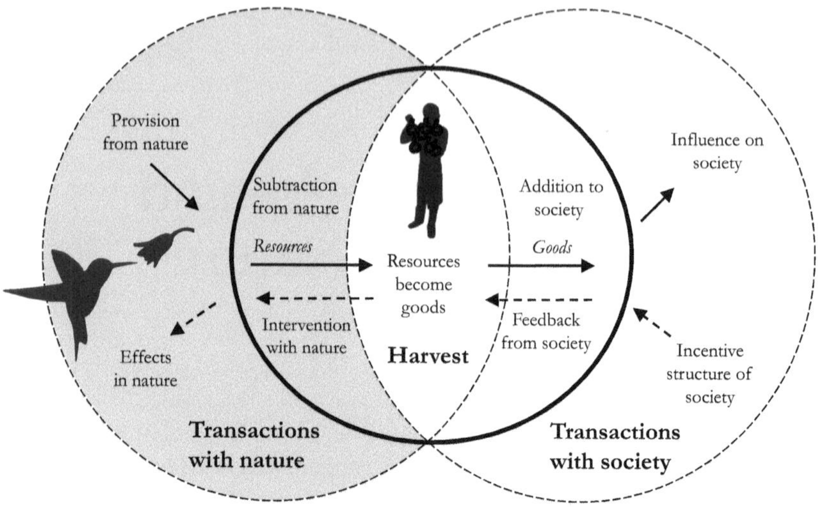

FIGURE 13.1 The harvester's multiple transactions with nature determine the sustainability of harvest from an ecological viewpoint. The subtraction of desired resources is based on the provision from nature, on which the harvest intervention, in turn, has direct and indirect effects. The harvester is one party in these transactions, whereas the other counterparts are the different components of the natural system that reciprocally interact with the harvesters. The areas in the graph shaded in grey indicate the main focus of the present chapter.

effects on any other species that somehow interact with the harvested species. The harvester's transactions with nature take place between the harvesters and different components of the ecological system and thus involve a myriad of ecosystem functions that are vital for wild species to exist and regenerate (Hooper et al., 2005) and and there is an infinite number of ways in which wild species can be impacted by the harvest itself (e.g. Ticktin, 2004; Ticktin and Shackleton, 2011). Although some of these transactions may appear self-evident, others are counterintuitive – and certainly many remain poorly understood or simply unknown. It is also important to bear in mind that some ecological consequences of harvest may not appear until after a time lag, or only after a certain threshold has been reached.

Additionally, in order to be sustainable, the harvest should not greatly alter the overall ecological system that provides what is harvested. From the ecological perspective, it is thus too narrow a viewpoint to see the provision from nature as consisting of the harvestable organisms only. We should rather include in the consideration the full variety of functions of the entire ecological system. This not only includes the assemblage of interacting populations of various species living together in the same geographical area but also the abiotic components of the ecosystem. Because it is difficult to set the exact limits to all of the components that are involved in the provision from nature, it should finally be considered as a holistic abstraction. Let us just define it as the functioning ecological system that makes the harvestable resource available in the first place.

To address the basic conditions of sustainability from a more theoretical perspective, we first and foremost need to understand the kinds and levels of direct and indirect effects of harvest on the wild species targeted, as well as their environments. Only then will it be possible to proceed to analyse how and to what extent the sustainability of harvest can be improved by adjusting harvest practices to fit increasingly well with the ecological boundary conditions. The first part of this chapter provides an overview of impacts of wild species harvest on the provision from nature and to the wider ecological system in which it is embedded. The second part of the chapter then proceeds to discuss how sustainability of wild species harvest can be theoretically evaluated, as well as the use and limitations of ecological models as tools to understand the dynamics of living resources.

ECOLOGICAL IMPACTS ON DIFFERENT LEVELS

The effects of wild species harvest take place on different levels of ecological organisation. To summarise what kinds of effects can be addressed, we will herein discuss the impacts of harvest at the levels of individual organisms, populations, communities, ecosystems, and the biosphere. Although these levels are not always easy to precisely define and delimit as clear-cut wholes, they are useful concepts that help to structure our thoughts.

An **individual organism** is commonly thought of as a contiguous body made up of cells with the same genome. Despite the apparent straightforwardness of this definition, at a closer look it turns out that not all of what we might think are individual organisms actually are, and vice versa. Many clone-forming plant species, including some trees, have so-called rhizomes, which are modified stems that grow underground and from which shoots can grow up to become what looks like, but are not, individual trees. Mushrooms easily look like separate individuals too, but many of them can be only different spore-bearing fruiting bodies of just one individual fungus. Each of these individual fungi consists mainly of mycelium – a dense mat of underground fungal threads that can cover several square metres or even square kilometres – and may produce a large number of fruiting bodies that can, in this sort of case, be seen to form parts of just one individual organism. On the other hand, although seeds may appear to be part of their mother plant, they are in fact – in the case of sexually reproducing plants – individual organisms with a genome that is different from that of their mother plant.

A **population** consists of all of the individuals of the same species living in the same area. How to define the 'same area' is by necessity an arbitrary decision, except for unusual cases such as island populations separated by expanses of water that form barriers for interactions between the populations on the respective islands.

An **ecological community** consists of all populations and all living organisms of the different species in a given area, and an **ecosystem** further consists not only of all of the living organisms but also the nonliving components of their environment. How to delimit an ecosystem in space is, again, basically a matter of choice for the observer. All of the ecosystems on Earth together, however, make up the **biosphere**, which refers to the relatively thin layer surrounding the planet, which is inhabited by living organisms.

Impacts on Individuals: Viability and Fertility

The most direct impact of wild species harvest is usually the impact on the individual plant or animal being subject to harvest. The degree of impact on the individual may vary, though. At one extreme, there is the case where the entire organism is either killed or removed alive from the ecosystem. This is typically the case for hunting (Figure 13.2) and for the capture of live animals for the pet trade or biomedical industry. Similarly, when trees are cut for timber, the tree dies, although often a large percentage of the tree biomass is actually left in the forest. Because each seed is an individual organism, the collection of fruits, nuts, or other seeds often implies such removal, too. This is, however, not necessarily the case for fruits that are consumed locally, whereby the seeds may be thrown away after eating the fruit without much negative impact on their viability – often even with the opposite effect.

In fact, humans may even increase the viability of some seeds. Plants that produce fleshy fruit actually often do so as an evolutionary adaptation to seed dispersal

FIGURE 13.2 Lethal harvest: Hunting normally implies that the animal is killed and removed from the ecosystem. However, only part of its biomass is directly consumed by humans. The paca (*Cuniculus paca*) is a rodent species commonly hunted in Amazonia.

by animal vectors, such that the seeds can pass safely through their digestive system and get deposited in some place far away from the mother plant. Passing through the digestive system of an animal may even speed up germination of the seed. This role as a seed disperser can be taken on by humans as well. By eating fruit and depositing the seeds on the ground when defecating, people may actually improve the viability of the seeds. Thus, although harvest usually implies death or reduced viability of the targeted organisms, the opposite can be true in certain cases.

In some cases, even removal of just a small part of the organism is enough to kill it. Many palm species are particularly susceptible because they grow only at one point – the growing bud located at the top of the stem. The so-called palm hearts, which are commonly consumed as food, consist of the growing bud and the surrounding tender leaf shoots and inner core of many palm species. When the growing bud is cut, the palm tree cannot produce any new leaves and thus dies after some time. A somewhat similar example among animals is the harvest of shark fins, which are considered a delicacy in the Chinese cuisine. Shark fishermen do not harvest the whole shark, but only cut off the dorsal fin and then release the mutilated fish into the sea, where it is doomed to die.

In other cases, however, a part of the organism can be harvested without killing it. From trees and other plants, careful harvesting of limited amounts of leaves, latex, or bark rarely kills the organism, at least immediately (Figure 13.5B). Even animals can be subject to this type of harvest. Vicuñas (*Vicugna vicugna*), a camelid native to the Andes Mountains, are nowadays sheared alive, instead of hunting them, in order to obtain their valuable wool (Bonacic et al., 2006). Rhino horns can also be cut from live individuals, even more than once, because they grow back, such that it could become unnecessary for rhinos to be killed by poachers seeking to sell the horns to Asian countries where they are used in traditional medicine.

Although the above kind of harvest is not necessarily lethal to the organism, it may nevertheless decrease its viability or fertility. All parts of an organism usually have their particular function and therefore one should always assume that removing a part of an individual does somehow affect it. Leaf harvest, for example, can create a physiological imbalance in the plant because its photosynthetic production of carbohydrates decreases, thus reducing the plant's overall viability (Duarte and Montúfar, 2012). Similarly, harvesting the bark of trees can increase mortality and reduce fruit production (Stewart, 2009). When harvest damages individuals, leaving open wounds, they may also become exposed to some opportunistic disease, such as fungal attacks. Even shearing the wool from vicuñas deprives them of their natural protection against the cold and may potentially increase mortality, particularly if these animals face extreme temperatures before the wool has grown back again (Sahley et al., 2007).

The same kind of harvest may have a different impact on the individual depending on the particular characteristics of the species in question. Most palm species have just single stems, and thus they invariably die if the stem or even just the top of the stem is cut. However, there are also palm species that have multiple stems and can produce new stems from basal shoots even if one or several stems are cut. There are also some palm species that are clonal, having horizontal underground stems that unite several above-ground stems. Cutting one such palm tree is usually not lethal because what we see as an individual tree is just a part of a palm individual consisting of many stems.

Conversely, harvest does not affect the viability of the organism at all if it is already dead when harvested. For example, indigenous people in Amazonia commonly use the wood from dead fallen trees as pillars for their houses. This is because the wood that is hardest and most resistant to rotting is the core wood found in the middle of the dead trunks of some particular tree species.

Impacts on Populations: Abundance and Structure

When many individuals of the same species in a locality are affected by harvest, the most obvious immediate effect at the level of population is that the number of individuals – that is, the population size – is reduced. This may lead to the extirpation of the local population or even to the extinction of the entire species. However, the population structure may also be affected. For example, when hunting leads to increased mortality throughout the population, few individuals live to old age and the age structure becomes skewed towards overall younger individuals. Similar effects can be even more marked if harvest efforts are particularly geared towards large individuals, such as is often the case in fishing. Alternatively, the harvest of fruits may reduce regeneration to such a degree that the age structure shows a deficit of young individuals. This is suspected to occur in some Brazil nut (*Bertholletia excelsa*) harvesting areas (Figure 13.3). A similar situation may result if fruit production is reduced because of, for example, repeated harvest of leaves or branches.

Changes in size structure of the population may also occur (Figure 13.4), often related to changes in age structure. In particular, those fish, reptiles, and trees that continue growing throughout their entire lifetimes are susceptible to this kind of population-level changes. It is, however, less apparent in the case of mammals that grow very little once they have reached their adult age, but equal-aged individuals may nevertheless display differences in size, which also leads the harvest towards particular size classes. This is how harvesting also exerts an evolutionary pressure that may lead to a selection process that favours certain genetic traits, often such that they are undesirable from the perspective of harvesters. Fishing that selectively targets large fish, for example, can lead to changes in the population's genetic structure; the fish may grow slower, thus reducing the provision of fish from the ecosystem – something that has happened, for example, with Atlantic cod (*Gadus morhua*) (Swain et al., 2007).

In theory, similar processes could affect many other harvested species as well. For example, there have been fears that selective logging may lead to deterioration in the genetic quality of timber tree populations when tree individuals that

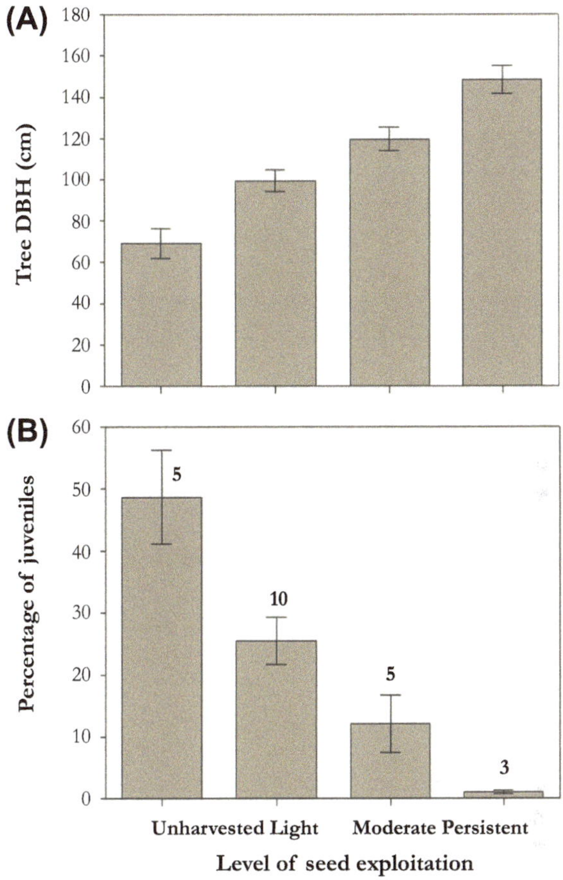

FIGURE 13.3 Relationships between historical levels of Brazil nut collection and (A) tree size measured as diameter at breast height (DBH); (B) percentage of juveniles (trees 10–60 cm DBH). Data are means ±SE. Numbers next to error bars in (B) indicate the number of populations studied. *From Peres et al. (2003).*

are deemed undesirable for harvest are left to reproduce while those with the most desirable characteristics are harvested. Research has shown, however that, at least for the different species of mahoganies (*Swietenia* spp.), this effect is quite insignificant (Cornelius et al., 2005). Other types of impact on the genetic structure of a population are the extirpation of genetically distinct subpopulations and the loss of genetic diversity that takes place when the number of individuals in a population is reduced to such a small fraction of what it once was that some gene alleles or genetic combinations that were not present among all individuals are completely eliminated; an example of the latter is the fate of the arctic fox (*Alopex lagopus*) in Sweden due to intense hunting in the early twentieth century (Allendorf et al., 2008).

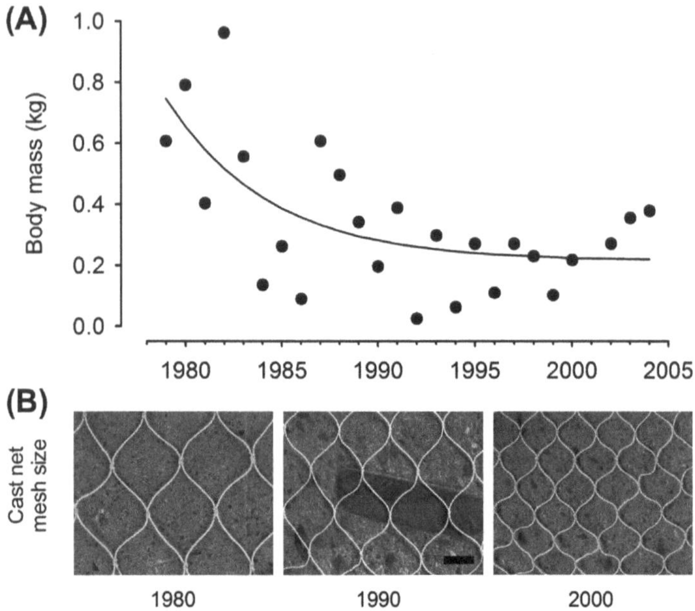

FIGURE 13.4 (A) Time trend of body mass for the migratory fish *Prochilodus mariae*. (B) Photographs of cast nets made over the past three decades by a fisherman in the community near the study site. Scale bar indicates 2.5 cm. *From Taylor et al. (2006).*

In addition, the sex structure of a population may be affected by harvest if either male or female individuals are particularly targeted. One example of this is the harvest of the fruit of the palm tree *Mauritia flexuosa*, known as *aguaje*, *moriche*, or *burití* in different parts of the Amazon Basin. This tree is dioecious, meaning that each tree bears either only male or only female flowers, of which only the female ones bear fruit. Although the fruit can be harvested from the ground or by climbing (Figure 13.5A) this is difficult and time consuming; therefore, it has been common to harvest the fruit by felling the tree. The result has been that large expanses of these palm forests have got heavily skewed sex ratios, with few female trees and therefore poor regeneration.

A skewed sex ratio is not necessarily an indication of overharvest, however. It can also occur for natural reasons. For example, the delicious cloudberries (*Rubus chamaemorus*) commonly growing on bogs in the boreal regions are dioecious too. The flowers grow on what look like decimetre-high plants, but in fact hundreds of these stems may be part of the same individual – a clone that is interconnected by underground rhizomes. These rhizomes can form extensive webs, so on a small bog the whole cloudberry population may consist of only very few individual organisms; sometimes, by chance, these individuals may therefore all be of the same sex. Comparing the cases of aguaje and cloudberry, what superficially looks like the same phenomenon is in fact due to entirely different causes.

FIGURE 13.5 (A) The fruit of the palm that in Peru is known as *aguaje* (*Mauritia flexuosa*) have commonly been collected by felling the huge palm trees. The adoption of climbing techniques such as this used in a village in the Pacaya Samiria National Reserve has made the collection of the fruit without killing the palm more feasible. (B) Collecting the leaves of the *irapay* palm (*Lepidocaryum tenue*) can be less harmful to the plant if not all of the leaves are cut but about one third of them are left intact.

Furthermore, animal populations subject to harvest often display behavioural changes. This can be seen as individuals becoming shyer or otherwise changing their habits in order to avoid getting killed. The mechanisms behind these kinds of changes are not well understood, but there are, in principle, three potential causes for changing behaviour: (1) genetic selection, such that individuals with a genetic predetermination for the shy behaviour experience a evolutionary advantage; (2) individuals that survive encounters with hunters learn to avoid them in the future; or (3) a process of evolution through social learning rather than genetic selection, so that young individuals learn by imitating the behaviour of the old (Avital and Jablonka, 2000). In particular, monkeys are known to be extremely clever in adapting their behaviour to the degree that they even can distinguish between known and unknown humans or between tourists, who are harmless, and locals, who are potentially dangerous because they sometimes hunt monkeys.

Some species are more vulnerable to overharvesting than others. Large size, conspicuous appearance and habits, and being slow or sessile facilitates harvest. When these traits are found in a species with a low reproductive rate, their capacity to withstand harvest is greatly reduced. The reproductive rate of a population, in turn, depends on the age at which the species reaches reproductive maturity, on the frequency of reproduction events, and on the number of offspring or propagules produced at each of these times. Harvest that targets individuals before their reproductive age can have more impact on the population than the harvest of individuals that have already passed their peak of reproductive capacity – an issue that can be particularly relevant to the management of fish and timber trees. Conversely, when young individuals have high natural mortality, their harvest is not necessarily as harmful for the whole population.

The impact of harvest also depends on whether the species in question shows **density–dependency** or not. Density-dependent population growth implies that the growth rate slows down as the population size grows in a given area – for example, due to increased competition for food or increased transmission of parasites or diseases – and increases when the population is reduced. Such density-dependence has been shown for several species of fish, affecting, for example, individual growth rates, mortality, and egg production (Rose et al., 2001). It has been similarly observed for many species of large herbivore mammals (Bonenfant et al., 2009).

Finally, harvest very often also affects the spatial structure of a population because its effects tend to be more intense near human settlements and transport routes than further away from these areas. Therefore, population densities of harvested species often increase with distance from settlements, rivers, and roads (more about this in Chapter 18). Similarly, the impacts on age, size, genetic, or sex structures may vary in space and be most marked where harvest intensity is highest, gradually diminishing with increasing distance.

Impacts at Community Level

Changes at the population level by definition also affect the ecological community. If multiple species are harvested, such as is typically the case for hunting, fishing, and logging, the species composition of the community changes as species that are of high value and vulnerable to overharvest decrease most, leaving the species of lesser value or the less vulnerable less affected. If such multiple-species harvest leads to resource depletion that proceeds over time, often this takes place through a species sequence, beginning with the depletion of the most valuable and vulnerable species, and then continuing until only the most resistant and low-value species are left. Large species often both have high value and are easier to find and harvest. Because they also usually occur at relatively low densities even in the absence of harvest, such large species are often the first to be lost. This kind of species sequence of harvest has been documented in many cases of wild species around the world, such as for fishery in the Oueme River in Africa (Figure 13.6). It has also been modelled for tropical rainforest hunting in Peruvian Amazonia and peninsular Malaysia (Figure 13.7).

When one species decreases in abundance, or disappears, due to excessive harvest, also other species are affected. Such effects may be mediated through a myriad of different sorts of interactions, including herbivory, predation, dispersal,

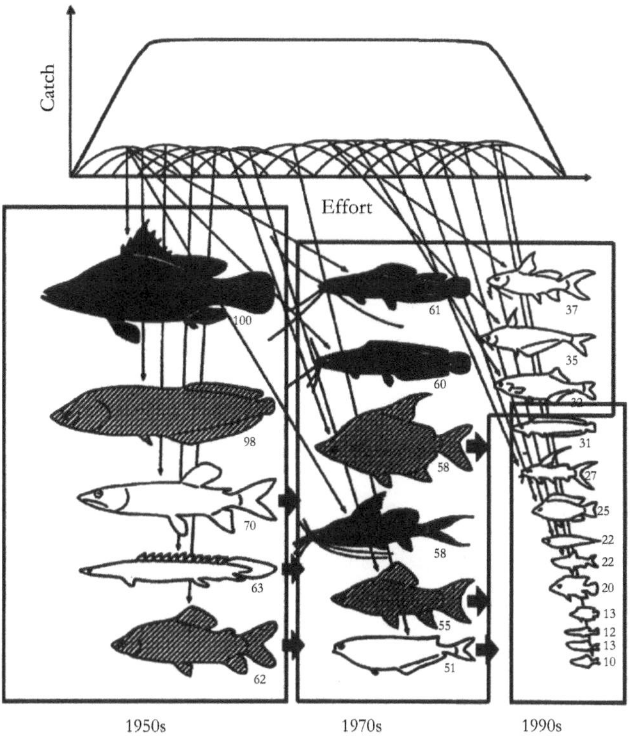

FIGURE 13.6 Increasing fishing effort leads to sequential depletion of species after species, although the total catch remains relatively stable over a wide range of effort levels. With time, due to increasing fishing effort in the Oueme River in West Africa, there was a change in the species composition in the catch. Large species became uncommon or disappeared altogether, whereas smaller species became more common. Black indicates species that disappeared before 1965, and grey indicates species whose numbers were seriously reduced. Black arrows represent species which reduced their breeding size. *From Welcomme (1999).*

competition, pollination, and the provision of shelter or microhabitats. In theory, any species that directly or indirectly interacts with the harvested species may also be affected by the harvest. Even entire ecological communities may change, particularly when harvesting depletes some species, thus opening the opportunity for other species to become more abundant or for completely new species to invade the community. This can take place because of a lack of predators or competitors or because the new species occupies the ecological niches of the depleted species.

Intense harvest of keystone species may cause drastic changes in ecosystems. A classical example of this is the hunting of sea otters (*Enhydra lutris*) in Alaska, which led to the proliferation of sea urchins, which had previously been controlled by predation by the sea otters. Sea urchins, in turn, feed on kelp (brown algae that forms dense underwater forests that can reach tens of metres

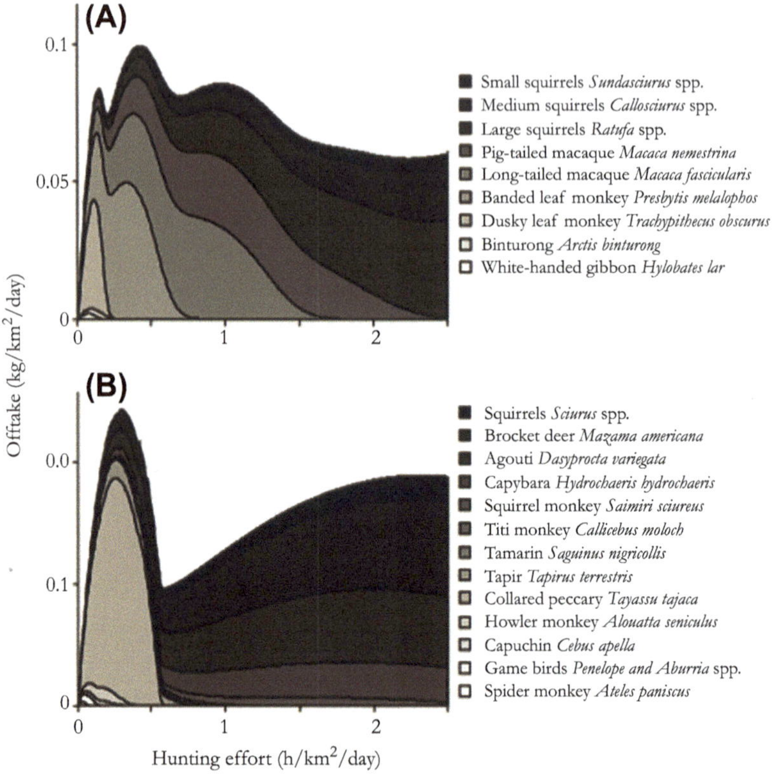

FIGURE 13.7 Result of simulation model of equilibrium harvest rates in multispecies hunting. (A) Blowgun hunting in peninsular Malaysia (B) Shotgun hunters in Peruvian Amazonia. *From Rowcliffe et al. (2003).*

of height) and the increase in sea urchins thus led to the collapse of the kelp forests (Jackson et al., 2001).

When harvest affects one species, it can alter the relationship through which two or more species (interspecific competition) or individuals of the same species (intraspecific competition) compete with each other for the same limited resource, such as food, water, or sunlight. A study of primate densities at 56 hunted and nonhunted sites in Amazonia and the Guiana Shield showed that where the abundance of large primates, such as howler monkeys (*Alouatta* spp.), woolly monkeys (*Lagothris lagotricha*), and spider monkeys (*Ateles* spp.), had been reduced by hunting, the abundance of nonhunted medium-sized primates, such as saki monkeys (*Pithecia* spp.) and uakaries (*Cacajao* spp.), had increased. Although most of these monkey species have quite a varied diet, they rely particularly heavily on ripe fruit whenever available, and the various monkey species are likely to compete over these scarce resources. Therefore, it seems, hunting of large monkeys favours the medium-sized ones by releasing them from competition over food – a mechanism known as density compensation (Peres and Dolman, 2000).

There are also many other kinds of relationships between different species of a community. In **mutualism**, two or more species benefit from each other. Some important examples of such relationships are those of one species relying on another for pollination or dispersal. In tropical forests, mutualistic relations are often highly specialised through the co-evolution of two parties, such as a plant species and its pollinator (see critique in Schleuning et al., 2012). One species may also benefit from another without affecting it, for example, a type of relationship known as **commensalism** by using the shelter or microhabitats it offers. For example, Brazil nut trees have complex interactions with many other species of the rainforest; bees, bats, and rodents contribute to its pollination and seed dispersal. The largest giant individuals of the Brazil nut tree can be centuries old, and their presence in the forest also provides shelter and habitat for numerous species of plants and animals. If long-lasting efficient harvest changes the population structure of the Brazil nut trees, possibly finally reducing the population overall, consequent changes will be likely to follow also in the populations of other species.

Sometimes rather unexpected consequences may occur through these types of mechanisms. As an example, the yellow-eared parrot (*Ognorhynchus icterotis*) lives in the cloud forests high up in the Andes Mountains, where it builds its nests in the hollow trunks of wax palms (*Ceroxylon* spp.). The populations of these palms, however, have been decimated not only because of habitat conversion but also because of the age-old tradition of the Catholic Church to harvest their fronds to use for decorations when celebrating the Palm Sunday, the Sunday preceding Easter. When the palms became rare, so did the parrots. Towards the end of the twentieth century, this species was thought to have gone extinct, and when it was rediscovered in 1999 in Colombia, a census revealed the existence of only 81 individuals that rarely nested. Since then, the population of these parrots has somewhat recovered as a consequence of successful protection of the wax palms (Quevedo-Gil, 2006).

Understanding the functioning of food webs through different trophic levels can help to distinguish some possible community-level response mechanisms to wild species harvest. Green plants are the primary producers with their photosynthetic products powering the consumers forming the rest of the ecological pyramid. The first-level consumers are plant-eating herbivores, which in turn support the group of secondary and tertiary consumers and finally the decomposers. To put it simply, the harvest of a species of a low trophic level will naturally induce a chain of change toward the upper trophic levels. In reality, however, the picture is rather multidimensional. For example, the harvest of some important species of primary consumers may not only cause the collapse of populations of some secondary or tertiary consumers, but it can also impact other primary producers through reduced herbivore stress. As herbivores usually consume only a selection of the community's available species pool, the conditions of competition among the different plant species may thus change. The consequences of this can be seen as changes in local vegetation.

There are a variety of different herbivore feeding habits, digestive systems, and other adaptive mechanisms. Folivores are specialised in eating leaves, nectarivores in consuming nectar from flowers, frugivors consume fruits, seed-eaters feed on seeds, and so on. Animals that are evolutionarily far from each other may have more or less similar feeding habits, and vice versa. Monkeys are a good example of a close species group in which a variety of herbivorous habits have evolved. Many monkey species are also flexible in their diets, changing their feeding behaviour from time to time, such as in response to seasonal changes (Peres, 1994). Most herbivores are free moving, and they may spend considerable times just searching for food. During the dry season, when only a few trees bear fruit in the seasonal rainforest, those trees that do so collect large groups of frugivore monkeys and birds to feed together. One may even hear the presence of such a tree from the noise the animals make. Selective logging of these keystone tree species can have devastating consequences in frugivore populations of the forest.

The fruit of the açaí palm (*Euterpe oleracea*) represent a case in which humans have assumed the role of an additional consumer of the wild fruit along with many species of birds and mammals to which the thin flesh of the açaí drupes constitutes an important food source. When humans harvest the fruit from the wild, there is less left for the animals; removing 41% of the açaí fruit in experimental plots has been shown to be enough to make howler monkeys stop visiting the stand. When 75% of the fruit were removed, tamarin monkeys also disappeared. The number of individual frugivore birds observed in the plots declined by 29%, and the length of their visits to the plots declined by 68% (Mogenburg and Levey, 2003).

Terrestrial ungulates and rodents eat fallen fruit from the ground or uproot seedlings. Sometimes animals may even trample seedlings to death. In the rainforests of Africa, elephants sometimes kill entire trees by debarking them, breaking the main stem, uprooting, or pushing them over. In Amazonia, there are no animals capable of doing such things (at least not anymore; see Chapter 3) but the animals feeding on plants nevertheless may have profound effects on the flora. Extirpation of large herbivorous vertebrates seems to lead to a defaunation syndrome of the forest understory characterised by 'seedling carpets, vertebrate damage-free understory herbs and seedlings, piles of uneaten rotting fruits and seeds' (Dirzo and Miranda, 1990). It is, however, difficult to firmly establish the net effect of any animal species on the vegetation because there are so many different kinds of interactions present.

Peccaries can be efficient dispersers of small seeds, but they destroy larger seeds, such as those of many palm trees. However, it also happens that palm seeds are too hard for them to crack open, such that they only chew off the fruit pulp and then expectorate the seed. Sometimes, the peccaries then trample such seeds down into the ground, thus protecting them from predation by insects

(Beck, 2006). Moreover, peccaries as well as some large rodents preferably eat palm seeds that are infested by larvae of certain beetle species. Thus, although they act as seed predators themselves, these mammals also suppress the populations of other seed predators, such as beetles. Different game animals act as seed predators and seed dispersers to different degrees, and they may also affect the abundance of other seed predators, thus making it very difficult to predict the net effect of hunting on seed mortality. Studies in Mexico and Panama, however, indicate that large-seeded plant species tend to become more abundant in forests subject to hunting, in comparison with nonhunted forests (Stoner et al. 2007a). Lianas and plant species that are dispersed by vectors not targeted by hunters (bats, small birds), as well as by water and wind, also seem to benefit from the hunting of larger seed-dispersing species of mammals and birds (Wright et al., 2007).

Particularly species-rich communities contain many co-occurring species that potentially could be negatively affected or even go extinct if harvest is carried out in excessive quantities or in a careless manner. On the other hand, high biodiversity may also imply increased **resilience** of the entire community because some particular species may be functionally substituted by others. For example, pollinators and seed dispersers may actually be less specialised for a particular plant species in the very diverse tropical areas; this strategy would at least be favoured by the fact that diversity is often accompanied by low densities of individuals of a particular species (Schleuning et al., 2012). The ecological interactions of species-rich communities are hard to comprehend, however, and innumerable kinds and amounts of interactions occur among the different species. This makes it also difficult to predict the consequences of an intended human intervention.

In Amazonia, species diversity is typically so high that just finding out what species of plants and animals actually exist even in a limited area of, say, 1 ha constitutes a long-term project for a multidisciplinary and well-funded research team. Because many harvested species occur at low densities, they are very difficult to inventory. As a consequence of all of these factors, it is hard to know what impacts harvest activities targeting one species will have on the status of others. This is how high species diversity in tropical forests complicates the scientific study itself (Figure 13.8).

Disturbances causing sudden but temporary changes in the physical environment or its biological interactions are inherent to nature. Natural disturbances include processes of biotic as well as abiotic origin. Species migrations and disease epidemics are examples of the former. Examples of the latter are, for example, storms that cause tree falls, exceptional rainfall and temperatures, drought, flood, landslides, and fire – all of which can destroy vegetation and kill animals, removing biomass even in very extensive areas. The harvester's transactions with nature mean that some additional disturbance is introduced into the community by human action. The extraction of species or their parts induces recovery processes with effects varying according to the type, magnitude, and frequency of human intervention. For example, the felling of a stand

FIGURE 13.8 The tropical rainforest in Amazonia hosts an enormous diversity of plant species.

of trees creates open space where natural regeneration processes by seeds and saplings can take place. The larger the human-made clearance is compared to typical natural gaps, the greater is the contrast with the natural conditions. Moreover, the frequency of such disturbance events is a critical factor to consider.

Irrespective of cause, recovery processes induced by a disturbance are called secondary **succession**. In these processes, a series of changes take place in the local species assemblages and with the ecological interactions, until a dynamic equilibrium state (climax) is reached in which more or less stable overall biotic characteristics again prevail. Because human activities tend to increase the degree of disturbance, successional stages with their specialist species become increasingly common in areas affected by humans, which also makes the variety of species of the climax community smaller and sparser. Knowledge about the local disturbance regimes and successional processes helps the assessment of the ecological constraints of wild species harvest. When harvest changes the environment or causes the death of individuals in a climax community, the risk of populations turning towards a decline is high. This is because climax species are in general more vulnerable to disturbance, their recovery is slow, and major changes in the prevailing environmental conditions may impede their survival. In turn, the harvest of successional species may be done in a way that reaches a dynamic equilibrium where the harvest induces disturbances that are favourable for the very species. The regeneration potential of the successional species tends to be high and, additionally, individuals may withstand considerable damage.

To a certain degree, the consequences of wild species harvest can be envisioned at a tentative level by intuition. However, the cascade effects on food webs and population structures are usually much more difficult to understand or predict. Generalisations and simple explanations should be applied with care, no matter how tempting they looked at the beginning. In fact, anyplace where some specific wild species is harvested at substantial levels, we should expect that it will indirectly affect other species too. The only way to really understand the interactions involved is to proceed by the means of scientific research. The character and magnitude of the indirect effects of harvest may often remain unknown because of the complexity of the problem at stake.

Impacts on Ecosystems and the Biosphere

Sometimes, wild species harvest has impacts not only on living organisms but also on the abiotic components of the natural world. However, because of the interwoven biotic and abiotic components of ecosystems, it is often difficult to clearly distinguish ecosystem level effects from those impacts that take place on the community level. For example, succession, as discussed previously, is a community-level phenomenon, but it typically occurs following a disturbance that also affects the abiotic environment. The effects of harvest on ecosystems can take place from very local to global levels, but most harvest interventions alter environmental conditions relatively locally, implying changes in microclimates, small-scale hydrologic and soil conditions, or local biogeochemical cycles.

Harvest can, however, also affect air, soil, or water and associated biogeochemical cycles at larger scales, up to changes at the level of the entire biosphere. Basically, these kinds of global effects are linked to harvest that implies large shifts in biomass and primary production, altering global biogeochemical cycles. The effects of harvest on abiotic components of ecosystems are often particularly seen in activities that involve the use of heavy machinery and/or imply the subtraction of large quantities of biomass from the ecosystem. Wild species harvest of such a magnitude that it would alone be able to alter the biosphere is uncommon, but together with other human activities some large-scale harvest activities certainly contribute to global environmental change (Hansen and Hoffman, 2011). Examples of this kind of harvest can be found at least in logging and fishing.

For instance, the opening of skidding trails and logging roads first removes vegetation locally, and the increased solar radiation due to canopy-opening and the use of trails and roads for transport then together cause soil compaction, increased surface water runoff, and erosion. These effects, in turn, often alter the regeneration of vegetation. Overall, logging can significantly reduce the carbon stocks found in the standing biomass of forest (Asner et al., 2010). Even when logging is very selective, targeting only a few species, it may lead to an increase in the amount of dead wood and litter on the ground. In combination with increased canopy openness, this may lead to an increased amount of

FIGURE 13.9 Tropical rainforest in Sabah, Malaysia, which first was subject to selective logging, and, soon after, was affected by forest fire. This sequence is fairly common in tropical forests around the world because logging operations leave large amounts of flammable material on the ground, while also the canopy opens up such that the microclimate on the forest floor becomes sunnier and drier.

dry material and associated elevated risk of forest fires (Holdsworth and Uhl, 1997) that further destroy vegetation and kill animals, leaving behind a strongly altered ecosystem (Figure 13.9). Careful application of reduced impact logging (RIL) methods can help to efficiently minimise long-term changes in the forest structure, hydrology, and microclimate, and thus contribute to a decreased risk of forest fire (Miller et al., 2011).

Ecosystem-level effects may also result from harvest in more indirect and unexpected ways. Still, these impacts can also be experimentally studied. *Prochilodus* is a genus of fish that feeds on detritus (dead organic matter) from the bottom of rivers and streams in the neotropics (Taylor et al., 2006). It is also a delicious fish to eat, and *Prochilodus* species belong to the economically most important fish species in Amazonia (one of the species is actually the *boquichico* caught by Ricardo and his crew on *Mi Marido* in Chapter 5). Experimental removal of *Prochilodus mariae* from one side of a 210-m stretch of the Las Marías River in the Orinoco Basin showed that the presence of this species has profound effects on the ecosystem.

Where *Prochilodus* fish were present, the pebbles on the bottom of the river appeared to be licked clean. On the other side of the enclosure where *Prochilodus* had been removed, the bottom of the river was completely covered by a thick layer of detritus (Figure 13.10). In quantitative terms, the amount of

FIGURE 13.10 An experimental study in the Orinoco River basin in South America tested the effects of removing the detrivorous fish *Prochilodus mariae* on the ecosystem. On the left, the fish fauna were left intact, whereas to the right all *Prochilodus mariae* have been removed. *From Taylor et al. (2006).*

particulate organic matter (POC) increased 450% in the part of river without *Prochilodus* fish. All of this was seen despite *P. mariae* being just one out of a total of over 80 species of fish present in the river. Smaller-scale experiments also revealed that *Prochilodus*, when present, decreased the amount of particulate matter in the benthic water and changed the composition of microbial biofilms from diatoms and heterotrophic bacteria to nitrogen-fixing cyanobacteria (also known as blue-green algae).

As seen so far in this chapter, the human interventions to subtract biological matter from nature can have multiple direct and indirect effects on all levels of ecological organisation. These effects, of course, do not alone determine what happens to the natural system that provides the harvestable resources, but rather they form one component in a multitude of other human activities and natural phenomena that together shape the environment and the world at large. Even in those cases in which the harvester's decisions seem to have little effect on resource dynamics in the face of larger forces, such as climate change or advancing land conversion, it is nevertheless important to understand the effects of harvest interventions on the provision from nature – even more so when the harvest is the primary driving force in the dynamics of the resource base. The next section will offer an overview of ways in which this kind of dynamic relationships between harvesters and the harvested resources can be studied.

MODELLING THE IMPACTS OF WILD SPECIES HARVEST

What is an Ecological Model?

The effects of wild species harvest were dealt with above in a purely descriptive manner. In this section, we will add some more analytical rigour to the analysis of the harvester's transactions with nature and present ways to explore them in a more formal way. Here we present a number of basic mathematical functions expressed as graphs and equations as representations of how the harvester's actions relate to the dynamics of the resource. We intend to keep this introduction to harvest models simple to such a degree that it can be rewarding reading also for those readers with little background or particular interest in mathematics. Reading the text and studying the graphs should be enough to grasp the main message and to appreciate the benefits of the mathematical approach, as well as to understand its limitations. We do, however, also recommend that the reader makes an effort to understand the mathematical formulas.

There are two basic types of ecological models; **analytical models** and **simulation models**. Analytical models include no empirical data but only symbolic expressions. Each analytical model is based on a set of specific assumptions, expressed in the language of mathematical formulas. By applying various algebraic operations to these formulas, one then obtains the results of the model. Such model results may provide explanations to empirically observed phenomena or lead to predictions that can later be compared with empirical data, in order to test the validity of the underlying assumptions in the model. Simulation models, on the other hand, perform numerical operations on numbers estimated based on empirical observations. Such models are used in order to make quantitative predictions or assess scenarios, rather than to provide logical explanations.

Ecological models are always simplifications of reality based on implicit assumptions that highlight some aspects of nature and disregard others. 'All models are wrong, but some are useful' is a common saying, originally coined by the British statistics professor George Box (1919–2013). Indeed, models can be useful, as long as one chooses to use a model suitable for the purpose and remembers to scrutinise the implicit assumptions of the chosen model to make a realistic judgment of its potential and limitations. Most ecological harvest models focus on the effects of harvest at the level of populations for one species at a time. Some models include two or even several species, but the multitude of the different types of interactions among species in real-life nature makes it very difficult to model the effects at the level of entire ecological communities.

The Logistic Model of Population Growth and Harvest

The harvest models we present here are of common usage and based on the **logistic model** of population growth, formulated in 1845 by the Belgian mathematician Pierre Francois Verhulst (1804–1849). The central idea of this model is that population growth is density dependent and that any species in a determined environment has an **intrinsic rate of growth**, which is the population's net rate of growth (births minus deaths) per capita when the population is very small. As the population gets larger, however, the per capita growth rate decreases linearly due to, for example, competition for food or other resources. At a certain point, the competition becomes so intense that deaths equal births, such that the per capita growth rate becomes zero. The population size (or density) at this point is called the **carrying capacity**; if the population is larger than this, the per capita growth rate becomes negative, as is graphically represented in Figure 13.11.

Mathematically, this concept can be represented using the following equation:

$$\frac{dN}{dt}\frac{1}{N} = r\left(1 - \frac{N}{K}\right) = r - \frac{rN}{K} \qquad (13.1)$$

where N represents the number of individuals at a particular time t, r is the intrinsic growth rate of the population, and K is the carrying capacity. Thus, dN/dt is the rate of population growth (change) at a particular moment in time, and $(dN/dt)(1/N)$ is the per capita population growth rate. The rest of the characteristics of this model can then be deduced from this simple assumption that net population growth per capita decreases linearly with population size.

We now turn the focus from population growth per capita to aggregate population growth. When the population size is zero, there are no individuals there to breed, so then there is of course no population growth at all. On the other hand, when the population equals the carrying capacity, the per capita population

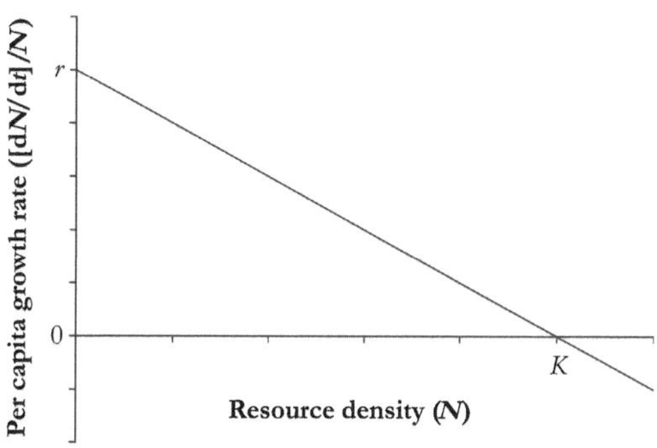

FIGURE 13.11 The relationship between population size and per capita net growth according to the logistic model. The letter r indicates the intrinsic rate of growth, and the letter K is the carrying capacity. Note that the function is undefined for negative population sizes: in real life, you cannot have a population of less than zero individual organisms. However, the growth rate can be negative if the population size exceeds the carrying capacity.

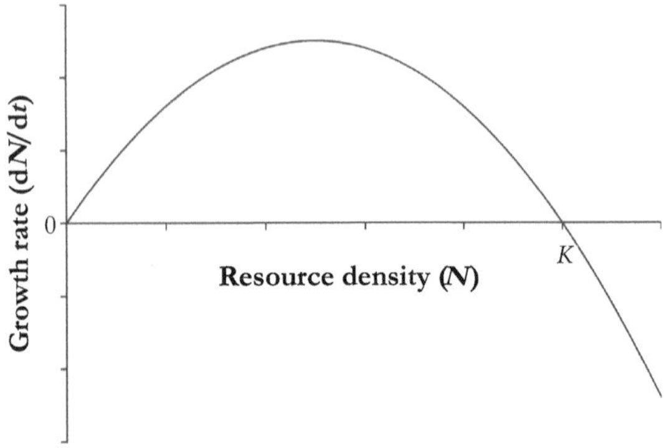

FIGURE 13.12 The bell curve showing the relationship between population size and absolute growth rate, according to the logistic model.

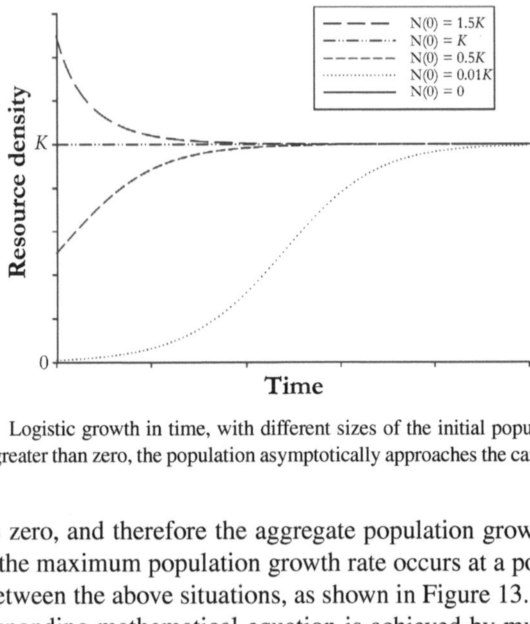

FIGURE 13.13 Logistic growth in time, with different sizes of the initial population. As long as the initial size is greater than zero, the population asymptotically approaches the carrying capacity, K.

growth rate is zero, and therefore the aggregate population growth also equals zero. In turn, the maximum population growth rate occurs at a population level somewhere between the above situations, as shown in Figure 13.12.

The corresponding mathematical equation is achieved by multiplying both sides of Eqn (13.1) with N, in order to move from per capita to an aggregate rate of population growth:

$$\frac{dN}{dt} = rN\left(1 - \frac{N}{K}\right) \tag{13.2}$$

Thus, according to this model, any population greater than zero will with time approach the carrying capacity (Figure 13.13).

The above-mentioned model captures a basic property observed for many populations in nature: they cannot grow for eternity. At some point, they reach a level at which further growth is limited and finally ceases due to the scarcity of food or other resources. Verhulst did not, however, provide any justification for the mechanisms underlying this model. Despite this shortcoming, the model became very widely used. It was only long after its description by Verhulst that other authors started to provide ideas explaining how the behaviour of individuals leads to mechanisms that explain population-level phenomena implied by the logistic model. For example, one mechanism could be that the birth rate is fixed, whereas the death rate is density dependent, such that it increases with increasing population density due to competition for scarce resources. According to Eqn (13.3), for example, the birth rate, b, is constant, whereas the death rate has one constant component, m, and one density-dependent component, αN, where N is the population size, and α is the coefficient of proportionality.

$$\frac{1}{N}\frac{\mathrm{d}N}{\mathrm{d}t} = b - (m + \alpha N) = (b - m)\left(1 - \frac{\alpha}{b - m}N\right) \tag{13.3}$$

Comparison of Eqn (13.3) with Eqn (13.1) then shows that $r = b - m$ and $K = (b - m)/\alpha$. Thus, the intrinsic rate of growth, r, and the carrying capacity, K, turn out not to be independent from each other, as the original logistic model suggests, but instead they both depend on the parameter b (i.e. the density-independent birth rate), as well as m (i.e. the death rate at low population density; Rueffler et al., 2006). A similar result can be obtained if one instead assumes a fixed death rate and a density-dependent birth rate. Because of its mathematical simplicity, we will continue using the logistic model in its original form. It is useful, however, to bear in mind this clarification of the biological meaning of the parameters r and K.

In order to further modify the model, **harvest** can be included so that the **net population growth** is the natural growth minus the harvest. This means that the population remains stable (at equilibrium) only if harvest equals natural growth (i.e. it is exactly on the bell curve shown in Figure 13.14). Here, the bell curve is the same as that in Figure 13.12, only that now also harvest is introduced. The graph shows that when the harvest is under the curve the population grows and when harvest is above the curve the population decreases.

Mathematically, we denote harvest with the letter H, such that

$$\frac{\mathrm{d}N}{\mathrm{d}t} = rN\left(1 - \frac{N}{K}\right) - H \tag{13.4}$$

Often, particular interest is in identifying an equilibrium situation in which harvest equals the natural growth:

$$\frac{\mathrm{d}N}{\mathrm{d}t} = rN\left(1 - \frac{N}{K}\right) - H = 0 \leftrightarrow H = rN\left(1 - \frac{N}{K}\right) \tag{13.5}$$

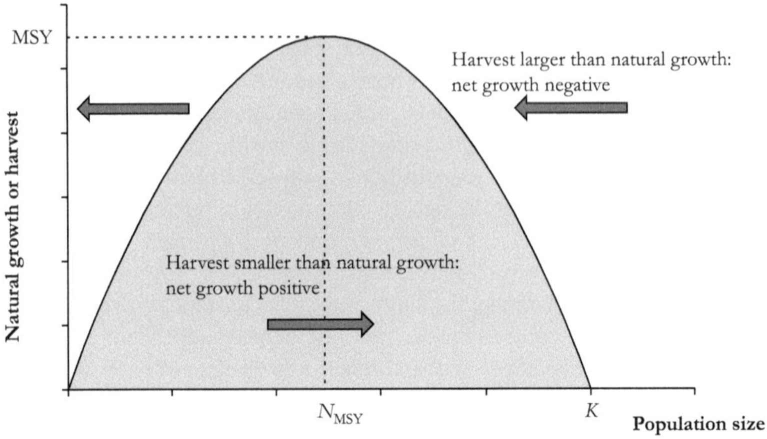

FIGURE 13.14 According to the logistic model, the population decreases if harvest rates exceed the equilibrium level represented by bell curve. In contrast, if harvest rates are under the curve, the population grows.

As seen in Figure 13.14, the model now starts to provide us with some important insights. For example, it can be noticed that the harvest that is initiated in a previously unharvested population found at its carrying capacity inevitably reduces the population size. This case presents a situation in which it is actually not possible to 'live off the interest and leave the capital intact', showing one of the limitations of that analogy. Moreover, the maximum level of harvest that can be sustained occurs at an intermediate population size – this being quite different from the case of interest on financial capital, too. This level of harvest is called the maximum sustained yield (MSY), and it can be mathematically shown that it equals the following:

$$\mathrm{H}_{\max} = \frac{rK}{4} = \mathrm{MSY} \tag{13.6}$$

Moreover, this is achieved when the population density is kept at the level of

$$\mathrm{N}_{\mathrm{MSY}} = \frac{K}{2} \tag{13.7}$$

Figure 13.15 helps us to further analyse the model visually. Each filled circle in the figure represents a possible combination of population size and harvest level, whose position can be compared with the bell curve showing the following equilibrium states:

A. Population is high and harvest is low, being right on the equilibrium curve. So, there is little reason to worry about resource depletion.

B. Harvest is above the equilibrium curve, so the population is decreasing. However, it will soon hit the equilibrium line, still at a relatively high population density, so there is little reason to worry.

C. Population is high, but also harvest is very high, being far above the equilibrium curve at a level that is definitively impossible to sustain in the long run. The population is decreasing and will eventually crash if the harvest level is not reduced.

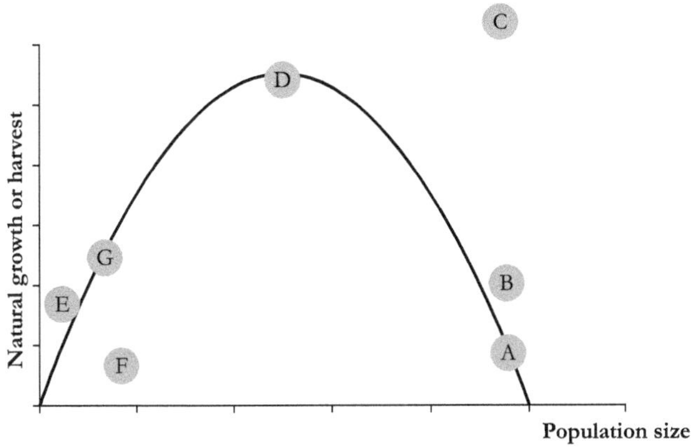

FIGURE 13.15 Some examples of possible combinations of population size and harvest. See the text for an explanation.

D. Population size is intermediate, and harvest is at equilibrium at the maximum sustained yield. This equilibrium is, however, unstable, so in fact, there might be reason for concern because the situation may easily become similar to that at point C, leading to the same problems although at a slower pace.

E. Harvest level is quite low but also is the population size. Harvest is therefore above the equilibrium curve, and the population is decreasing even further and will soon approach zero.

F. Population is small but also the harvest level is low, being well below the equilibrium curve. The population is recovering.

G. Both population level and harvest are low, with harvest being near the equilibrium curve. The system is near equilibrium and any change that might be occurring is probably proceeding relatively slowly.

Comparing the above harvest situations to how we defined sustainability in Chapter 2 reveals that some harvest situations can be considered more sustainable than others. There is little doubt that those situations that are clearly unsustainable are undesirable from most viewpoints. However, it is worth asking whether all of those situations that can be considered sustainable are also desirable from the harvester's viewpoint or from the perspective of some other actor involved.

Extensions of and Alternatives to the Basic Logistic Model

It is important to note that the logistic model is implicitly based on several assumptions. For example, it assumes that all individuals are equal in all significant respects, the environment is stable, and there is a stable supply of energy, nutrients, and anything else that the organisms may need. Furthermore, it is assumed that population growth is limited only by competition among individuals belonging to the same species. The fact that all of these assumptions are never perfectly met in real life does not render the model useless, but rather it means that the model is useful for answering some questions and not others. There are also ways in which the model can be modified in order to better fit reality when the situation diverges too much from the basic logistic model in all its simplicity.

For example, the assumption of the basic model that all individuals of a species are equal in size is fairly well met for hunted mammals that after a rather short immature period reach reproductive maturity and more or less stop growing. This assumption fits much less, for example, fish or trees that continue growing during their entire life span and harvesters who are interested not in maximising the number of fish or trees harvested but rather in the biomass or volume. For such cases, the logistic model can be used with the modification that the number of individuals, N, is replaced by the biomass, B.

$$\frac{\mathrm{d}B}{\mathrm{d}t} = rB\left(1 - \frac{B}{K}\right)$$

Another modification to the basic model is motivated by the fact that when the population density gets very low, many species cease to display maximum per capita growth as assumed in the basic logistic model. Instead, they may display reduced per capita growth, a phenomenon known as **depensation** (Figure 13.16). If per capita growth becomes negative in low densities, this is known as **critical depensation**. A wide variety of causal mechanisms may lie behind this phenomenon, including the difficulty of encountering mates for breeding (the so-called Allee effect) and beneficial group dynamics becoming impaired (e.g. for highly social animals that need a minimum group size in order to maintain successful reproductive behaviour). A wide variety of different modifications of the basic logistic model have been elaborated in order to represent such different causal mechanisms (see Lierman and Hillborn, 2001).

It has been empirically observed that the maximum population growth for large, long-lived organisms, in fact, often occurs when the population density is quite near the maximum possible equilibrium population density (i.e. the carrying capacity). For very small, short-lived organisms,

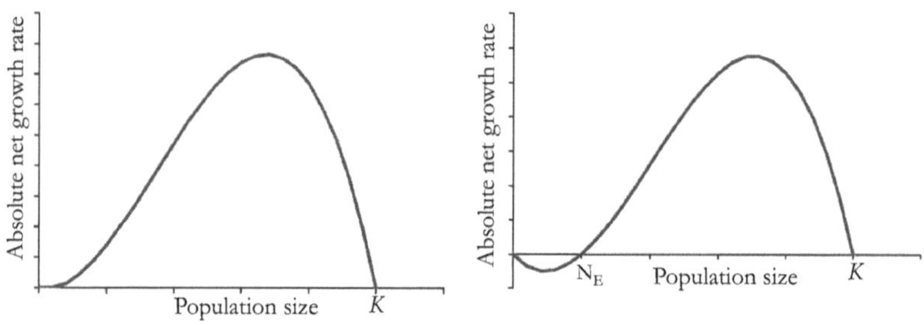

FIGURE 13.16 Depensation (left) and critical depensation (right). Critical depensation implies that if the population falls below the level N_E, it will inevitably go extinct.

the maximum population growth instead tends to occur at quite low population densities. Thus, the assumption of the basic logistic model that per capita population growth decreases linearly with population density, and that MSY therefore is achieved at $N=K/2$, may be quite far from reality in many cases. If this is important for the harvest question being analysed, the model can be modified accordingly, as shown in Box 13.1.

In some real-life cases, there are also important deviations from the assumption of stable environmental conditions that imply no fluctuations in time. Many species have a distinct breeding season, and then it may be more appropriate to use a **discretized logistic model** (Box 13.2). Population oscillations in real life may also be caused by chaotic and unpredictable environmental fluctuations. An example of this is the natural year to year variability in rainfall in many dry areas of the world. In such cases, the logistic model, or other deterministic models, may not be of much use at all.

The basic logistic model does not include any interactions among species (Figure 13.18). Therefore, alternative models or modifications of the logistic model have been made so as to couple the dynamics of, for example, predators and their prey (e.g. Berryman, 1992), a herbivore population and the plants they feed on (e.g. Caughley, 1982), species that compete with each other (e.g. Ayala et al., 1973), and of species that mutually benefit from each other (e.g. Dean, 1983). Such models, however, usually contain no more than two species, as they otherwise would become excessively complicated to handle and very difficult to interpret. All of these models can, moreover, be modified to incorporate the effect of harvest in a similar way as presented above for the basic logistic model. The impacts of harvest on multispecies ecological communities, however, are very difficult to model.

Modelling Structured Populations

The harvested population of a wild species can sometimes be strongly structured according to the age or life stages of individuals it consists of. Among fish, for example, the young individuals often rely on completely different food sources than the old. The young fish also typically have much higher mortality and they do not reproduce at all until they have reached a certain age. In addition, humans are typically more interested in harvesting large fish than small fish. Nevertheless, also for such cases, the logistic model can often be useful for analytical purposes. However, for the purpose of simulation modelling, it is often necessary to take into account the distribution of individuals in different classes of age, size, or life stage. This can be done using so-called matrix models.

To use a matrix model, a number of classes based on, for example, age, size, or life stage must first be defined. A field inventory of the species of interest can then be carried out, classifying each individual into one of these classes. Ideally, each individual is marked and observed again at regular intervals during

Box 13.1 The Theta-Logistic Model

The shape of the typical bell curve representing growth as a function of population density can be modified by introducing the parameter θ into the model:

$$\frac{dN}{dt} = rN\left(1 - \left(\frac{N}{K}\right)^{\theta}\right)$$

When $\theta > 1$, the maximum population growth occurs at some level of $N > K/2$, which may be more realistic for many hunted game mammals, for example. However, if $\theta < 1$, the maximum population growth occurs at some level of $N < K/2$. Setting $\theta = 1$ equals the basic model (Figure 13.17).

Box 13.2 The Discretized Logistic Model

The discretized logistic model has the form

$$N_{t+1} = N_t + rN_t\left(1 - \frac{N_t}{K}\right)$$

This implies that the population density, N, changes in discrete steps, and the population density N can therefore sometimes overshoot the carrying capacity K such that population growth in the next step of time becomes negative. Thus, this model often produces fluctuations in the population density, and the character of these fluctuations depends on the intrinsic growth rate, r. At low values of r, these fluctuations gradually dampen such that the population approaches an equilibrium state, whereas higher values of r instead lead to cyclic oscillations. At even higher r, the oscillations become chaotic instead of cyclic.

Again, this model can be further modified in order to include a time lag, as the net population growth may depend more on the previous than the current population density, such as for herbivores, whose grazing depresses the availability of feed. Another common modification is to include for each year a prereproduction (spring) population, as well as a postreproduction (autumn) population (e.g. Kot and Schaffer, 1984).

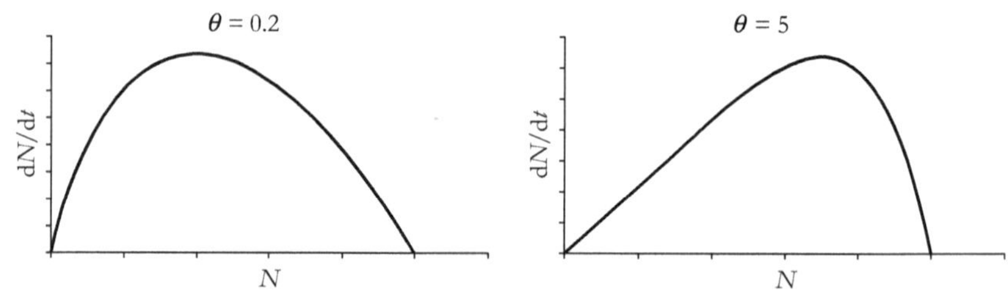

FIGURE 13.17 The theta-logistic model implies that the peak growth does not necessarily occur at $N = 0.5K$ but may be skewed to the left or the right, depending on the value of the parameter θ.

FIGURE 13.18 A reindeer (*Rangifer tarandus*) killed by a lynx (*Lynx lynx*). Such interactions are not included in the simple logistic model, but they can in real life have very significant ecological effects. *Photo: Olof Johansson.*

a number of years. This is how estimates can be obtained of the proportions of individuals in each class during a given period of time, as well as of their mortality, transitions from one class to another, and reproduction. These values can be used to analyse whether the population is increasing or decreasing, as well as to assess various harvest scenarios or design schemes for sustainable harvesting (Box 13.3).

USING HARVEST MODELS IN PRACTICE

Applying the Logistic Model to Hunting in Amazonia

Harvest models can be used to assess the status of resource stocks or the sustainability of the current harvest, and to analyse how harvest practices can be optimised. Choosing what model to use depends not only on the characteristics of the resource system and the purpose of the analysis but also on the feasibility of acquiring the needed data. For example, the unified harvest model (Bodmer and Robinson, 2004) was developed to assess the impact and sustainability of hunting of Amazonian game animals. It follows the basic logistic harvest model but has undergone some important modifications. One of them is that the model assumes that maximum sustained yield is not necessarily achieved at a population density of 50% of the carrying capacity, but rather this level varies according to the longevity of each respective species (Table 13.1), similarly to the theta-logistic model described in Box 13.1.

Box 13.3 Example of Using an Age-Structured Matrix Model

Figure 13.19 shows an example of an age-structured so-called **Leslie matrix model** using four age classes. The vector N_t represents the number of individuals in each respective age class at time t, and the vector N_{t+1} similarly represents the number of individuals in each respective age class one unit of time (e.g. a year) later. The symbols R_0, R_1, R_2, and R_3, respectively, indicate the rate of reproduction – that is, the number of offspring produced by each individual – in each respective age class. The symbols S_0, S_1, and S_2 indicate the rate of survival in each age class – that is, the proportion of individuals that from one year to another pass into the next age class. No individuals are assumed to survive beyond the age class 3. The matrix is multiplied with the vector N_t in the following way. First, 'tilt' the vector N_t to the left, so that it becomes horizontal, and lay it over the first row of the matrix. Then multiply the value of each cell in the matrix with the value in the overlying cell in the vector N_t and take the sum of the resulting four products. This is the entry in the uppermost cell of the vector N_{t+1}. Repeat the procedure for the second row of the matrix, and put the resulting sum of products in the second cell from the top in the vector N_{t+1}, and so on until filling all of the cells of vector N_{t+1}. This whole procedure is then repeated several times for N_{t+2}, N_{t+3}, N_{t+4},... etc.

Depending on the initial number of individuals in the respective age classes, these may show considerable fluctuations from one time step to another. Gradually, however, the fluctuations tend to smoothen out. The λ (lambda) value of the matrix is a measure of the long-term population growth, representing the average ratio between the population size at one year and the population size the year before. Thus, $\lambda>1$ implies that the population is increasing, and $\lambda<1$ implies that it is

$$\begin{array}{c|cccc} \text{Age} & 0 & 1 & 2 & 3 \\ \hline 0 & R_0 & R_1 & R_2 & R_3 \\ 1 & S_0 & & & \\ 2 & & S_1 & & \\ 3 & & & S_2 & \end{array} \times \overset{N_t}{\begin{bmatrix} N_{0_t} \\ N_{1_t} \\ N_{2_t} \\ N_{3_t} \end{bmatrix}} = \overset{N_{t+1}}{\begin{bmatrix} R_0N_{0_t}+R_1N_{1_t}+R_2N_{2_t}+R_3N_{3_t} \\ S_0N_{0_t} \\ S_1N_{1_t} \\ S_2N_{2_t} \end{bmatrix}}$$

FIGURE 13.19 The calculations involved in a Leslie matrix model. The letter R represents the rate of reproduction at different ages, indicated by the subscript. The letter S, in a similar manner, represents the age-specific survival rates, and the letter N represents the population size in each respective age class.

Box 13.3 Example of Using an Age-Structured Matrix Model—cont'd

decreasing. Leslie matrices are often used only in order to determine the current population trend, with no ambitions of predicting population changes in the long term. In order to increase the realism, some matrix models, however, do include some sort of density dependence – that is, the reproductive rates diminish as the population gets larger, thus prohibiting that the population grows in eternity (e.g. Jensen, 1996).

A further development of the matrix models are so-called **integral projection models**, which are based on the same principle. However, instead of dividing up individuals into distinct classes, they model demographic rates as continuous functions of the individuals' state (e.g. Ramula et al., 2009). A serious limitation of all such models is, however, that they require quite detailed quantitative field data. Although it is fairly simple to count, measure, and even mark individual plants, this is much more difficult to do for animals. And although it is still possible to do for fairly common animals living in open habitats, it is in practice unfeasible for animals occurring at low population densities in dense tropical forests with limited visibility.

TABLE 13.1 Modifications of the Basic Logistic Model Used in the Unified Harvest Model

Life History Strategy	Sustainably Harvestable Proportion of Production (In Basic Logistic Model: 100%)	Estimated N_{MSY} as a % of K (In Basic Logistic Model: 50%)
Short-lived	60%	50%
Medium-lived	40%	60%
Long-lived	20%	80%

N_{MSY} is the population size, or density, at which maximum sustained yield is achieved. K is the carrying capacity.
Source: *Modified from Bodmer and Robinson, 2004*

Another modification in this model is incorporated in the way in which the natural population growth is calculated. In the basic logistic model, the rate of natural population growth is a function of the parameters of the carrying capacity (K), intrinsic rate of growth (r), and the variable population density (N) Eqn 13.2. The rate of natural population growth decreases linearly with population density, but it is not defined whether this is because of decreased fertility or increased mortality. The unified harvest model, however, instead uses species-specific estimates of fertility based on data from a wide variety of sources. This model then assumes that the natural mortality differs from species to species due to longevity, and thus the whole production of the populations cannot be harvested sustainably but only a fraction.

Similar to the theta-logistic model (Box 13.1), the unified harvest model also assumes that the population density that yields the maximum production varies between different species (Table 13.1).

Finally, the unified harvest model requires data on the population density of the studied species. These data are needed not only from the study area itself but also from an unharvested control area in order to get an estimate of the carrying capacity, *K*. This implies doing an inventory in the field using the **distance sampling method**, which requires walking along transects; each time an animal is observed, one measures the distance between the transect and the place where the animal was first spotted. A minimum of some 20 observations is needed to generate a reasonable density estimate using the mathematical procedures described by Buckland et al. (1993). Because there are many game animal species that are rare and difficult to come across, not to mention the general difficulty of observing any animals in dense forests, obtaining good estimates for the carrying capacities of the hunted fauna may require walking thousands of kilometres (Carrillo et al., 2000). This is not impossible if one has patience and good funding, but it is a serious limitation for the feasibility of using the unified harvest model for practical wildlife management.

The model was applied to analyse hunting in the Pacaya Samiria National Reserve in Peruvian Amazonia (Bodmer and Robinson, 2004). The results shown in Figure 13.20 reveal that, for example, for red brocket deer (*Mazama americana*), the population density was found to be near carrying capacity and the harvest well below the estimated level of sustainable yield. The situation was quite different in the case of the South American tapir (*Tapirus terrestris*), whose population density was at 60% of the carrying capacity and the harvest was well over the estimated sustainable yield. Thus, whereas the red brocket deer seemed to be increasing, the tapir appeared to be decreasing. The authors, however, suspected that the high harvest of tapirs was sustained by dispersal of animals into the study area from surrounding unhunted areas.

As the authors also point out, however, the output of this model are products of a number of crudely estimated parameters and variables, so one cannot expect it to produce very precise results. It may be added that, given that the

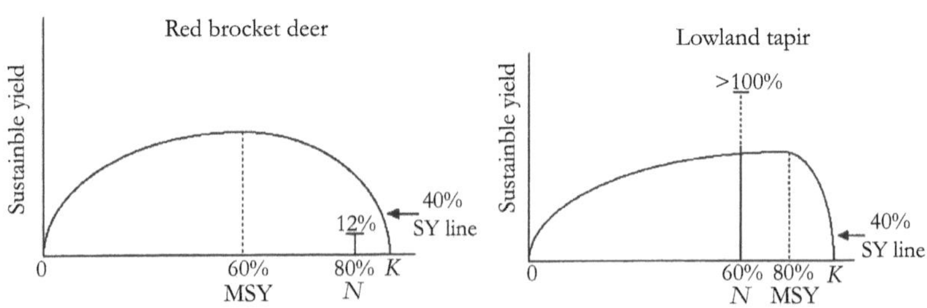

FIGURE 13.20 Results of using the unified harvest model in order to evaluate impacts and sustainability of hunting in the Pacaya Samiria National Reserve, Peru. It can be very instructive to compare these graphs with Figures 13.14 and 13.15. *From Bodmer and Robinson (2004).*

area in question has been subject to continuous hunting during at least several decades, one could expect that the populations of the hunted species would be fairly near an equilibrium state; that is, they ought to be very near the respective bell curves in Figure 13.20, similar to the points A, D, or G in Figure 13.15.

Studying Harvest of Palms Using Matrix Models

Often, matrix models have been used to study the sustainability of palm harvest, both for leaf harvest and harvest of entire trees. Such models usually differ in a number of ways from the example shown in Box 13.3. To start with, rather than being age-structured, these models are stage- and size-structured (so-called Lefkovitch matrix models). Typically seedlings, i.e. plants that do not yet have an above-ground stem, are divided into one or two classes, and the trees with stem are divided into a number of classes based on either height or diameter at the base. In this type of model, any individual may, from one time step to the next, not only either die or pass to another class, but also stay in the same class where it was. This implies that the matrix is somewhat different from the example in Box 13.3, having non-zero entries also along the main diagonal. This kind of model often also includes a density-dependence function, in particular for the youngest stages, because it has been shown that the higher is the density of individuals of the same species, the lower is the survival and growth of seedlings.

Field work involves measuring and tagging every tree within the study plots. After a year, the plot is visited again in order to tag new seedlings and measure all previously tagged trees a second time. Based on this, the probabilities of an individual of any particular class to either die, stay in the same class, or pass to the next, are calculated and entered into the matrix. In addition, fecundity is estimated by counting fruiting bodies or inflorescences. Germination, in turn, is assessed by sowing a known number of seeds in seed plots, as this is needed for the entries in the first row of the matrix (see Box 13.3). A matrix multiplication is then performed with the vectors containing the current number of individuals in each class, as shown in Box 13.3. This is repeated a large number of times so that the value of λ can be calculated and the long-term trend can be estimated. Ideally, field data should be collected over a number of years, as short-term fluctuations in survival and growth otherwise may obscure the long-term trends.

This kind of a model was used, in combination with experimental leaf harvest, in order to compare the effects of different leaf harvest regimes on the understory palm *Chamaedorea radicalis* in the El Cielo biosphere reserve in Mexico. The harvest of the leaves of this palm is an important income-generating activity for the local people and most of the harvest goes for export, which in 1999 amounted to over 2000 tons and a value of $20 million. The results of the study showed that sustainability could be improved while maintaining current harvest levels, but this required slightly altered harvest practices, harvesting leaves four times a year but only one leaf per tree per time, and, in addition, not harvesting reproductive individuals (Endress et al., 2006). This study spanned

FIGURE 13.21 *Iriartea deltoidea* is a common palm tree in neotropical rainforests, but it is in many locations heavily exploited for its wood, which is black, hard, and durable.

over 4 years, and it highlights the importance of doing this type of study over several years because the λ values for the different treatments varied quite considerably from year to year. One year indicated a population decrease, and the next year indicated an increase, and so on. Therefore, a study spanning only one year easily could have provided misleading results.

A similar approach was used to study the demographics of the dioecious palm *(Mauritia flexuosa)*, known as *aguaje* in Peru and commonly felled to harvest its delicious fruits that are sold at local markets. The results suggested that if felling the trees instead of climbing is seen feasible, a sustainable harvest regime would be to fell 22% of the female trees every 20 years. Field data was collected during only 2 years, however, so there is quite a risk that year-to-year fluctuations affected the result in one way or another (Holm et al. 2008).

A somewhat simplified and more rapid method was used to study the sustainability of extraction of the palm *Iriartea deltoidea* (Figure 13.21), used in construction in an extractive reserve in Acre, Brazil (Pinard, 1993). The field work was carried out within a short period of time, so it was not possible to tag individual palms and return to measure their growth one year later. Instead they were divided into six classes based on their height, and the probabilities of transition from one class to another was based on the current distribution of individuals in the respective classes and an estimate of the average age span of each class.

The age span was assessed by counting leaf scars on a selected sample of individuals in combination with published data on leaf production of the same species in another location. The study concluded that the palm populations were relatively insensitive to harvest of individuals higher than 20 m, given that they were then near the end of their life span. Because the palms start to reproduce long before they reach such a height and harvest actually does primarily target high trees, as these offer the best quality material, it was concluded that harvest did not pose any serious threats to the local populations of the species.

Models like those described in this chapter easily lead the reader to think that maximising the sustainable harvest would be something desirable. This is, however, erroneous. For one part, these single-species models do not take into account all the myriad of potential indirect ecological impacts that may occur when the abundance of one or more species are reduced. Such indirect impacts cannot easily be incorporated into general models such as those described in this chapter, and must therefore be considered specifically for each particular case. For another thing, rational management of harvestable wild species should not only take into account the benefits acquired by harvesting but also the costs involved in doing so. This is the theme of the next chapter.

Chapter 14

Costs and Benefits Weighted by Harvesters

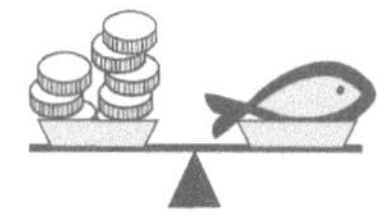

RESOURCES BECOMING GOODS

Provision from nature means that there are resources that can be harvested from the wild, but this does not mean that all resources actually get harvested. A condition for harvest to actually take place is that somebody perceives that the expected **benefits** from harvest outweigh the **costs** involved. In practice, the harvester first has to calculate or estimate how much time, effort, and material assets (tools, equipment, money, and the like) he or she must invest in the intervention with nature in order to extract resources from the wild. In addition to the extraction itself, other costs may be caused by, for example, planning, transporting, storing, processing, and marketing the harvested goods. The benefits that can be obtained from the addition of the harvested goods to society can range from the pure pleasure of the harvest itself to the fulfilment of subsistence needs and the generation of monetary income. Often, the net benefit is compared – at least unconsciously – to alternative options of using time, effort, and assets. Figure 14.1 illustrates how the harvester's calculations of costs and benefits take place at the interface of the natural and social systems.

Wild species resources provided by nature become **goods** when humans harvest them. These goods can be consumed as such (e.g. many fruits) or they may be processed further. The addition of these goods to society can have different consequences, from satisfying individual or collective subsistence needs to influencing the larger society through commodity chains involving processing and trade. The influences of harvest on the social system largely depend both on the effects of harvest in nature and on the specific utility and value of the harvested goods. The more value people generally see in the harvested goods, the more trade, processing, and societal arrangements emerge and are devised to regulate the harvest itself, as well as the ways to distribute its contributions for the well-being of people (Shackleton et al., 2011c). The same is true for the ecological effects of harvest; when the effects are perceived to be significant for the integrity of ecological systems, there is also likely to be more social pressure to regulate harvest interventions, something seen commonly, for example, in how the use of forests is organised. The societal arrangements that emerge as a consequence of these concerns regarding the transactions with nature and with society directly and indirectly influence the costs and benefits of harvest.

Diagnosing Wild Species Harvest. http://dx.doi.org/10.1016/B978-0-12-397204-0.00014-0

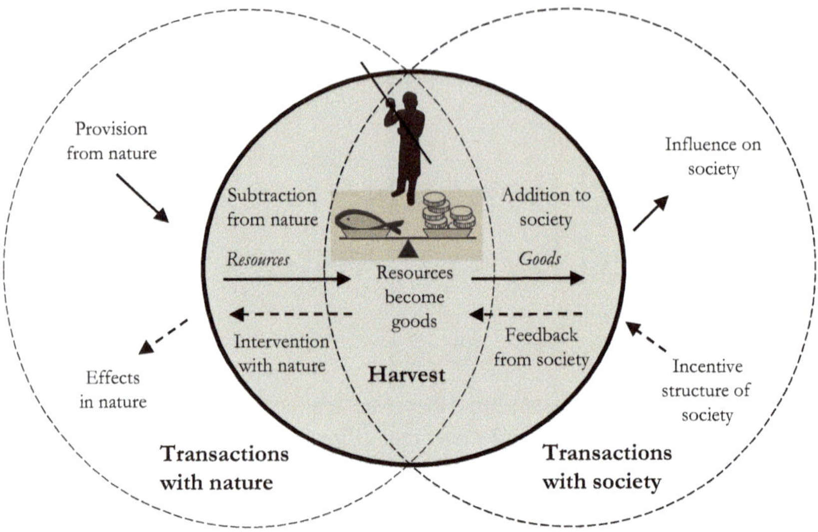

FIGURE 14.1 The utility of a biological organism or its part makes it a potential resource that can be extracted from nature and added as a good to society. When deciding whether to carry out the harvest act, the weighting of costs and benefits takes place in the interface between the transactions with nature and with society.

UNDERSTANDING HARVESTERS' DECISIONS

Understanding how potential harvesters value costs and benefits and make decisions about wild species harvest is crucial for one to be able to assess how harvest behaviour may change in response to changes in socioeconomic conditions, policy, or management measures. There is, however, no single and universally valid way of studying such decision-making. Instead, different scientific approaches are based on different specific assumptions; thus, they highlight some aspects of reality while disregarding others. One such approach, often used in the study of the harvest of wild foods in subsistence economies, has been to measure costs in hours and benefits in the energetic content of the foodstuffs procured; it has been predicted that harvesters strive to maximise the return in terms of energy per unit of time, according to so-called **optimal foraging theory** (e.g. Beckerman, 1983; Hill et al., 1987).

Another approach is to use money as a measure of costs as well as benefits, assuming that harvesters are well-informed and rational actors who attempt to **maximise profit**. This approach is, of course, very appropriate for cases of commercial wild species harvest, when the costs and benefits do consist of monetary transactions. Nonetheless, even when transactions are mainly nonmonetary, money can still be used as a measure of how people value the costs and benefits involved. An advantage of this approach is that it can be used for all resources, not only foodstuffs. Nowadays, it is also increasingly relevant because the monetary economy has penetrated almost the entire world, such that even people in some of the most remote places on Earth have gotten used to value costs and benefits in monetary terms. The benefits of harvest may then consist of either the monetary **revenue** acquired when selling the harvested goods; nonmonetary benefits, such as those acquired through direct consumption of the goods; or nonmonetary exchanges. In any case, nonmonetary benefits can be measured in monetary terms.

Although it is obvious that large harvests, or revenues, are something desirable for harvesters, one may easily forget the role of the costs, believing that harvesters simply try to maximise revenues. This is not so and it is very important to take into account the role of the costs. In fact, one could say that much, if not most of the potentially useful wild plants and animals that still exist out there do so only because harvesting them has been, in one way or another, expensive enough to dissuade potential harvesters. For analytical purposes, it can be useful to categorise these costs into different types, and any of them may be either monetary or nonmonetary.

Fixed costs are unrelated to the quantity of the resource units harvested; for example, they may be payments for permits or long-lasting equipment. The **search cost** per harvested resource unit is typically related to the abundance of the resource. The less abundant a resource is, the more time and other resources it takes, on average, to be able to harvest one resource unit. The search cost is a per-unit-of-harvested-resource cost that, on average, decreases with increasing resource abundance. **Handling costs** are related to the amount of resource units harvested, consisting of the cost of the actual action of converting a resource into a good, or product, after having located the resource. For example, this cost may be the time it takes to cut a tree and saw it into pieces small enough that they can be transported from the forest to the market. It may also include the **pursuit** of a spotted animal in order to hunt it down, and, of course, eviscerating it afterwards. The cost of consumables, such as shotgun cartridges or gasoline for chainsaws, also belongs to this category (Box 14.1, Figure 14.2).

Transport costs are related to the amount harvested as well as the distance that must be covered in order to get to the place of the resource and back to the place where it will be consumed or sold. The cost of transport often involves a nonmonetary dimension, such as the time and physical pain and exhaustion of carrying a heavy load a large distance. Finally, there may also be other costs that are specific to the type of harvest activity. These costs may include taxes paid according to harvest levels or, in the case of illegal harvest, the costs that are associated with the risk of detection and any associated fines or other punishment. These costs may even consist of the social rejection one may face, for example, for transgressing harvest rules set by the community where one resides. Technological development often contributes to increased efficiency and thus reduced costs, and therefore it has profound impacts on harvest behaviour (Figure 14.3).

Costs of harvest fall on the harvester before any benefits can be obtained from the harvested matter. Particularly in commercial harvest, it is common that

Box 14.1 The Effects of an Increase in the Price of Consumables

In early 2008, the government of Ecuador imposed a 300% tax on ammunition, as part of a general policy to impose stricter control on the trade with firearms, with the objective to curb violent crime. Whether this particular tax actually had any effects on crime rates is doubtful, but it did have a marked effect on subsistence hunting in Amazonia, as revealed by a study in the community of Sarayaku (Sirén, unpublished data). After the price rise, it was not worthwhile to shoot such relatively small animals whose meat was no longer worth the cost of buying the cartridge. The number of annually hunted prey therefore dropped to just half of what it had been a few years earlier, whereas the average biomass of hunted prey doubled, such that the total biomass hunted annually remained stable. With time, there was also a change in the technology used for hunting, as many hunters switched from cartridge shotguns to muzzle-loaders, which were less effective but for which the ammunition was considerably cheaper.

FIGURE 14.2 The costs for consumables, such as shotgun cartridges, form part of the handling costs. In the photo, a Baka ('pygmy') man in Cameroon receives elephant bullets from a meat trader. *Photo: Karl Ammann* (www.karlammann.com).

harvest takes place based on credit. This can be in the form of **advance payment** or through provision of harvest gear, food, or other items important for the life and work of the harvester. Advance payments and possible consequent **debt bondage** can, however, significantly affect the harvesters' subsequent bargaining positions with the buyers, as was seen in the case of the Brazil nut (*Bertholletia excelsa*) harvesters in Chapter 8 (see also Boscolo et al., 2010; Medina et al., 2009).

FIGURE 14.3 Chainsaws and outboard motors are among the most ubiquitous items in hardware stores in many Amazonian towns, such as this one in Puyo, Ecuador.

When analysing harvesters' decisions, it is useful to consider not only **explicit (accounted) costs**, such as the value of actual monetary transactions, but also the alternatives that were rejected each time one course of action was chosen in favour of others. The cost of not choosing the best available alternative allocation of one's time or capital, is called the **opportunity cost.** For example, if two persons go out in the forest in order to spend a day picking wild fruits or berries, the opportunity costs may consist of the money they could have gained instead if they had spent the day doing something else. Thus, the opportunity cost of taking an unpaid day off from work and spending it in the forest is much higher for a businessman with a high daily income than for a subsistence farmer or an unemployed person, who forgo little or no income by spending one day in the forest. The sum of the accounted cost and opportunity cost is called the **economic cost**, and, accordingly, the **economic profit** is the revenue minus the economic cost.

Costs as well as benefits may be affected by the market conditions (Figure 14.4). **Perfect market** conditions imply that no sole actor can influence any prices, such as for the harvested goods, labour, or other inputs. This is often a useful approximation of real-world situations of wild species harvest, in particular when it takes place for commercial purposes and there are many individual sellers as well as buyers. In other cases, the concept of perfect market conditions can serve as a benchmark against which to assess the presence of **market imperfections** and their implications. In the context of wild species harvest, several significant types of market imperfections may be in place. Much wild species harvest takes place in the context of subsistence economies where the harvested goods are consumed locally, often within the harvester's own household, rather than being sold on the open market. In such a situation, the assumption of an invariable price is obviously erroneous, not because of the absence of monetary transactions but because the amount of any resource harvested by one single individual may significantly influence how the consumers value each unit of the resulting goods. An input of one extra kilogram of wild meat into the family economy is valued much higher if there is nothing else to eat than if there is already more meat stored than what the family possibly is able to eat.

Under such conditions, rather than assuming profit-maximising behaviour in a perfect market, it may be more reasonable to assume that harvesters make a **tradeoff between consumption and leisure**, implying that the higher the net benefit obtained from the harvested goods, the less willing people are to give up leisure time in order to increase the net benefit even more. Alternatively, one may even assume that their objective is to **minimise the labour** employed to reach a **subsistence target** (cf. Angelsen, 1999). This would mean a case in which the members of a household need a given quantity of goods to subsist and do not aspire to more beyond that point, actually stopping to work once the needed quantity of goods has been acquired. Even when harvest is carried out for commercial purposes and the number of end consumers is large, significant market imperfections may result if the harvesters can only sell their produce to one or a few middlemen. The middlemen then will have the power to set the price based on their own strategic considerations, such as has been common in Amazonia in the cases of rubber (Chapter 3), Brazil nuts (Chapter 8) and timber (Chapter 9).

Other imperfections may have to do with the labour market, such as the ease of switching from harvesting a certain wild species resource to other economic activities and vice versa. Wage labour may be scarce, or engaging in wage labour may imply that one must move to urban areas; therefore people may tend to continue harvesting wild species even in spite of an apparent lack of profitability. On the other hand, even if there are wild species that could be profitably harvested in remote and sparsely populated places, there may not necessarily be enough people to do the job. This may be the case especially if there are institutional barriers to entry into wild species harvest, such as on indigenous peoples' lands, where outsiders may not be allowed to take up residence or engage in harvest activities.

FIGURE 14.4 Market conditions as well as final destination varies between different harvested wild species products. (A) Subsistence use: Palm heart extraction in Sarayaku, Ecuador. (B) Local trade: Bushmeat sold outside the railway station in Yaounde, Cameroon *(Photo: Karl Ammann)* (www.karlammann.com). (C) International trade: Brazil nuts from Amazonia in a grocery store in Finland.

When assessing the potential effects of different management and policy interventions on wild species harvest, it is important to be explicit about one's assumptions regarding market conditions and the economic objectives of harvesters. Often, such assumptions are made only implicitly and without deeper reflection about their validity, and then there is a risk that interventions intended to improve harvest situations do not lead to the desired results. For example, it is not uncommon that conservationist organisations attempt to provide alternative sources of food to local people such that they should not *need* to hunt anymore. This may work in cases where subsistence economy conditions prevail and households strive to reach a subsistence target while minimising labour. However, if households are immersed in an open economy without many market imperfections, and aim to maximise their profits, it is quite likely that they will continue to hunt and sell any portion of their harvest that they do not consume within the household.

Similarly, providing salaried jobs to harvesters can help people move away from wild species harvest that is considered harmful. But, unless there are significant imperfections in the labour market, this may just mean that someone else takes up the harvesting activity instead, such that the impact on wild species remains unchanged. Protected areas in the tropics are commonly surrounded by buffer zones on which public agencies and nongovernmental organisations carry out activities that aim to provide the local population with sustainable alternatives for making a living so as to reduce the pressure on the resources of the protected area. It has been suggested that flows of money and new services to the buffer zones can potentially attract new migrants, in the long term even increasing the pressures instead of diminishing them. To what extent this is really happening has been a matter of intense debate (e.g. Joppa et al., 2009; Wittemeyer et al., 2008).

DETERMINING LABOUR INPUT

Profit maximising means that the harvester tries to maximise the difference between total **revenues** and total **costs**. The revenues consist of the **harvested amount** of resource units multiplied by the **price** per resource unit. The costs of wild species harvest activities vary greatly from case to case based on the type of resource that is targeted, the logistics, and the different kinds of legal requirements that the harvester has to take into account, to name a few classes of cost variables. For example, although informal harvest of palm leaves or fishing can involve considerable handling, transport, and salary costs, the harvest of Brazil nuts or timber from concessions often additionally require investment in planning, administration, and sometimes even processing. On the other hand, wild animals in tropical forests typically are inconspicuous and occur at low densities, such that search costs are the major costs involved in harvesting the resource.

To harvest any particular quantity of a wild species resource, it is necessary to incur certain costs. The harvester thus attempts to choose an optimal level of harvest, such that the difference between total revenues and total costs can be maximised. As an example of how the implications of such behaviour can be assessed, we will here focus on a stylised case where costs include only the cost of the time spent searching for the resource and where market conditions are perfect, such that there is a fixed price for the harvested products. These assumptions are fairly reasonable for the case of commercial hunting near a city consuming large quantities of game meat, for example. Under such conditions, each harvester will attempt to choose the amount of time devoted to searching for the resource, in such a way that the profit is maximised. Typically, the more time one dedicates to the search labour (also called **harvest effort**), the greater is the quantity one harvests. Commonly, it is assumed that this relationship is linear.

Hence, on one hand, given a certain resource density, doubling the time of labour devoted to hunting should lead to a doubling of the harvested quantity of wildlife (Figure 14.5A). On the other hand, the harvest per unit of search labour time also depends on the resource density. Again, it is often assumed that this relationship is linear, such that doubling the resource density, given a certain time devoted to harvest labour, would also lead to a doubling of the harvested quantity. (In real life, this assumption of linearity may be a more or less accurate representation of reality, depending on the particular case-specific conditions, cf. Maunder et al., 2006; Rist et al., 2010). In the long run, thus, increased aggregate search labour leads to decreased resource density

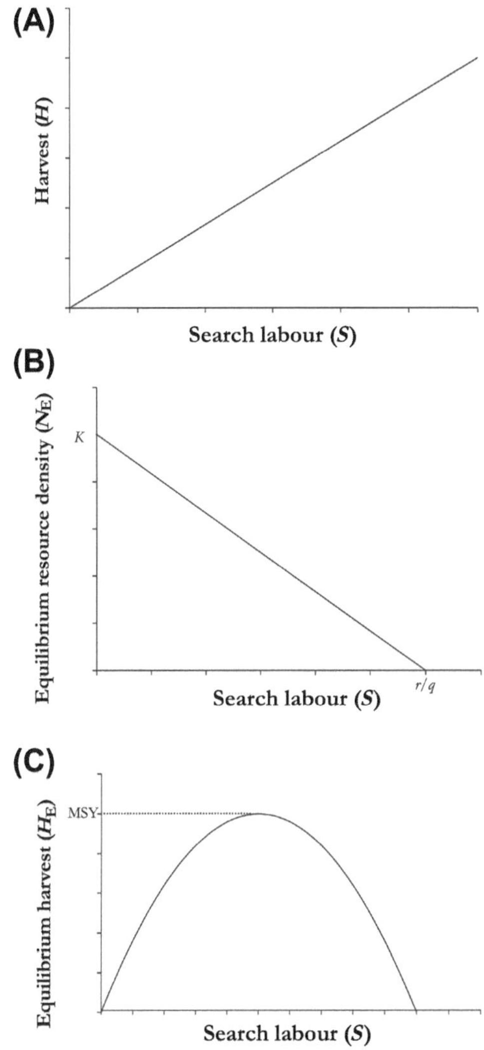

FIGURE 14.5 Function graphs showing the basic properties of the harvest model. (A) At a given resource density, i.e. in the very short run, or for one single harvester among many, harvest is directly proportional to search labor. (B). In the long run, and at aggregate level, however, resource density decreases linearly with search labour (C) Therefore, at equilibrium, harvest increases with aggregate search labour only up to a certain point, and then decreases again, forming the characteristic 'bell curve'.

(Figure 14.5B). Therefore, at **biological equilibrium**, the amount of harvest is a bell-shaped function of the amount of aggregate search labour (Figure 14.5C)

Because we assume that hunters face a fixed price for their output, then revenues depend on the amount of labour according to a function with the same

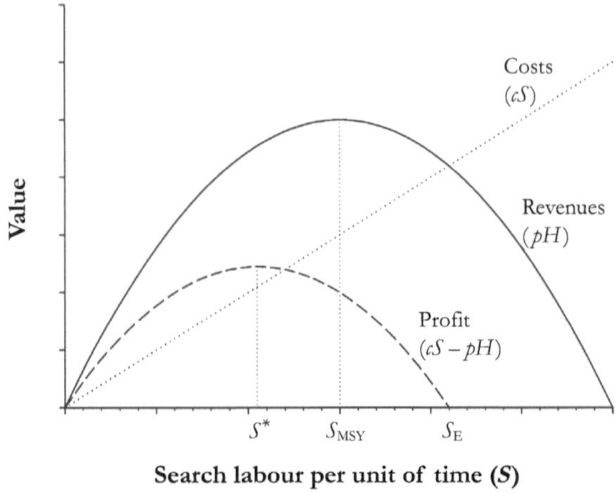

FIGURE 14.6 When taking into account not only the revenues of harvest but also the costs of search labour, it turns out that the optimal amount of search labour is lower than the amount of search labour that yields the maximum sustained yield. In this graph, the profit curve is acquired by subtracting the cost curve from the revenue curve. S_{MSY} represents the amount of search labour that yields the maximum sustained yield. S^* is the optimal amount of search labour and S_E is the level of search labour at economic equilibrium, where profits are zero.

bell shape as does the harvested quantity itself. If there is a sole owner of the resource or a so-called social planner who is making decisions in order to maximise the collective benefit, this decision-maker would therefore choose a level of labour input that would keep the resource density somewhat higher than that which would lead to the maximum sustained yield, as this would maximise the difference between revenues and costs (Figure 14.6).

If there are many individual harvesters acting independently, however, each one of them in this situation can potentially increase his or her profit by harvesting a bit more, thus reaping a larger share of the total profit. However, if many harvesters do this the total profit decreases to the detriment of them all. In fact, this can continue up to the point where revenues equal costs and thus the profit equals zero. Only when this point is reached are there no more incentives for the individual harvesters to attempt to reap a greater share of the total profit; this is the **bioeconomic equilibrium** of the system. **Rent dissipation** is a term commonly used for this mechanism in which the self-interested behaviour of many harvesters, although rational from the strictly individual perspective, ultimately leads to a situation where the potential economic profits of wild species harvest are forgone for everybody.

SUPPLY AND DEMAND OF WILD SPECIES

In the previous section, we assumed a fixed price for the harvested products. In an open economy, any individual harvester faces a fixed price that he or she

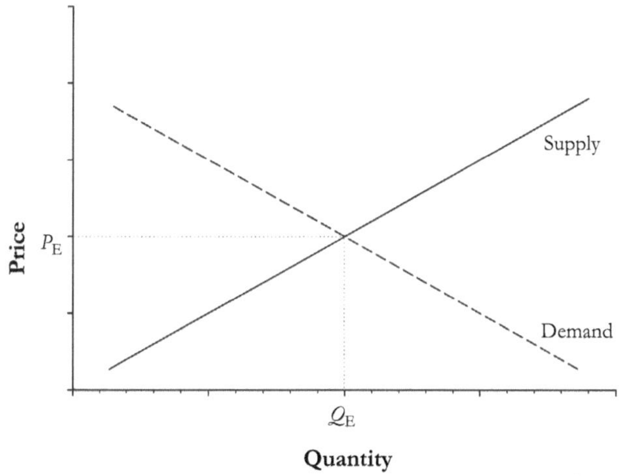

FIGURE 14.7 The standard textbook model of supply and demand. The equilibrium quantity produced, Q_E, and the equilibrium market price of the product, P_E, are determined by the crossing of the supply and demand curves.

cannot affect. At an aggregate level, however, prices depend on the interplay between supply and demand. The principles of **supply** and **demand** in standard economics are quite simple: when consumers demand more, the prices tend to increase; when producers supply more, the prices tend to decrease. Price fluctuations also influence the behaviour of both consumers and the producers accordingly: when prices increase, then consumer demand tends to decrease but the producers are willing to produce more. These mechanisms together lead to an equilibrium at the price level where supply equals demand (Figure 14.7).

For wild species harvest, however, the supply curve is very different. If prices increase from a low level, the harvest (or 'production' in economic terms) increases, just as in the standard case. This can, however, only continue up to the point where the Maximum Sustained Yield is reached. Further price increases instead lead to that the equilibrium harvest decreases, due to the reduced resource density. The long-term aggregate supply curve of a wild species subject to unrestricted and uncoordinated harvest therefore increases only up to a certain point and then bends downwards again (Clark, 1990) (Figure 14.8). This supply curve can be described mathematically as a bioeconomic model (Box 14.2).

If harvest decisions are made by a **monopolist** (sole owner or social planner), however, another supply curve may result (Figure 14.8). In the monopolist case, the harvester can consciously select harvest levels such that he or she maximises the long-term profit and does not exhaust the resource base completely. In real life, however, monopolist control does not guarantee that the resource stock will not be exhausted due to intense harvest. A monopolist may also choose to harvest a large part of the resource today even if that reduces future harvests. This is because people often, if not almost always, value benefits obtained in

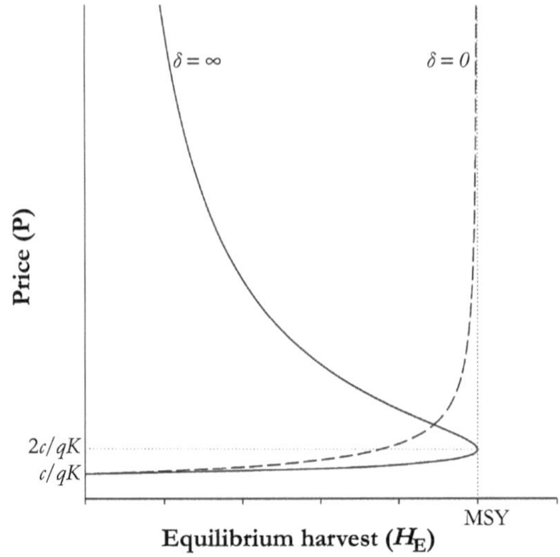

FIGURE 14.8 Supply curves for a harvested wild species product, according to the model described in the text. The solid line represents the supply that results when many self-interested individuals harvest from the same resource pool in an uncoordinated manner. The dashed curve shows the supply that results when a monopolist owner (or so-called social planner) harvests the resource so to maximise sustainable profit. These two curves can also be understood as representing different discount rates. The solid line represents an infinite discount rate, whereas the dashed line represents a zero discount rate. Any discount rate between these two extremes yields a supply curve that does bend back, but less than the solid line shown. *Modified from Clark (1990) and Milner-Gulland and Mace (2009).*

Box 14.2 A Bioeconomic Harvest Model

Economists have developed a set of theoretical concepts that allow mathematical modelling of economic behaviour at individual or household levels. Such modelling is a useful tool for understanding the complex phenomena that form the basis of such decisions. One central concept through which this kind of decision making can be evaluated is the concept of so-called *Homo economicus*, a theoretical representation of people as self-interested, well-informed, and rational actors, who as producers strive to maximise their profit, and as consumers strive to maximise their well-being (or, in strict economic terminology, their **utility**). *Homo economicus* has no sense for altruism or reciprocity and is thus a quite gross simplification of real human beings, but nevertheless it is a useful concept for specific analytical purposes.

Combining such economic models with population ecological models can be very useful in order to explore potential impacts on harvest behaviour and resource stock of changes in costs and prices, infrastructure and market access, policy and macroeconomic variables, or management interventions, for example. The models can also be used to assess the sustainability of harvest, particularly in the absence of detailed biological data. There are many different types of bioeconomic harvest

Box 14.2 A Bioeconomic Harvest Model—cont'd

models, such as dynamic versus equilibrium, stochastic versus deterministic, spatial versus nonspatial, and analytical versus simulation. The following model, originally created by Clark (1990), is an equilibrium model, meaning that it shows an end state in equilibrium rather than changes over time. It is deterministic because it does not incorporate any randomly occurring events, nonspatial because it does not take into account any spatial variability, and an analytical model because it uses only symbolic expressions and no real numbers.

To start with, the harvested amount is linearly proportional to resource density, search time, and the catchability coefficient:

$$H = qSN \tag{14.1}$$

where H is the harvest, q is the catchability coefficient, S is the search labour per unit of time (also called 'effort'), and N is the resource density. The parameter q, the catchability coefficient, is here a measure of the risk for any single individual of the resource population to be harvested due to one unit of search labour. Its value depends both on the species in question and on the technology employed. Large-bodied animal species living in the canopy are, for example, much easier to spot and hunt than small-bodied species living in the dense underbrush, and thus the former have higher value of q.

Resource growth is modelled using the logistic equation described in Chapter 13:

$$\frac{dN}{dt} = rN\left(1 - \frac{N}{K}\right) \tag{14.2}$$

At biological equilibrium, harvest equals natural growth, such that net growth is zero:

$$0 = \frac{dN}{dt} = rN\left(1 - \frac{N}{K}\right) - H \tag{14.3}$$

Combining Eqns (14.1) and (14.3) provides the equation for the linear relationship between equilibrium resource population density and search labour shown in Figure 14.5(B):

$$N_E = K\left(1 - \frac{Sq}{r}\right) \tag{14.4}$$

Profit equals revenues minus costs:

$$\Pi = pH - cS \tag{14.5}$$

where Π is the profit, p is the market prize for one harvested resource unit, and c is the cost of one unit of search labour.

Combining Eqns (14.1) and (14.5) yields

$$\Pi = S\,(pqN - c) \tag{14.6}$$

It is furthermore assumed that many self-interested individuals harvest from the same resource pool in an uncoordinated manner, resulting in an **economic**

Continued

Box 14.2 A Bioeconomic Harvest Model—cont'd

equilibrium where profit equals zero. Accordingly, substituting $\Pi=0$ into Eqn (14.6) and reorganising yields the formula for the resource density at economic equilibrium:

$$N_E = \frac{c}{pq} \tag{14.7}$$

This implies that if $N<c/pq$, no harvesting of the resource takes place because it would be unprofitable. On the other hand, if $N>c/pq$, harvesting yields a positive economic profit or rent – that is, a higher return than spending the same time in any other alternative economic activity.

Rearranging and substituting Eqn (14.7) for economic equilibrium into Eqn (14.3) for biological equilibrium and, finally, rearranging and substituting Eqn (14.1), we get the following equations for the bioeconomic equilibrium:

$$H_E = \frac{rc}{pq}\left(1 - \frac{c}{pqK}\right) \tag{14.8}$$

$$S_E = \frac{r}{q}\left(1 - \frac{c}{pqK}\right) \tag{14.9}$$

Plotting H_E against p and switching the axes, we get the backwards-bending supply curve shown by the solid line in Figure 14.8.

If we instead assume that the resource has one single monopolist owner who intends to maximise sustainable profit, the model becomes slightly more complex. In short, as can be deduced graphically from Figure 14.6, the optimal level of search labour is then half of its level at bioeconomic equilibrium:

$$S^* = \frac{r}{2q}\left(1 - \frac{c}{pqK}\right) \tag{14.10}$$

Substituting Eqn (14.4) into Eqn (14.1), and then substituting Eqn (14.10) into the result of that, yields the equation for how the optimal harvest level varies with price:

$$H^* = \frac{rK}{4} - \frac{rc^2}{4Kp^2q^2} \tag{14.11}$$

Studying this equation and comparing it with Eqn (13.6) of Chapter 13, one can easily see that with increasing price, also the level of harvest increases, ultimately approaching maximum sustained yield, as also shown graphically by the dashed line in Figure 14.8.

Box 14.2 A Bioeconomic Harvest Model—cont'd

Unlike *Homo economicus*, real people are not entirely egoistic, but quite on the contrary have a sense of altruism and are able to cooperate and sacrifice personal gains for the common good. Also, real people are not perfectly informed, and lack of knowledge of the potential effects of different actions may have important implications on people's behaviour. In real life, it is quite possible that a group of several harvesters coordinate their harvest such that they act as one single monopolist owner, thus maximising their aggregate profit. Similarly, a sole owner who is unaware of the impacts of his or her actions on the resource base may deplete the resource to the point of rent dissipation instead of maximising profits.

the present more than the same benefits obtained in the future – a phenomenon discussed at the end of this chapter.

As shown in Box 14.2, the equilibrium density of a harvested wild species is directly determined by the price it is given on the market. Thus, it is also important to understand the factors that affect this price as a reflection of the consumer demand. First of all, this helps one to predict how changes in policy, management, or socioeconomic conditions of harvesters may affect their harvest practices and consequently the wild species resources. One important question is whether the demand for products acquired from wild species increases or decreases when incomes increase. It has been shown, for example, that as incomes increase, the consumption of wild plants often goes down, whereas the consumption of wild meat goes up (Figure 14.9). Such issues can be analysed using the concept of **income elasticity of demand** (Box 14.3). Another important aspect is how the demand for wild species products is affected by the availability of **substitutes** (Figure 14.10). It has, for example, been shown that the demand for wild meat in central Africa is negatively correlated with the availability of fish from the sea (Brashares et al., 2004). It has also been suggested that the development of synthetic medicines may lead to decreased exploitation of plants and animals used for traditional medicinal practices. Such relationships can be studied using the concept of **cross-price elasticity of demand** (Box 14.3).

MULTISPECIES HARVEST

So far, this chapter has mostly discussed cases of harvest of a single species. Multispecies harvest differs from single species harvest in some important respects; this can be exemplified using the case of tropical forest hunting. Unless there are effectively enforced restrictions on hunting, once hunters are out on a hunting trip, they tend to attempt to maximise their return in terms of yield per unit of time. Species preferred by hunters are therefore those that provide a large benefit for a small extra cost – that is, the handling cost including pursuing, shooting, and eviscerating the animal. Therefore, when encountering a prey

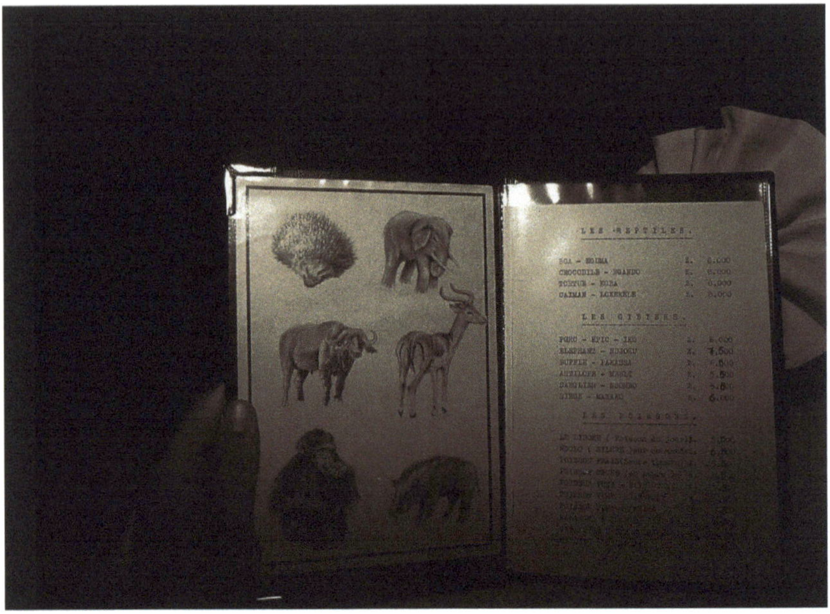

FIGURE 14.9 Wild meat is sometimes available even in fancy restaurants, such as this one in Kinshasa, Kongo. The consumption of wild meat often increases with increased income, but can also decrease, depending on the specific socio-economic context. *Photo: Karl Ammann* (www.karlammann.com)

Box 14.3 Price Elasticities

In order to analyse how demand and supply of a certain good vary with changing circumstances, the concept of *elasticity* is used. There are several different kinds of elasticities, but we will here concentrate on the **own price elasticity of demand**, the **income elasticity of demand**, and the **cross-price elasticity of demand.**

The own price elasticity **coefficient** of demand is defined as:

$$\varepsilon_p = \frac{dQ/Q}{dP/P}$$

where Q is the quantity demanded, P is the price of the good, and d represents an infinitesimally small change. A somewhat simplified version of this definition, which works well for small changes is

$$\varepsilon_p = \frac{\text{\% change of quantity demanded}}{\text{\% change in price}}$$

Because demand typically decreases with increasing price, the numerical value of elasticity tends to be negative. A price elasticity coefficient of demand $\varepsilon_p = -1$ is

Box 14.3 Price Elasticities—cont'd

called **unit elasticity**. It implies that if price increases with a certain percentage, the demanded quantity decreases with the same percentage. The more the demanded quantity changes with increased price, the more **elastic** is the demand said to be, and the higher is the absolute value of the elasticity coefficient (Figure 14.11). A study among indigenous peoples in Honduras and Bolivia showed that the consumption of wild meat clearly decreased with increasing prices, particularly among the more well-off households; among low-income households, the price elasticity of wild meat was $\varepsilon_p = -0.4$, whereas among high-income households it was $\varepsilon_p = -2.6$ (Wilkie and Godoy, 2001).

Similarly, the income elasticity coefficient of demand can be defined as:

$$\varepsilon_i = \frac{\%\ \text{change of quantity demanded}}{\%\ \text{change in income}}$$

Most goods are so called **normal goods**, meaning that $\varepsilon_i > 0$, which implies that the more income people have, the higher their demand for the goods. If this increase in demand is less than proportional to the change in income (i.e. if $0 < \varepsilon_i < 1$), the good is sometimes called a **necessity**. If the increase in demand is more than proportional to the change in income (i.e. if $\varepsilon_i > 1$), the good is called a **luxury** or **superior good**. On the other hand, **inferior goods** are such that $\varepsilon_i < 0$, implying that the demand decreases if peoples' income increases. Typical inferior goods are low-quality foods, which people substitute with better food if their incomes increase. Whether products from wild species are inferior, necessity, or luxury goods can have important implications for predicting how income changes may affect harvest and thus the conservation status of the species in question. A study in Gabon revealed that wild meat was a necessity ($\varepsilon_i = 0{,}17$), meaning that the higher income people have, the more wild meat they consume (Wilkie et al., 2005). Elasticities may, however, vary between different segments of a population. The study in Honduras and Bolivia showed that among low-income households, wild meat was a necessity good ($\varepsilon_i = 0.04$), whereas among high-income households it was an inferior good ($\varepsilon_i = -0.14$), suggesting that demand for wild meat may follow an inverted U pattern with income (Wilkie and Godoy, 2001).

Finally, the cross-price elasticity coefficient of demand shows how the demand of one product depends on the price of another, defined as:

$$\varepsilon_c = \frac{\%\ \text{change of quantity demanded}}{\%\ \text{change in price of another good}}$$

A positive cross-price elasticity of demand, $\varepsilon_c > 0$, implies that the two types of goods are substitutes – that is, that one can substitute for the other. A negative coefficient, $\varepsilon_c < 0$, on the other hand, implies that the two goods are complements – that is, that the two goods preferably are consumed together. In particular, the concept of substitutes is relevant for questions about wild species harvest. Studies in tropical forest regions have shown that fish as well as chicken were substi-

Continued

Box 14.3 Price Elasticities—cont'd

tutes for wild meat, whereas, for some reason, the same was not true for livestock (Wilkie and Godoy, 2001; Wilkie et al., 2005). In accordance with that, a study in Ghana spanning three decades revealed that years of poor supply of fish from marine fisheries coincided with increased hunting in nature reserves and decline of hunted wildlife species, in general and in particular near the coast (Brashares et al., 2004).

Similarly, the elasticity coefficients of supply tell how the quantity supplied varies with, for example, changes in price (e.g. Morra et al., 2009). For harvested wild species, however, one must be careful to distinguish between the short-term and the long-term supply. In the short term, an increase in price is likely to increase the supply, such as in Figure 14.7. But in the long run, supply of wild species products is highly dependent on ecological factors and excessive harvest may lead to resource depletion and a back-bending supply curve, such as in Figure 14.8.

FIGURE 14.10 Canoes are indispensable for people living in remote Amazonian villages without road connection. Glass fibre canoes and canoes made of hollowed logs can be substitutes for each other. Those made of glass fibre last longer and weigh less, but they are much more expensive than the traditional hollowed log canoes.

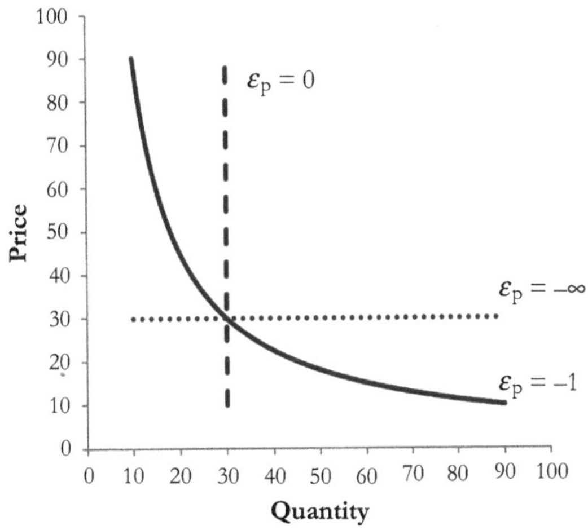

FIGURE 14.11 The price elasticity of demand may, in theory, range from 0 to $-\infty$. Perfect inelastic demand (dashed line, $\varepsilon_p=0$)is a theoretical concept implying that the quantity demanded remains constant regardless of the price level, whereas perfectly elastic demand (dotted line, $\varepsilon_p=-\infty$)conversely implies that if the price is above a certain level, the quantity demanded is zero, whereas if the price is below that same level, the quantity demanded is infinite. Unitary elasticity (solid line, $\varepsilon_p=-1$), finally, implies that any increase of the price leads to an equal proportional decrease of the quantity demanded.

animal, hunters decide whether to attempt to hunt it or not based on the expected ratio of the value of the prey over the handling costs (i.e. the cost of pursuing, shooting, and eviscerating the animal), in comparison with the expected ratio of the value of the hunted prey over the total costs, as well as whether to continue searching for prey of species higher ranked in terms of hunter preference.

This implies that a certain set of species, in particular large-bodied ones, constituting the **hunting set** (also called diet breadth) will almost always be pursued if encountered, whereas the rest of species almost never will be pursued. If the abundance of species with high value to handling cost ratios declines, the ratio of total harvest value to total cost drops, and then additional species are added to the hunting set. However, no matter how rare a species within the hunting set becomes, it will continue to be hunted whenever encountered, as long as the presence of other, less vulnerable, species makes it worthwhile to go out hunting at all. This is the mechanism behind the rapid decline of apes in central Africa. They are so rare that it would be unprofitable to hunt specifically for these species. However, as hunters go out to hunt other, less vulnerable species, such as duikers (*Cephalophus* spp.), the hunters may, on rare occasions, also encounter gorillas or chimpanzees and then hunt them too (Wilkie and Godoy, 1996).

Similar, although slightly different, mechanisms are at work for other kinds of multispecies harvest. Fishing methods are often unselective in terms of species. Although some species may be driven close to extinction, fishing with unselective gear may continue being profitable because of the presence of other, less vulnerable species. Also, selective logging targets a variety of species with highly variable market prices. For example, in Amazonia only few species are so valued that timber sales alone cover the cost of opening forest roads. One such species is the big-leaf mahogany (*Swietenia macrophylla*), which alone has been able to motivate the opening of the logging frontier to hundreds of kilometres from sawmills (Veríssimo et al., 1995). When the road is in place, other timber also can be more profitably logged, not to mention the extraction of other kinds of lower-value resources.

DISCOUNTING THE FUTURE

One factor that commonly contributes to a wild species being rapidly harvested to exhaustion is that people tend to prefer receiving a benefit today rather than getting the same benefit tomorrow or after 10 years. To some extent, this is simply a psychological phenomenon; the desire for immediate satisfaction seems to be something very fundamentally human. But it also has to do with expectations regarding returns on investments. For example, a decision regarding the felling of trees for timber has much to do not only with the current prices but also with the growth potential of the resource stock in relation to possible price changes in the future, as well as concerns of expected returns to alternative financial investments.

Either you can invest in letting the trees grow another few years and thus be able to sell a larger amount of timber, or you can fell the trees now and invest the money in something else. In rapidly growing economies, the expected returns from such capital investments are often much higher than the maximum rates of growth of slow-growing resources, such as stocks of valuable timber (Rice et al., 1997). Therefore, even a monopolist owner may find it more profitable to drive a resource to extinction than to manage it sustainably. Such issues are often represented as the maximisation of the **discounted present value** (DPV) of the resource, which decreases with time according to the formula

$$\mathrm{DPV} = FVe^{-\delta t}$$

where FV is the future value, t is time until the future value will be received, and δ is the **Discount Rate** (Figure 14.12). A zero discount rate, $\delta=0$, implies that DPV=FV – that is, future harvests, or future revenues in general, are valued just as much as present ones. Whereas a negative discount rate, $\delta<0$, means that a future harvest is valued more than the present one, a positive discount rate, $\delta>0$, on the other hand, implies that future harvests are valued less than present ones, leading to a more or less short-sighted behaviour. An infinite discount rate, $\delta=\infty$, implies that only the present is valued and that the future is not valued at all.

To manage wild species sustainably, it may seem preferable to promote views that assign the resources a discount rate of zero. Unfortunately, this is often only

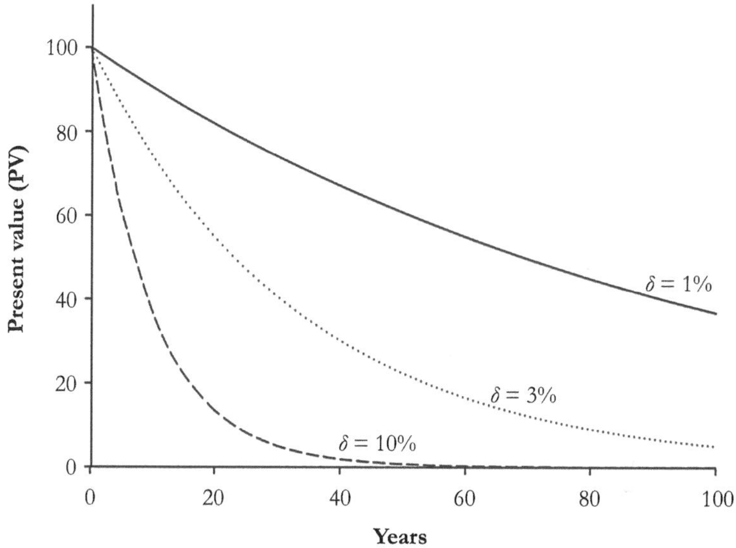

FIGURE 14.12 How future revenues are valued in the present, depending on the discount rate. Even at the relatively modest discount rate of 3% annually, a revenue 25 years into the future is considered to be worth less than half of the value of receiving the same amount of revenue today.

wishful thinking. In particular, immediate survival always has supreme priority. People that actually risk dying unless they satisfy their immediate needs for food or other basic requirements for physical survival will not think about long-term sustainability. However, people who already have their basic needs satisfied will usually still put a higher preference on additional immediate satisfaction than on benefits in a distant future.

However extreme this sounds, many wild species are, in fact, harvested in ways consistent with an infinite discount rate. This is a consequence of unrestricted access to harvested resources. In this kind of a situation, there are no incentives for an individual harvester to refrain from harvest. Even if everyone is aware that the resource stock could grow and thus provide more harvest in a few years if just left alone, the individual harvester has nothing to gain from refraining from harvest: 'If I do not cut the tree today, somebody else will do it tomorrow' is the reality that they face. Therefore, in conditions of unrestricted harvest, each harvester easily behaves according to an infinite discount rate in practice – only the present counts, and no importance is given to tomorrow. Avoiding such behaviour is one central challenge for the management and governance of harvested wild species.

Chapter 15

Management of Resource Systems

NOT ONLY TAKING AWAY

Wild species harvest, in its simplest form, involves just taking away from nature without any action to mitigate adverse impacts on future resource availability, the environment, or other fellow human beings. Often, this leads to reduced resource abundance and other undesired environmental impacts. Taking note of this, however, people often take action to avoid such negative impacts and maintain the abundance of the resource at a desired level, possibly even above natural levels. Much scientific research about wild species harvest aims at finding ways to optimise harvest systems to support the long-term goals of increasing harvests, reducing costs, and conserving biodiversity and ecosystem functions. There are also, however, many cases where local people, without any support from modern science, have developed such systems of **management** for harvested wild species (Aswani et al., 2012; Berkes, 2009; Weiss et al., 2012).

The word 'management' itself has a number of more or less ill-defined meanings depending on whether it is applied to human organisations, technical systems, or natural resources that are managed. In relation to wild species harvest, we define management as any **purposeful human behaviour** carried out to **achieve specific goals** of improved harvest outcomes (Figure 15.1). Acts that are not purposeful are, thus, not management, even if they have positive effects, such as the unintentional dispersal of seeds of fruit trees or other plants. Finally, **management planning** considers the overall conditions where wild species harvests takes place, answering to overarching harvest- and management-related questions. Planning may, for example, have the intention of establishing ways to regulate the conditions in which harvest takes place, thus having no direct linkage to any specific intervention made in the field.

When game, fish, and other renewable resources are readily available and harvest levels remain low, no specific management measures are necessarily required. Under these circumstances, resilient ecosystem functions are sufficiently able to maintain continuous provision from nature. This is the way in which many small-scale extractive uses are still practiced in different parts of the world. Mushroom or berry picking in Finnish forests is a good example; people just go out to the forest and take what they find, neither affecting the fungi or the berry-bearing plants, nor the future provision of mushrooms or

Diagnosing Wild Species Harvest. http://dx.doi.org/10.1016/B978-0-12-397204-0.00015-2

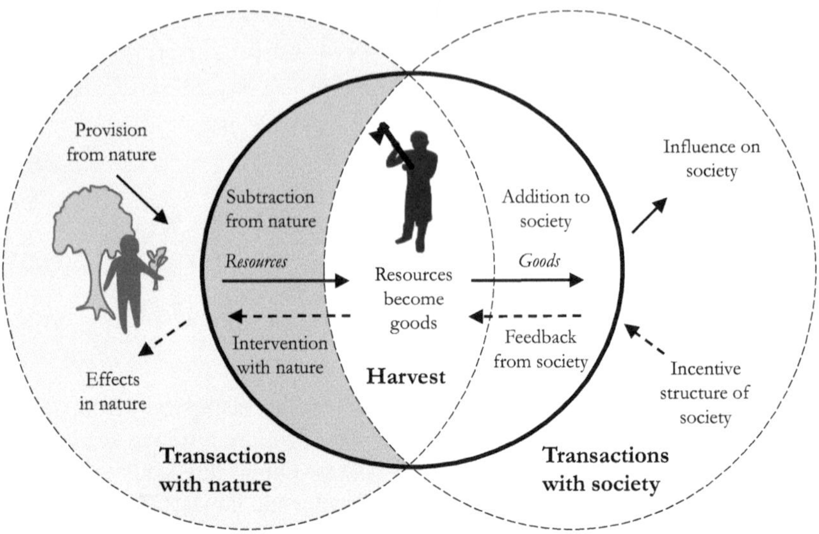

FIGURE 15.1 Harvesters can purposefully act to improve the outcomes of harvest. These acts of management can aim at increasing the provision of subtractable resources or improving their quality, decreasing the costs of harvest, or mitigating the negative effects of harvest interventions with nature. The area shaded in darker grey indicates the main focus of the chapter, and the area shaded in lighter grey indicates where effects of management are also seen.

berries in any significant way, and leaving the entire forest ecosystem unaltered. These resources are so plentiful that the harvest made by humans is insignificant for the species concerned. At the other extreme, intensive management of forests for wood production can significantly change tree species assemblages, diversity, and ecosystem functions. It is sometimes also hard to define a clear line between the harvest of managed wild species and the farming or raising of semicultivated or semidomesticated species. An example of a borderline case is reindeer herding as practiced in northern Finland, Sweden, Norway, and Russia. The reindeer *(Rangifer tarandus)* is a native species to the area, and they live most of their life roaming freely in forests and mountains, but they are private property, and a few times per year the herds are rounded up in pens to, for example, mark the calves or select animals for slaughter. Often, their owners also aid them in their seasonal migrations, sometimes even transporting them on trucks between summer and winter grazing lands (Figure 15.2).

The examples given above are from relatively species-poor boreal environments in which the few existing species can nevertheless offer surprisingly many harvestable resources. Yet, in species-rich areas, such as tropical rainforests, the number of potentially useful species can be orders of magnitude higher than that provided by the boreal forests. Managing ecosystems for the harvest of multiple species can be very challenging. A management activity

FIGURE 15.2 The semidomesticated reindeer of northern Scandinavia have some characteristics of wild as well as domestic animals. A few times a year, they are gathered in pens, for example to mark new calves for their owners and to separate animals destined for slaughter. *Photo: Olof Johansson.*

that benefits one harvestable species may have the opposite effect on other species. High diversity also implies that there are complex interactions among species, which makes the consequences of harvest and management interventions increasingly difficult to predict, as described in Chapter 13. When the individuals of the target species occur at relatively low densities, the mere distance between one individual and the next makes their identification, observation, monitoring, or harvest costly. Unless management practices are devised to overcome these kinds of problems directly related to the provision from nature, management easily becomes unprofitable, harmful to the environment, or both.

Management interventions generally aim to enhance the growth of the resource stock, decrease the costs of harvest, improve the quality of the harvested matter and resulting goods, diminish adverse effects in the environment – or a combination of these. Basically, this can be done in two different ways. One way is to simply refrain from harvest, not necessarily completely but to some extent, thus allowing the harvested resource to maintain its stock or even recover if it was previously reduced by excessive harvest. Another way is to actively enhance the resource stock by, for example, planting, thinning, pruning, fertilising, or providing shelter and feed.

REFRAINING FROM HARVEST

Management can take the form of refraining from harvest either completely or, more often, partly. There are a myriad of different strategies for doing this, which can be summarised by the words how much, which, when, where, and how to harvest and are related in different ways to specific management goals, such as maximising benefits from harvest, conserving biodiversity and ecosystem functions (Figure 15.3). Different specific ways of refraining from harvest may be part of a single harvester's self-chosen management strategy to maximise long-term benefits, or they may be the basis for rules that restrict or disincentivize harvest for everybody.

Thinking in terms of the logistic harvest model, presented in Chapters 13 and 14, a logical strategy to employ to avoid resource depletion, and optimise harvests in the long run, is to limit **how much** of the resource is extracted in a particular area during some period of time. Indeed, in many cases this is the main guiding principle of wild species management in practice. One example of this is bear (*Ursus arctos*) hunting in Sweden, for which the authorities each year, based on inventories of the stock, set a maximum number of individuals that

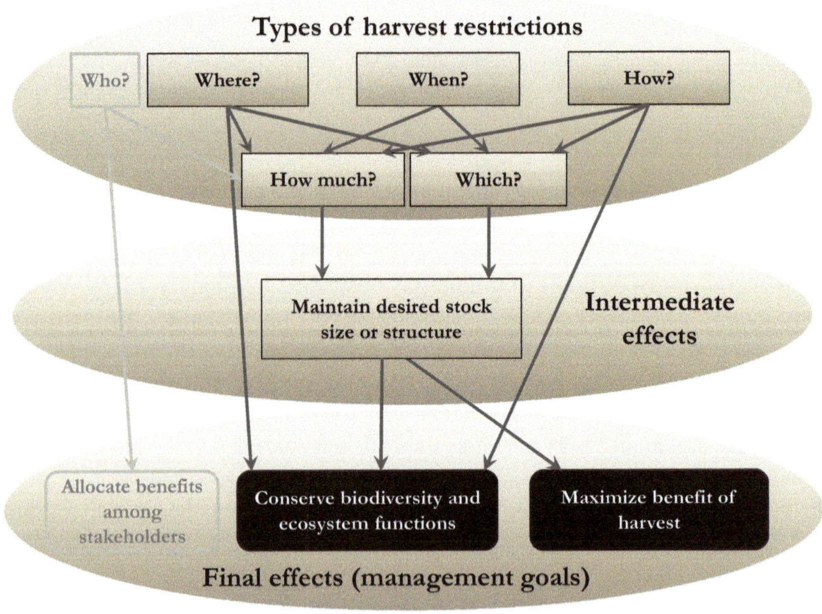

FIGURE 15.3 The logic of different management strategies that include refraining from harvest can be illustrated using the words **how much**, **which**, **where**, **when**, and **how**, as further explained in the text. In different ways, illustrated in this figure, these strategies serve management goals such as maximising the benefits obtained by harvest or conserving biodiversity and ecosystem functions. Many strategies for wild species governance, treated in Chapter 16, also include restrictions on who may harvest.

may be hunted in a certain area. Anybody who hunts a bear must then immediately report this to the authorities, and when the limit is reached, bear hunting is then prohibited until the next year's hunting season opens (Figure 15.4). Restrictions on how much to harvest are ideally defined in terms of time and area, such as the annual harvest within an area of land or water with clearly defined limits. Restrictions that do not fulfil these criteria may not always be efficient. For example, the traditional rules that oblige the harvester to leave four leaves on the tree when harvesting palm leaves (see Chapter 7) are not very effective alone as they do not provide any restrictions on the frequency of such harvest.

In taking into account the structure of the harvested population, however, it may turn out to be important to define **which** individuals may be harvested and which not. For example, harvest can be restricted in order to save individuals that are about to reproduce or are taking care of offspring. This is sometimes the goal of aquarium fish capturers in Peruvian Amazonia. The highly valued discus fish (*Symphysodon aequifasciatus*) has a peculiar behaviour in relation to their offspring: both of the parents stay with the young for one to three weeks

FIGURE 15.4 Bear hunting in Sweden. To avoid risking the long-term persistence of the species, only a limited number of bear are shot each year. *Photo: Tomas Nybrand.*

after hatching and even feed them with a nurturing substance from their skin (Juvonen and Salo, 2004). The fishers that collect discus fish often work during the reproductive season and purposefully remove only one of the parent fish and leave the fry with the other. However, what commonly follows is that the single parent, in order to maximise its success in the next breeding event, devours the young instead of caring for them (Sanna-Kaisa Juvonen, personal communication, April 2013).

Harvesting only those individuals that have already carried out reproductive cycles can also be one way to **optimise the age structure**. The harvest of tropical timber is often subject to legal regulations stating minimum sizes, in the form of diameter at breast height, so as to let the trees grow old and bear seeds before they are felled. Similarly, fishery regulations in Amazonia and many other parts of the world often define minimum lengths for each fish species, or alternatively minimum mesh sizes to be used, all of which have the same effect: fish are left to grow until they have reached reproductive maturity. However, concerns have been raised that this strategy may be ineffective or even counterproductive. With fish, large individuals have been shown to have more reproductive success than small fish, and, in addition, targeting large fish may lead to a process of genetic selection favouring slow growth. Therefore, it has been suggested that fishery management instead must focus on the protection of large and old fish (Birkeland and Dayton, 2005). Optimising the age structure for mammals is quite different. Mammals have much fewer offspring than fish and lower natural mortality rates. Old individuals have stopped growing and do not have more offspring than younger ones, but they compete with them for food. Here, it may be optimal to harvest a fairly large proportion of the young, such that relatively few grow to old age. This is, for example, the strategy that has been adopted for reindeer herding in Fennoscandia over the last few decades (e.g. Kumpula, 2001).

In order to **optimise the sex structure**, female individuals are often spared because one male can fecundate several females. If taken too far, however, this strategy can backfire. In elk (*Alces alces*) populations where the proportion of males has been reduced far below 50% of the adults, this has led to a deficit of old males. As female elk preferred to mate with old males, they mated with the younger individuals only as a last resort and very late in the breeding season in the autumn. The consequence of this delayed mating, was that the females gave birth to their young only late in spring. This, in turn, implied that the elk calves had shorter time to feed and grow during their first summer, which inhibited their growth, reducing their size as adults as well as their reproductive success throughout life (Lavsund et al., 2003; Saether et al., 2003). **Improving the genetic structure** can be made by harvest that spares individuals with particularly desirable characteristics, such that they pass on these traits to the next generation. For example, it is common among elk hunters in the Nordic countries that they voluntarily refrain from shooting females with twin calves, as they wish to increase the occurrence of twin births in the local elk populations (Lavsund et al., 2003).

Another common management strategy is based on defining **when** to harvest and when not. This can, indirectly, affect how much is harvested or which segments of the population are harvested. In some rivers in Sweden, for example, fishing is allowed only on certain days of the week, for example from Monday to Wednesday. This is, of course, not because fishing on a Thursday would be more harmful than fishing on a Wednesday, but just a way to indirectly reduce the total quantity of fishing effort and, thus, fish harvest. Other restrictions on when to harvest are, however, related to biological cycles. For example, fisheries may be closed during spawning to specifically protect breeding individuals and reduce disturbances during their reproduction.

The practice, common in many northern countries, of having hunting seasons only after the reproductive season (e.g. in the autumn and winter) is based on the idea that the animals should be protected during the critical seasons for migration and breeding, and instead should be hunted when there are many more individuals than will survive until the next summer anyway, due to natural mortality. Implementing this kind of management strategy can, however, be difficult if harvest traditionally coincides with seasons that are important for resource regeneration. For example, there has been a tradition to hunt migratory waterfowl in the Finnish Archipelago during the spring months as the birds pass through areas where they used to be a very welcome contribution to people's restricted diets after the long months of winter, and legal restrictions on springtime hunting are still a conflictive issue in these areas (Vuorisalo and Laihonen, 2000).

Management strategies may also be spatial in nature by defining **where** harvest takes place. For instance, certain areas may be assigned as so-called **no-take areas**, where nothing is harvested (Figure 15.5). Such areas should be large enough to maintain a breeding population producing a surplus that disperses out of the area to where harvest is allowed, the **harvest zone**. This is attempted, for example, in Sarayaku (Chapter 4). Such no-take areas do not necessarily need to be located in any specific type of habitat. However, if the ecosystem is heterogeneous in terms of habitat quality for the target species, it is more effective to locate no-take areas in places of high habitat quality and importance for the regeneration of the species in question. Management by no-take areas has many advantages, but their efficiency from an economic perspective may vary. If harvesters respond to the setting aside of a no-take area by increasing the time they spend harvesting outside its borders, it may happen that just a new bioeconomic equilibrium is reached, where rents again are dissipated and net profit is zero (see Chapter 14).

Spatially based management strategies may also serve to maintain ecosystems and the population of the species in question in their natural state, unaffected by harvest, thus conserving biodiversity and ecosystem functions. This is the basic idea of many protected areas and other kinds of areas reserved for

FIGURE 15.5 The reserves created by the people of the Sarayaku community (Chapter 4) are examples of so-called **no-take areas**.

the conservation of nature or natural resources. In more detail, harvest bans can be devised to protect particular habitats that fulfil some specific ecological role in the life cycles of the harvested species, such as with the spawning areas for fish. The culturally grounded disdain of hunting at salt licks in parts of Amazonia is an example of this kind of restriction of harvest in certain specific habitats. Important resting areas of migratory birds are in many parts of the world considered places where hunting should not take place. In practice, when national parks and other protected areas have been originally established for the purpose of the strict protection of nature, they can also effectively function as no-take areas providing benefits for the hunters in the surrounding lands.

Management strategies may also combine a spatial and a temporal component (i.e. the 'where' and 'when' questions) to create a system of rotational harvest. For example, as was described in Chapter 9 regarding the Peruvian forest concessions, harvest rotation times can be adjusted to the total harvest area in such a way that after one harvest, the same area is left to recover for a time estimated to be long enough for stock regeneration.

Wild species management may also involve refraining from certain methods and technologies for harvesting (i.e. putting limits on **how** to harvest). This typically implies that the use of efficient harvest methods and technologies is avoided in favour of less efficient ones. For instance, in the Manu National Park in Peruvian Amazonia, park authorities allow the indigenous population to hunt but not with firearms. In the Nordic countries, hunting from snowmobiles is prohibited, and fishing with nets often faces more restrictions than fishing with hook and line. These measures aim to ensure the viability of game and fish populations; for example, it is prohibited to close an entire river cross-section with a chain of nets. Recreational harvesters of wild species may even voluntarily choose not to use the most efficient technology available, for example hunting with a bow and arrows instead of firearms, or fishing with rods instead of nets. When it comes to harvest for commercial or subsistence purposes, however, restricting the use of efficient harvest methods and technologies has the drawback that it increases the costs for harvesters and thus often is inefficient from an economic point of view.

Restrictions on harvest methods and technology may, however, serve not only to reduce the quantity harvested but also to avoid the other adverse collateral ecological effects of harvest. Some harvest methods not only affect target species but also can damage entire ecosystems. One such method, commonly used in commercial fisheries, is trawling. This fishing method is regulated on the high seas in order to protect other marine species from being killed as side catch, but in the case of bottom trawling the aim is also to protect the integrity of the seafloor. This is particularly important for coral reefs, where trawling may destroy the reefs that constitute the habitat of the targeted fish species as well as many other species. In this situation, management measures work against the short-term profitability of harvest but can efficiently protect the ecosystem from excessive harm and may therefore, in the long run, be economically beneficial to the harvesters.

Other examples of restrictions on harvest methods and technology in Amazonia are found in the struggle of the Sarayaku people to stop fishing with explosives or pesticides, and also the legal ban in Peru on fishing with the poisonous root of the *barbasco* plant (*Lonchocarpus* spp.) (see Chapter 5). The effects of such restrictions may, however, be more complex than thought. An example of this is the Peruvian ban on the use of chainsaws to transform logs into boards in the forest. This practice may imply a waste of timber, and thus the ban works in favour of the overall efficiency of timber transformation, but simultaneously the prohibition of the use of chainsaws can also indirectly imply more collateral damage to the forest vegetation. This is because transporting entire logs through the forest involves more damage to the forest than transporting boards.

Managing harvested wild species by refraining from harvest can, ideally, be a very cost-efficient way of management, as the direct costs can be very low. In fact, it may even imply a reduction of the direct costs of harvesters, if they spend less time and resources on harvesting activities than what they would do otherwise. Unfortunately, however, often the implementation of such management strategies requires costly monitoring and enforcement activities. Another drawback of these strategies is that they can never raise the long-term sustainable harvest of the target resources above the limits set by the natural dynamics of the species in question. In order to do that, other kinds of management are needed that more actively enhance the resource stock.

ENHANCING RESOURCE STOCKS

The long-term benefits of harvest may also be increased through active manipulations in order to enhance resource growth, so as to counteract the impact of harvest. In this way, the stocks of the target species may sometimes increase even above their natural levels. Similarly, active manipulations can be applied with the objective of improving the resource stock in terms of quality, be it in the long term through genetically based selection or more in the short term by promoting the development of certain traits that are not inheritable. In Chapter 13, it was explained how the unintentional and often negative impacts of harvest can be perceived at different ecological levels, such as individuals, populations, ecological communities, or ecosystems. In a similar way, management manipulations can be directed towards improvements at any of these levels, although the distinction between them is not always clear-cut.

Actions at the level of **individuals** serve to directly promote the growth or quality of each individual. The fertilisation of açai palms (*Euterpe oleracea*), mentioned in Chapter 11, in order to promote growth is one example of this, and another is the removing of branches from the lower trunk of a tree to promote log quality, something that is common in plantation forestry in many parts of the world and sometimes also takes place in managed seminatural forests. More commonly, however, actions are taken in order to enhance growth at the level of **populations**. This may involve the promotion of the establishment of large numbers of young individuals (Figure 15.6). In its simplest form, this may imply, for example, just planting seeds by burying them in the soil to protect them from predation and provide optimal conditions for germination. But often it involves rearing the plants or animals outside of their natural habitat to ensure survival during their earliest and most vulnerable life stage, and then returning

FIGURE 15.6 (A) The fibres obtained from the *chambira* palm (*Astrocaryum chambira*) in Peruvian Amazonia are an important raw material for handicrafts, including hammocks and bowls; and (B) in the Tahuayo River basin, the local women manage stands of *chambira* by enrichment planting. *Photos: Natalia Aravena.*

them to nature. The planting of tree seedlings grown in nurseries is a common example of this, and another example is the river turtle management, based on relocation of eggs, described in Chapter 6. Similarly, stocking of fish fry into natural bodies of water is common in North America, Europe, and China. Actions at the level of populations may also target adult life stages, such as the winter feeding of deer and other game animals, which reduces winter mortality and is common in many northern countries.

One special case is when species are introduced to areas where they were previously absent. Very commonly, all around the world, people have often taken high-yielding or otherwise popular game and fish species and introduced them to places where they do not occur naturally and where they have become **exotic** species. Examples of this are the introduction of rainbow trout (*Oncorhynchus mykiss*) into the waters of the Andes Mountains and Patagonia in South America, and the introduction of Nile perch (*Lates niloticus*) to Lake Victoria in Africa, which have greatly increased the productivity of fisheries but also, especially in the latter case, have had dire consequences for biodiversity, as much of the native aquatic fauna has become extinct. In many freshwater habitats of the American continent, species of tilapia fish (*Tilapiini*) fish, originating from Africa and having escaped from aquaculture ponds, also have become invasive and have outcompeted much of the native fish species.

Such dire consequences are, however, much less likely if a species is introduced just very near its natural range, such as the planting of *wayuri* palms (*Pholidostachys synanthera*) in Sarayaku, as described in Chapter 7. Although this palm does not occur naturally at the site, natural stands are just 40 km away, and there are no natural barriers to dispersal. The reason why this species does not occur naturally in the Sarayaku area seems to have to do with soil characteristics, so there is hardly any risk that this species would become invasive in the area where it has been introduced; more likely, its long-term survival in the area where it has been introduced will require continuous management inputs by humans. Very different from the introduction of exotic species, the **reintroduction** of useful native species that have gone extinct locally may have positive consequences both from an ecological and economic point of view. The vicuña (*Vicugna vicugna*), a camelid with very valuable wool living in the high Andes Mountains, had become extinct in Ecuador but was then reintroduced. In Peru, the wool is nowadays harvested by shearing live animals instead of hunting and killing them, and there are plans to begin to allow shearing and commercialisation of such vicuña wool in Ecuador, too.

Actions at the level of ecological **communities** usually involve the removal of plants or animals whose presence somehow is detrimental to the growth of the target species. Predators that compete with humans for the same prey species are often hunted to reduce losses to predation and thus increase the provision of harvestable resources available for humans (Figure 15.7). Liana cutting is commonly proposed, and to some extent practiced, for timber management in tropical rainforest, partly because lianas compete with the trees, reducing

FIGURE 15.7 The nature protection board of Sweden authorised the hunting of these wolves *(Canis lupus)* in order to protect semidomestic reindeer from predation. Other predators that are subject to limited hunting for similar reasons are bears, lynx, wolverine *(Gulo gulo)*, seals *(Pusa hispida, Haliochoerus grypos)*, and great cormorants *(Phalacrocorax carbo)*, the latter two because they eat fish and cause damage to fishing equipment. *Photo: Olof T. Johansson.*

their growth, but perhaps more importantly because cutting lianas significantly reduces logging damage to young growing trees.

Felling canopy trees to increase sun insolation and weeding competing undergrowth are other common management practices, such as those described in Chapter 11 regarding the management of the açaí palm. Similarly, in some places in Ecuadorian Amazonia, competing vegetation has been removed in order to establish monospecific stands of the *fibra* palm (*Aphandra natalia*), which provides fibres that are used for elaborating brooms, and also bears tasty and nutritious fruits (Figure 15.8).

Clearing to create gaps in the forest may be favourable for the germination of the seeds of certain useful species, for example big-leaf mahogany (*Swietenia macrophylla*) or Brazil nut trees (*Bertholletia excelsa*) (see Chapter 8). If such measures are taken on a larger scale, they may lead to significant impacts also at the level of entire **ecosystems**, including changes in abiotic conditions such as hydrological cycles and microclimate. The abiotic conditions of the resource system also can be manipulated in many other ways to enhance the growth of desired species, for instance by fertilisation of entire ecosystems, or by ditching in order to drain wetlands and thus promote tree growth, both of these techniques being common practices in forestry in the Nordic countries.

FIGURE 15.8 An almost monospecific stand of the *Aphandra natalia* palm in Ecuadorian Amazonia, near the village of Villaflora (Chapter 10). The underbrush and much of the canopy trees have been cleared in order to favour the production of leaf fibres from this species.

Although these different ways of enhancing resource growth benefit the target species, and may increase the sustainability (in a narrow sense) of harvest, this is typically at the expense of other species. Thus, the goals of increasing sustainable harvest may come in conflict with the goal of conserving biodiversity and ecosystem functions. As the intensity of manipulations increases, it becomes more and more questionable whether the target species are genuinely wild at all. The intensively managed açaí palm forests described in Chapter 11 are a borderline case between wild and cultivated (Figure 15.9), as are those *wayuri* palm stands described in Chapter 7, which are planted in primary forests slightly outside of the natural range of the species. An ever-larger share of the timber harvest in the world comes from heavily managed forests. Sometimes these can be considered seminatural, if they are made up of native species and are established through natural regeneration. Others, however, consist of trees planted by humans, often using exotic species, and they bear so little resemblance with natural ecosystems that they rather should be considered wood plantations.

One additional issue that is very central to the management of harvested wild species is the question of who should be allowed, or obliged, to take part in the harvest as well as management activities. Restrictions on who may harvest a particular wild species resource can be an efficient way to reduce the

FIGURE 15.9 The stands of the açaí palm are often managed to increase the density of the palm and to improve growth and fruit production. The density can be increased (e.g. through enrichment planting). Growth and fruit production can be enhanced (e.g. by weeding competing species or using fertilisers).

amount harvested, and thus it can contribute to the maintenance of the resource stock. However, such restrictions inevitably also influence the allocation of the benefits of harvest among stakeholders, including some and excluding others (Figure 15.3). Often, this is done in ways that promote social equity, but it has also often happened that the right to harvest wild species resources is reserved for the most privileged segments of society only, as in many European countries where the right to hunt large game species was long reserved for the aristocracy. Similarly, although there are many ways that resource growth can be enhanced through active manipulations, this will unlikely get done unless there are incentives that motivate somebody to do it. How people organise in order to govern wild species harvest is the topic of the next chapter.

Chapter 16

Governance Shaping Incentive Structures

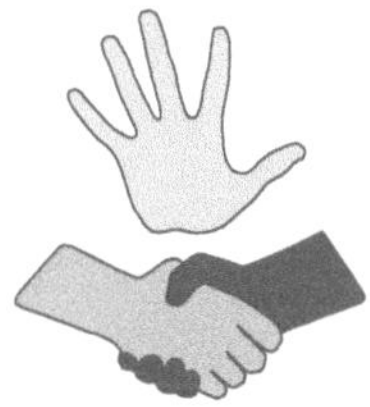

REGULATING TRANSACTIONS WITH NATURE AND SOCIETY

Chapters 13, 14, and 15 explored wild species harvest, focusing on the harvester's transactions with nature and society, which together form the basis for continued harvest. Extraction can continue only if the interventions do not compromise the provision of the targeted resources in the long term. On the other hand, continued harvest also requires that the harvester is left with the experience of a positive net benefit, when taking into account the costs involved. In order to improve the balance of these transactions, harvesters commonly engage in management of the resource system. However, the transactions of the harvester with nature and society are also embedded in a case-specific societal context that further forms part of a larger human system. How different societal relationships and arrangements regulate the harvest system is the topic of the present chapter (Figure 16.1).

Almost all harvest of wild species is in one way or another, subject to rules and other conditions specified by society. Typical questions that arise in the minds of harvesters include, for example, the following: Who is the owner of this resource? Who has the right to use it? Whose interest might be violated when a wild species is harvested? Is anyone controlling if I now choose to harvest here? Questions such as these are critical in the **governance** of renewable natural resources.

Rules that regulate human behaviour can be personal and unwritten precepts based on each individual's values and ethical considerations. Most often, however, there are also collective rules that range from informal customs and habits to formally decreed laws that are legally binding and enforced by authorities – even to the point of using violence against those breaking them. The basis of harvest and resource use regulation lies on fundamental structures of social organisation. **Institutions** – or, in other words, collectively agreed rules and their enforcement – are critical for governance of any human system. The 1993 Nobel laureate in economics, Douglass North, calls institutions the 'rules of the game' in a society:

Diagnosing Wild Species Harvest. http://dx.doi.org/10.1016/B978-0-12-397204-0.00016-4

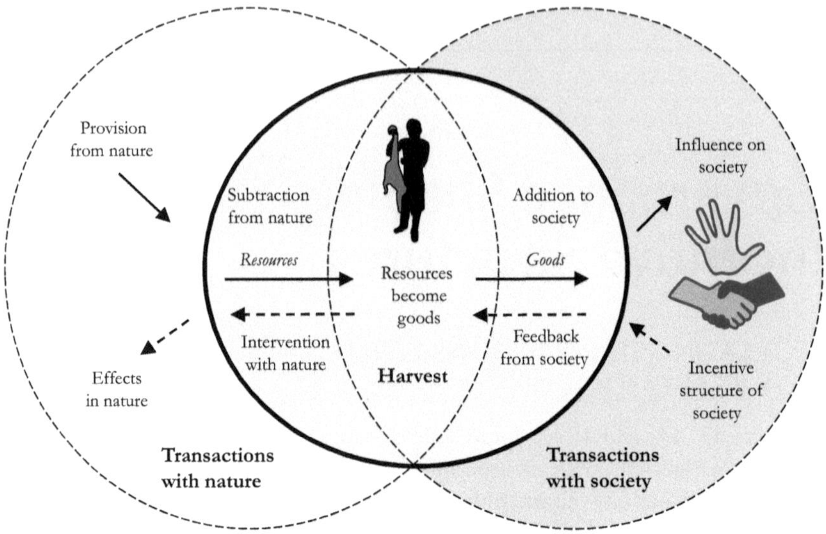

FIGURE 16.1 Wild species governance is about how society takes conscious action in order to influence harvest and management so to achieve some defined goals. The area shaded in darker grey indicates the main focus of the chapter, and the area shaded in lighter grey indicates where governance is seen from the harvester's direct viewpoint.

Institutions are humanly devised constraints that structure human interaction. They are made up of formal constraints (e.g., rules, laws, constitutions), informal constraints (e.g., norms of behaviour, conventions, self-imposed codes of conduct), and their enforcement characteristics.

North (1994).

Different kinds of institutions are interrelated, and, as North (1994) wrote, together 'they define the **incentive structure** of societies and specifically economies'. An incentive structure includes the whole of the institutional mechanisms that guide the decisions of people in relation to their acts and behaviour. **Incentives** motivate people to take some certain course of action, by providing either some reward for doing or a punishment for not doing so. Negative incentives, which discourage people from some action, are often called **disincentives**. The rewards and punishments can take any form from moral rewards or condemnation to economic benefits or losses. Thus, incentives and disincentives may include, for example, subsidies, taxes, credits or other funding, resolution and enforcement of property rights, capacity building, or even outright prohibitions and obligations. Sometimes, incentives can unintentionally work against their intended purpose. For example, in Madre de Dios in Peru, the high concession fees levied for timber concessions in relation to any other areas form an incentive for the loggers to extract timber from Brazil nut (*Bertholletia excelsa*) concessions subject to lower fees and of easier accessibility (Giudice et al., 2012), as was described in Chapter 8. Such incentives with counterproductive outcomes are called **perverse incentives**.

TABLE 16.1 The Four Main Forms of Incentives in the Governance of Harvested Wild Species

	Prohibit/discourage	Oblige/encourage
Harvest	A	B
Ecosystem manipulation	D	C

In the context of wild species harvest, several different incentive mechanisms concern either the harvesting activity itself or such actions that aim to enhance resource growth and quality. Their role in governance can be classified into four broad categories represented by the respective cells A-D in Table 16.1. Disincentives to harvest (cell A) may include, for example, harvest bans on certain species, harvest quota, closed seasons, fees for harvesting licenses, or anything that makes people refrain from harvest in one way or another. As many wild species are subject to excessive harvest, there are often good reasons to provide such disincentives.

Sometimes, however, incentives are provided that, on the contrary, encourage harvesting (cell B). An example of this are the subsidies that many countries in the European Union provide to their fishing fleets (e.g. Cardinale et al. 2013). Another example is the law that was introduced in Sweden in 1983, and abolished in 1997, which obliged forest owners to extract, in each decade, a certain minimum proportion of the timber growth on their property (Appelstrand, 2007, pp. 67, 232). As is often the case with incentives to wild species harvest, these are prime examples of efforts with perverse consequences. Much of the waters where the EU fishing fleet operates have been subject to overfishing, and in order to promote the fishery in the long run, the best approach would be to, on the contrary, discourage fishing until fish stocks have recovered. Obliging forest owners to harvest timber, on the other hand, implied that the wood and pulp industry could buy raw material at a price lower than the equilibrium price between supply and demand, thus implying a net transfer of capital from forest owners to the wood and pulp industry.

There are, however, also cases where it may be justified to provide incentives of this kind. The State government of Acre in Brazil has supported the collection of wild rubber through direct subsidies as well as capacity building, as a way to promote rural development as well as to protect the forest from conversion to pasture, with mixed results (Sills and Saha, 2010). In Finland, the government subsidises fishing of a number of fish species that have little commercial value, because this has turned out to be a relatively cheap way to remove phosphorus from many lakes and the Baltic Sea, which are heavily eutrophicated due to sewage effluents and leakage from fertilised agricultural fields (Jyväsjärvi et al. 2013).

Incentives are also sometimes provided to encourage actions that enhance resource growth (cell C). Participation in the nest relocation and release of juvenile turtles described in Chapter 6, which is obligatory for members of management groups, is an example of this type of management. Another example are the laws in many countries that oblige replanting trees after clear-cutting. In other cases, however, disincentives are provided to such actions that, although they would enhance resource growth, also would have other, negative effects (cell D). For example, although the hunting of predators could enhance the growth of stocks of wild game and fish, it is strictly regulated in the Nordic countries, with the intention to reach a balance between production and conservation interests.

Taxes or fees are often used to influence the way people behave. They may act as disincentives for some types of harvest and make alternative methods more attractive in relative terms. However, fees and taxes are collected not only to make certain types of harvest more attractive than others but also to raise public funds. Their collection can be directly based on harvest rights or harvested material, or indirectly on other economic activities generated by harvest. The funds they raise can in turn form an incentive for authorities to channel more effort towards the regulation and monitoring of the harvest activity in question. Harvest fees and taxes may not generate much revenue directly, but when cumulative benefits and revenues created in processing and trade are also included, wild species harvest can form an important sector of the economy and significantly contribute to public funds that can sometimes be further reinvested in developing the harvest economy as well as for use in monitoring and control.

Other examples of institutional arrangements that can provide incentives for environmentally or socially desirable harvest are different kinds of fair trade and sustainability **certification schemes** that build on the consumers' conscience and aim to guarantee a stable and economically reasonable price paid to the harvesters. These kinds of incentives for harvest that fulfil determined criteria can make such harvest more attractive in the face of low market prices or high market uncertainty.

In broader terms, abundant natural resources (renewable or nonrenewable) have been described as both a blessing generating revenue and a curse dooming other sectors of the economy to underdevelopment, depending on how societies have been able to capture the value of such resources (e.g. Brunnschweiler, 2008; Sachs and Warner, 2001). Although this kind of phenomenon is most often attributed to such resources as minerals and oil, it also has relevance for wild species harvest, one of the famous examples of a 'resource curse' being found in the Amazonian rubber boom that generated much wealth and even substantial public funds but in the end was not able to contribute to durable development (Barham and Coomes, 1994). Why harvestable resources can sometimes become a blessing while in other cases they seem more like a curse is no doubt a result of many contributing factors, of which institutions and their quality are important ones; Botlhole et al. (2012) suggest an intermediate position that 'natural resources are neither a curse nor a blessing *per se*, but that they become a curse or a blessing depending on the underlying quality of institutions'.

The incentive structure of wild species harvest is further embedded in an underlying overall economic, social, ideological, and cultural context that determines how different institutions function. The societal setting of wild species harvest in general is often a complex bundle of impacts, including indirect consequences like the generation of employment or cultural heritage. The character and intensity of wild species harvest largely depend on general characteristics of societies at the local, national, and global levels. Economic development leading to increased incomes may increase or decrease the demand for wild species harvest, depending on the specific circumstances (Chapter 14).

Human population growth, at the local, national, and global levels, has profound impacts on the relation between humans and nature but its consequences are not always direct and straightforward. In general, wild species harvest is heavily dependent on the local human population density if it takes place for the purpose of subsistence and local consumption, but it is much less so if harvest is for commercialisation on the open market. In the latter case, the direction of causality can even be reversed: the presence of wild species for which there is a sufficient market demand may attract people to an area, such that wild species harvest causes human population growth instead of the other way around.

Human migrations due to wild species resources are not a very common phenomenon in the modern world but it has been important in the past. For example, in the Amazonian rubber trade (see Chapter 3) as well as the fur trade in Siberia and Alaska during the eighteenth and nineteenth centuries, led to a significant influx of people to these respective regions (e.g. Gibson, 2011). In a broader sense, an incentive structure finally encompasses all these general characteristics of societies that, in one way or another, influence the costs and benefits facing any potential harvester of wild species resources. All of this needs to be taken into account when devising policies to guide harvest and management planning, and when building regulatory instruments that can steer people's harvest behaviour towards the desired goals expressed in the policies.

POLICY INSTRUMENTS: HOW TO TURN GOALS INTO ACTION

Although there is no universally accepted definition of what the term **policy** means, policies can be seen as stated intentions that express commitment to defined outcomes and principles of how these outcomes are to be pursued. Thus, policies concerning wild species harvest are guidelines that steer resource decision making towards politically chosen objectives. How such policies are formulated depends on the specific characteristics of the wild species resources in question, as well as on the involved political entities and their power relations (Reed et al., 2009). Policies can be presented through anything from global UN declarations to personal principles stated by individuals. The level of **legitimacy** that resource policies bear depends on their scope and the mandate of their creators. The legitimacy of both the decision-making processes and their outcomes is important; while the former depends on the levels of **participation** and **transparency**, the latter is manifest in the practical applicability of the resulting rules. For the legitimacy of policies and rules, it is important that their outcomes are considered distributionally fair, consistent with other rules, and coherent in relation to other policies present in society (Ramcilovic-Suominen and Epstein, 2012). For example, to be widely accepted and endorsed, policies promoting sustainable wild species harvest should guarantee a distribution of costs and benefits that is considered fair, should not be in contradiction with other rules, and should be in line with more general societal goals and values.

In order to be efficient, policies are promoted through different **policy instruments**. For analytical purposes, these are generally divided between command-and-control policy instruments, on one hand, and market-based ones, on the other. In the first category, the policy is endorsed by enactment and enforcement of rules stating what to do and what not to do, whereas in the second category, the choice between different options is left open, but the desired policy goals are pursued by rules that grant benefits to the actors that choose to take certain actions instead of others. In practice, incentive structures work through the existence and application of a multitude of mechanisms that may combine elements of both command-and-control and market-based or other voluntary instruments.

Command-and-control policy instruments include laws and other rules that establish strict limits to what is allowed and what is not, and those who break the rules are subject to direct sanctions. Such sanctions form a disincentive to resource users due to the costs that the sanctions imply. However, rules in place are, of course, no guarantee of success. There are two basic ways in which the outcomes of command-and-control policy instruments can cause failures. For example, the failure may be caused by flaws in the regulations themselves, or the cause can be tracked down to noncompliance to rules that in themselves are adequate in the pursuit of defined ends.

Often, failures result from a combination of these two problems. Thus, any regulatory reforms should not only address the flaws in existing rules that have been put in place but also carefully address such customary codes of conduct or habits that clearly are adaptations to local conditions (Smith et al., 2006). **Compliance** with rules can be studied in various ways, of which three main types of models are emphasised in the literature. These are focused on (1) instrumental (or economic) incentives, (2) institutional arrangements, and (3) legitimacy as a basis of decisions about compliance (Ramcilovic-Suominen and Epstein, 2012).

The models based on **instrumental incentives** lie on the assumption that rational actors will choose between the alternatives of either complying with the rules or breaking them on the basis of the expected costs and benefits resulting from each alternative. This would mean that if breaking the rules is likely to benefit more than following them, the choice is for noncompliance (see Chapter 9). Although such mechanisms undoubtedly are present, this perspective alone provides only part of the picture. Even if an actor's behaviour is guided by economic self-interest, rules representing social **institutions** among individuals can have weight on the actor's decisions because of fears of loss of social status as a consequence of noncompliance. These kinds of rules can be purely personal and be based on individual ethics, or they can be shared with a larger group of people. The third perspective to be explored is compliance based on legitimacy. This means that when those who are supposed to follow the rules accept the authority of those who make them, the system has gained more legitimacy, which, in turn, is likely to increase compliance (Ramcilovic-Suominen and Epstein, 2012).

How rules are followed can be seen as a combination of many different and sometimes conflicting motivations, which are largely based on instrumental incentives, institutional arrangements, and considerations of legitimacy. These motivations, however, are greatly influenced by a series of context-specific circumstantial factors, including regulatory constraints, enforcement capacity, corruption, property rights, markets, culture, conflicts, and poverty (Ramcilovic-Suominen and Epstein, 2012). All of these factors can contribute to continuous conflicts between legal and extralegal economies, and they often contribute to the problems caused by overly complicated and bureaucratic regulations that are difficult and costly to follow. For example, the Peruvian forest law reform described in Chapter 9 was perceived by many involved actors to establish regulations that were overly complicated and costly to follow, were in contradiction with locally adapted norms, and lacked legitimacy. Many of those actors that criticised the new law on these grounds also expected it to lead to more illegal logging (Salo et al., 2013).

Illegal harvest beyond permitted areas or times, or exceeding quotas, is often a serious problem, as the effectiveness of harvest restrictions and other management schemes rely on efficient monitoring such that transgressions can be detected and sanctioned. Apart from regulatory measures, cultural values, social control, and even market mechanisms can also provide important means to combat illegal harvest. In the absence of buyers, there would hardly be any significant problem of illegal logging, and the demand would increase for timber coming from better controlled, and probably also better managed, sources. This, of course, requires that there is a way to verify the source of timber. As another example, just a few decades ago, in spite of being illegal, it was socially more or less acceptable among some Finnish hunters to eliminate eagles and hawks from their hunting grounds as these animals were considered to be competitors to humans over the region's valuable game; nowadays, the awareness of the importance of biodiversity conservation has increased, and such acts are generally repudiated.

There is a continuum of ways in which rules intended to regulate wild species harvest can be taken by those who are supposed to follow them. First, the rules can be taken as such, their letter as well as spirit being respected in full compliance. This often requires that the actors share the underlying values and agree upon the motives behind the rules, such as restricting harvest in order to conserve biodiversity or guarantee long-term sustainability. Often, however, rules are followed strictly only as far as the compliance can be efficiently monitored and enforced, whereas otherwise harvest activities are just adjusted according to the enforcement capacity of the authorities, sometimes disguising noncompliance as compliance, for example through bribes. This easily takes place when harvesters perceive that following the letter of the rules is the most convenient path but they do not genuinely share the goals of the regulatory apparatus. At another extreme, local resource users can openly reject the rules and resist their implementation – in addition to disagreeing about the very goals behind the rules (Hiedanpää et al., in preparation).

Different kinds of reactions of harvesters and other actors in the face of institutional change and other acts of governance are described in various chapters of Part 2. Yet there are also other ways to receive rules with criticism; when harvesters agree with policy goals, such as harvest restrictions, but perceive the means as inadequate, they sometimes engage in creative self-organisation to devise new institutional arrangements (Hiedanpää et al., in preparation). For the allocation of present and future benefits that arise from wild species harvest, having rights to make decisions about such resources is of primary importance.

Market-based policy instruments, in contrast to command-and-control instruments, are built upon the assumption of a freedom of choice among a variety of alternative options. The choice between different options is then regulated by rules that make some options more attractive than others. This can be attained through different mechanisms, including **tax rises** for undesired activities or **tax cuts** for desired ones, as well as by establishing direct **subsidies** and **credit programmes** that promote activities deemed desirable. Using these kinds of financial sticks and carrots, a society may direct development towards the desired goals.

The tax imposed on gunshots, as described in Box 14.1, is an example of a way to make it more expensive to use a specific harvest technology. As noted there, this tax had quite some effect on hunting behaviour, although it was not the original purpose of the measure. When applying this kind of method to restrict harvest, there can be reasons to evaluate not only the ecological and economic effects from the perspective of the harvester, but also the distributional effects on the entire society. In this particular case, in some areas where hunting is important, the tax on gunshots implied a net annual transfer of money to the government of over US $100 per family, quite a significant sum in rural areas where wages are low and access to salaried jobs is limited.

Incentives can also be created through other, more indirect, mechanisms based on the behaviour of market actors. Such mechanisms include, for example, **certification** of harvest and management methods to supply the markets with special products for which changes in consumer consciousness create demand. Certificates may be about organic production, fair trade, child labour-free production, or any other such ethical principle that producers, traders, and consumers favour. Examples of regulation arising from product quality concerns are also common when wild species products become global commodities (Laird et al., 2011). For example, the Brazil nut trade is very dependent on the **quality requirements** set by the importing countries, which want to guarantee the safety of this product, known to sometimes contain carcinogenic aflatoxins. Several kinds of incentives can also be bundled together in order to promote sustainable harvest. For example, harvest of nontimber forest products (NTFPs) can support such approaches as reduction of emissions from deforestation and forest degradation (REDD and REDD-plus, or REDD+), which aims to combat climate change by decreasing carbon emissions (REDD) and combines this with the larger goals of biodiversity conservation, forest management, and carbon stock enhancement (REDD+). For example, Brazil nut concessions can couple certified NTFP production with REDD, as seen in Chapter 8 (see also Kalliola and Flores, 2011).

When the role of wild species products widens from local use to trade, rules that regulate transport, processing, and markets become more important. These rules may be of either the command-and-control or the market-based policy instrument type. For example, decisions affecting transport costs can be of vital importance for the supply of harvested wild species products. When investments are made to improve transport infrastructure, the transport costs decrease, and it becomes cheaper to get products to markets. On the other hand, subsidies or taxes on transport, for example through fuel prices, can either facilitate or hinder the marketing of wild species products. Transporting products from one place to another can also be subject to direct regulation in the form of specific permits or licenses. For example, in Peru and many other countries, timber transported from the forest to the sawmill for transformation has to be accompanied by documents that define its legal origin (Chapter 9).

Globally traded wild species commodities potentially become subject to **international treaties**. The most comprehensive of these agreements is the Convention on International Trade in Endangered Species of Wild Fauna and Flora (CITES), which regulates cross-border trade on products derived from endangered species. This convention can either ban trade altogether (as is the case of the species listed in Appendix I of CITES) or restrict trade on the basis of the state of the resource in question (species listed in Appendices II and III). CITES targets both the compliance of the actors involved in harvest and trade and the harvest levels. First, all traded shipments of a species under CITES regulations have to fulfil the requirement of legality. The **lawful acquisition finding** means that the products to be traded have been harvested and processed according to the regulations laid by the relevant national legislation. Lawful acquisition alone is not enough, however. The shipments also have to meet the requirement of nondetrimentality. This means that a **nondetriment finding** can be conceded only if the harvest has not endangered the species as part of its environment.

Environmental policies are increasingly international, which has implications for national legislation. Sometimes, international agreements can be specific and binding, such as CITES, while in other cases the international policies only set a framework for national application. Furthermore, the Nagoya Protocol of the Convention on Biological Diversity will, when it enters into force, regulates the access to and benefit sharing of genetic resources (Buck and Hamilton, 2011). Other national laws and policies apart from those directly regulating wild species harvest also have important implications. For example, agricultural and foreign trade policies can affect competitive situations between wild species and agricultural products. Labour legislation and social welfare policies may also have similar links to wild species harvest. Equally important is that, many times, regulation takes place through local community-level agreements or wider societal arrangements that are prepared so that they actually fit the characteristics of the resources for which they have been devised and thus address the practical on-the-ground challenges of resource governance.

FOUR TYPES OF RESOURCES

A number of basic characteristics of wild species resources have important implications on their harvest and use from an institutional perspective. For example, on a very fundamental level, all natural resources (as well as other resources) can be described along two axes: their subtractability and excludability. In this section, we explore these concepts in the context of wild species harvest, concentrating on resources (provision from nature) rather than on goods (addition to society), which, from an institutional viewpoint, behave somewhat differently (e.g. compare the fish in the river as a resource and the fish on your plate as goods). For the purpose of this book, the difference between wild species resources and wild species goods is that resources are in their natural state as part of the ecological system and goods have already been harvested by humans.

Of the two basic characteristics of resources, **subtractability** describes a resource in terms of its exhaustion as a consequence of extraction. A highly subtractable (rivalrous) resource is such that for one person to enjoy the benefits from the resource, he or she must withdraw some of it such that less is then left for others. In principle, harvestable wild species resources are always highly subtractable: one animal, fish, or fruit extracted means one less for other harvesters to extract. A resource with low subtractability, on the other hand, is such that it can be enjoyed without withdrawing any of it from nature (Figure 16.2). Beautiful scenery is a commonly used example of such a resource. However, in practice, subtractability can also be expressed in terms of abundance in relation to the harvest pressure. Some resources can be so plentiful that it makes no real difference in practice even if everybody withdraws from the resource to satiate their own needs, because there is still enough for everybody else. This can be the case, for example, with oxygen molecules in the air we breathe. However, this can also be true for highly abundant harvestable resources that, in practice, even many harvesters at an aggregate level cannot affect the density of in any

FIGURE 16.2 Nature when used for tourism is usually a resource with low subtractability. Quite a large number of tourists can enjoy the beauty of nature without leaving less nature behind for others. (A) Local guide with tourists searching for gorillas in the cloud forests on the slopes of the Virunga volcano in Rwanda. (B) Young male gorilla watching back at the photographer. *Photos: Benedict von Wright.*

perceivable way (e.g. mosquitoes killed for feeding a pet lizard). These examples, however, are relatively rare and insignificant in relation to wild species harvest in most situations.

The second resource characteristic is **excludability**, the degree to which users can be excluded from harvesting a wild species resource. Excludability largely depends on the inherent characteristics of the resource; sessile organisms have higher excludability than mobile, let alone migratory, species. But excludability also depends on case-specific social and technological conditions that affect the capacities of surveillance and enforcement.

Combining these two basic features, subtractability and excludability, a matrix can be generated in order to define four basic types of resources (Figure 16.3). Those that we deal with in this book tend to show high subtractability, being primarily **common-pool resources** or **private resources**. The difference between these two resource types is that common-pool resources show low excludability, whereas private resources have high excludability. Those resources that have low subtractability but are relatively easily reserved for exclusive use are called **club resources**. Such resources are uncommon in the context of wild species harvest, although fish in the case of catch-and-release fishing in a small water body can sometimes come close. Resources whose characteristics make them both difficult to subtract and to exclude from the use of others are called **public resources**. Beyond perhaps fisheries subject to catch-and-release fishing in the sea, this resource type is not easy to imagine as subject to wild species harvest.

One and the same species can actually constitute different types of resources depending on their use (Figure 16.4). Gorillas or whales as hunted game constitute highly subtractable resources, but the same species can also be used for tourism in so-called gorilla or whale safaris, during which tourists just watch the animals without reducing the number of them available for others to watch, thus constituting public or club resources rather than common-pool or private resources.

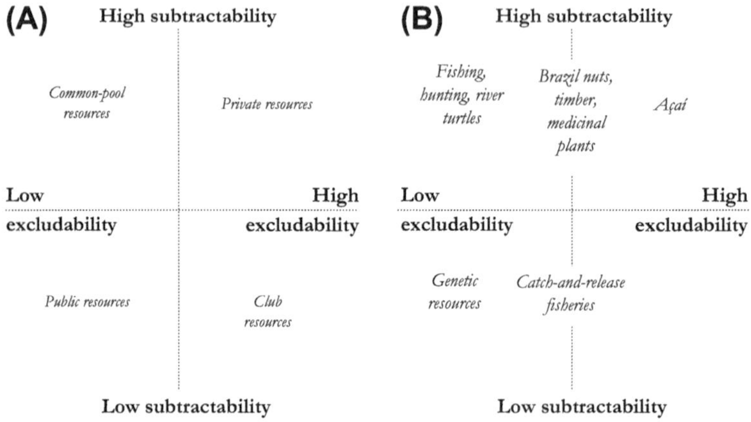

FIGURE 16.3 (A) Classification of natural resources into different resource types. As can be seen, harvested wild species are typically on a continuum between common-pool and private resources. When used for some specific purposes not covered in depth in this book, however, such as catch-and-release fishing or making use of their genetic material, wild species may also be public or club resources. (B) A representation to illustrate different types of harvested wild species according to their subtractability and excludability.

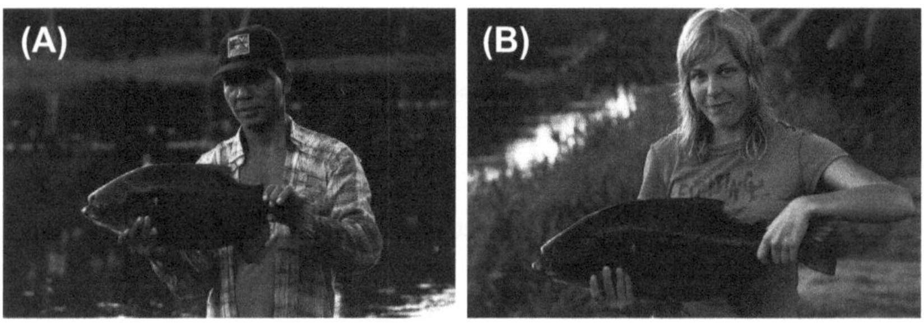

FIGURE 16.4 Subsistence fishing or catch-and-release fishing? The same kind of fish in the hands of two different people can turn from a common-pool resource into a public resource, depending on whether the idea of fishing is to subtract the fish or not.

To use examples described in our stories from the forest floor, highly mobile game resources (Chapter 4), turtles (Chapter 6), and migratory fisheries (Chapter 5) are typical common-pool resources, as excluding potential harvesters from appropriating these resources is quite difficult, although far from impossible. On the other hand, trees and other plants do not run away, and thus it may be easier to protect them from potential competing harvesters. Palms for roof thatch (Chapter 7), Brazil nuts (*Bertholletia excelsa*) (Chapter 8), timber (Chapter 9), and fruit trees such as açaí (*Euterpe oleracea*) (Chapter 11) are somewhat more similar to private resources. This, however, also depends on how the feasibility of surveillance varies over a given space. In remote places, where surveillance is difficult, resources such as these may also resemble common-pool resources more than private ones.

RIGHTS TO RESOURCES

When many people harvest the same resources or work in the same areas their interests may become conflicting. Property rights and use right arrangements arise from this basic. This means that a property right is an enforceable authority to take a specific action within a particular domain (Commons, 1968), or, to put it more simply to fit the topic of this book, the authority to decide about one's actions related to a given wild species resource. Ownership, property, and tenure rights are terms that are often used interchangeably to describe the rights that define the relationship between the owner and the owned property (e.g. Ostrom and Hess, 2007). The basis of these arrangements typically arises and evolves through long periods of time and along with associated socio-cultural development.

An important question could be, then, to whom the mandate and authority to decide about resources is assigned – who is able to take that authority? Again, a great variety of actors can take or be assigned property rights related to specific wild species resources: the continuum spans from a single individual to the whole of humankind. Between these extremes, property rights can be assigned to any subgroup of the whole of humanity, say, to a household formed by a family of six, to a rural community, to a transnational company, or to a municipality or a state as a representative of its citizens.

The continuum of different levels of exercising **property rights**, as discussed here, can be divided into private property, common property, and nobody's property (rights left unassigned). Strictly speaking, private property means that there is only one clear owner who exercises authority. Common property implies that several people, on equal grounds, share this authority, while in the case of nobody's property there is no one who is forbidden to make independent decisions about resource use. A special case of private property is that of private companies, which sometimes is called corporate property.

In many cases, private ownership or tenure may facilitate rational management, as the problem of coordination between many harvesters is eliminated. Private owners may, however, also deplete resources. For example, even if only one single individual had exclusive use of a resource, this individual can deplete it because he or she prioritises immediate gains (i.e. uses a high discount rate; see Chapter 14) or because he or she is unaware of the consequences of harvest actions (see Chapter 17). The actions that are profitable for the owner may also be detrimental to the interests of other segments of society (i.e. they cause negative **externalities**); for example, managing a forest exclusively for timber production may harm its biodiversity or its ecosystem functions related, for example, to hydrology and climate.

Sometimes, common property is further subdivided into government property and collective property. This division depends on the nature of the social organisation on which the groups exercising property rights are based; 'government property' refers to entities of formal political governance, such as national governments or municipalities (e.g. publicly owned forests), and 'collective property' to groups of people who are organised in some other way, such as rural communities (Starrett, 2003). In several of the Amazonian countries, indigenous people in some parts of the region hold large extensions of land as the collective property of communities or entire ethnic groups.

Commonly, communal property rights may lead to further internal regulation of resource use through, for example, hereditary use rights or other arrangements similar to private property rights. In such situations of collective property rights, the property regime can thus be internally further divided into private properties allocated among the members of, for example, a community (as described in Chapters 7, 8, and 10). This may imply an even-handed allocation of resources, or some additional criteria can be followed to distribute the benefits, as was seen in the case of the Brazil nut stands in the community of Sonene (Chapter 8).

It is important to note that the types of resources (private, common-pool, public, and club resources) and the types of property rights (private, common, or nobody's property) are different concepts. Natural resources that have the characteristics of private resources are often private property, but by no means always. They can also be common property or nobody's property, as resources become property only when property rights are assigned. This is often problematic not only because of questions of distribution but also in terms of enforcement, chiefly when a particular type of property right is assigned to a resource whose characteristics make the assigned right difficult to enforce in practice. For example, assigning private property rights to a resource of the common-pool resource type can imply very high costs for surveillance and enforcement of property rights, as has been the case with timber in many concessions in Peru (Chapter 9).

Property rights, however, are not only about who has the authority to exercise them but also about other questions, namely, about the degree of control the property rights holders have (Starrett, 2003). All property rights do not assign absolute dominion over resources. The characteristics of property rights define what exactly the actors holding rights over a specific resource are authorised to do with the resource itself and also with the property right. Commonly, five characteristics of property rights are considered as central constituents that define the specific nature of property in different contexts. Together, these five characteristics form the 'bundle of rights', and they are the rights to **access**, **withdrawal**, **management**, **exclusion**, and **alienation**.

Although access is often understood as more than this, in the bundle of rights the **right to access** means simply that the right holder is authorised to enjoy the existence of the resource by approaching it in one way or another. This can mean entering a forest or paddling on a lake. The right of access combined with the **right of withdrawal** forms the basis for any property rights that authorise the extractive use of resources; a right holder that has authorisation for access and withdrawal can, for example, fell trees or catch fish. The further **right to management** authorises the user to make more decisions about the resource base itself. This means that the user can carry out activities that alter the resource or its environment in one way or another. The right holder, then, is not a mere user of the resource anymore but can also choose to act in order to change the course of the future and in so doing hope to improve the benefits that the resource generates. This could be, for example, the right to thin a stand of pines or the right to dredge a lake.

The **right of exclusion** makes the property right even more complete as now the right holder has the authority to make decisions about how others are allowed to deal with the same resource, often including the option to close the use for others. For a full property right to emerge, the right holder also has to exercise the **right of alienation** by which the right holder is authorised to transfer the property right to another actor. This can be either completely or partially, as well as permanently or temporally. The right of alienation enables the owner, for example, to sell forest land altogether (permanently transfer all the five rights) or to lease fishing rights (transfer the rights to access and withdrawal for a period of time). Many countries (e.g. Peru) have constitutions that define all natural resources inalienably as the property of the state. So, full private property ownership of these resources is not possible, but some combination of the five rights can be transferred to different actors, often being further transferable from one actor to another. National legislation usually defines which of the five rights are considered transferable in different cases, and harvesting rights are, in many cases, transferable. This means that by paying a fee to the owner or the authority in charge of the administration of the area one may acquire the rights to harvest (e.g. hunting or fishing permits).

The right to use land and its resources without altering them significantly is called **usufruct**. States often grant such usufruct rights to indigenous peoples or rural communities over resources that have relatively little economic value, while reserving for themselves the rights to more valuable resources, such as timber and minerals. In Peru, for example, forest land is mostly owned by the state, and whereas the forest legislation explicitly grants usufruct rights to local communities for hunting and the use of other forest resources for subsistence purposes on public land, it has reserved to the state the right to lease out these lands as timber concessions. In the Nordic countries, the indigenous Sami people have been granted no property rights to their traditional lands, but they have usufruct rights to grazing lands for reindeer and to hunting, fishing, and timber extraction. Except for the rights to grazing lands, these usufruct rights are for subsistence purposes only.

Resolution of unclear property rights and efficient enforcement of the established rights are often proposed as actions that can improve the sustainability of resource use. This view is based on the assumption that property rights guarantee the right to future benefit flows and thus form an incentive to long-term **commitment** to sustainable levels of resource use. Different property rights situations combined with varying resource and user characteristics no doubt are important factors influencing the development of different resource use situations. It is important to keep in mind, however, that no form of property right alone guarantees sustainable use (see e.g. Ostrom and Cox, 2010).

It is very common that property rights and associated 'bundles of rights' vary considerably among different sites or areas where the same wild species resource is harvested, or even in one place among different actors. Examples of this kind of variation are found in several of our stories from the forest floor in Part 2 of this book. Moreover, how the property rights situations have come about varies as much as the rights themselves. Sometimes, wild species are harvested on the basis of an **age-old customary right** that has not been formally granted, thus taking place beyond the formal property rights regime. Nowadays, such informal rights are, however, sometimes becoming formalised (Figure 16.5), as was seen in the case of Brazil nut harvest in Chapter 8.

Formalised usufruct rights range from those granted to traditional peoples, as described here, to the **public right of access** or 'everyone's right'. The latter is enjoyed by anyone who wants, for example, to pick berries in Finland, no matter what the nationality of the picker is or whether the picking takes place on public or private lands. This kind of public right of access formally exists in a few countries only, including Finland, Sweden, Norway, and Scotland. In the Finnish case, the right is not directly written in any particular law, but it is rather based on several formal laws in combination with customary institutions that have developed as part of people's behaviour through the centuries. These kinds

FIGURE 16.5 *Primeiros povos* – first peoples. An entrance to an indigenous reserve in Brazil. Customary land and resource rights have been formalised in many countries. *Photo: Aili Pyhälä.*

of informal norms are called customary law, and they exist and have legitimacy as well as practical implications by virtue of a shared history and perceptions among the group of people that they concern.

Rights to resources can be granted by formal institutions through a variety of mechanisms based on **legally binding contracts**, such as ownership titles, permits, authorisations, concessions, licenses, or easements. These contracts transfer specific formal property rights from the holder of such rights to other parties. For example, in a **concession contract**, the owner of the resource can transfer rights related to access, withdrawal, management, and exclusion to a concession holder (concessionaire). These rights come with associated obligations related to extraction levels and methods, and they often also oblige the concession holder to pay concession fees, manage the resource, and guarantee the exclusion of other users (Figure 16.6). Concession fees can be based on, for example, the harvested volumes or the size of the concession area (see Chapters 8 and 9). The right of alienation can also be transferred in such a way that the concessionaire can trade the obtained rights further to third parties, always according to the original contract terms. Permits, authorisations, and licences are other kinds of formal agreements in which specific resource rights are transferred between the resource owner and another party.

In the **formal assignment of rights** to resources, one approach is to select the actors through prescribed procedures. In the **direct awarding procedure**,

FIGURE 16.6 Forest concessions, such as this one in Loreto, Peruvian Amazonia, are one way in which property rights to state-owned timber resources can be assigned to private actors.

anyone willing to gain rights to a resource expresses his or her willingness to the owner of the resource, and after negotiation a contract can be signed transferring defined rights under determined terms. As opposed to direct awarding, **competitive allocation mechanisms** are based on the assumption that more than one actor is interested in the resource in question, and it is in the owner's interest to choose the most suitable of those actors. This kind of allocation procedure aims at selecting among the interested parties the most appropriate, keeping in mind the overall benefits that are expected from the transaction. These benefits can be seen in relation to the policies established by the resource owner, such as policies of sustainable forest use or those of economic growth through private enterprise in the case of many states.

The exact design of competitive allocation procedures depends on these policies. Thus, allocation can be predominantly based on economic offers made by the interested parties, on one hand, or, on the other, the process can put more weight on other characteristics of the interested parties, such as their technical capacity, experience, size, social role, or others. In cases in which the economic offer (i.e. how much the actor is willing to pay directly for the transferred rights) has a central role, the allocation is often called an **auction**. Even in auctions, the winner normally has to fulfil determined qualification standards, but among those who do, the highest bid wins. Many times, other factors, as described here, have an essential role in the process, however. In these public or competitive **tenderings**, as they are often called, the resource owner thus selects the winner of the contract based on a variety of criteria. Many times, allocation processes are hybrids with characteristics of both auctions and tenderings, and there is also great procedural variation in relation to how the applicants enter the process (e.g. open, selected, or through a prequalification round).

These forms of rights allocation to wild species resources are mostly applied when the resources are found in specific areas under clear administrative jurisdictions. The case is somewhat different, for example, when the resources in question are highly mobile and not tied to specific political jurisdictions. Many (ocean) fisheries are examples of such a situation (Figure 16.7). But for those circumstances, there are also mechanisms for harvest rights allocation. Harvest rights under the total allowable catch limit can be assigned spatially, as is the case of the Territorial Use Rights of Fisheries, which are normally applied to those fisheries that are not very mobile (Arnason, 2012), or through other **catch share** mechanisms such as individual transferable quotas (ITQs). The initial allocation of quotas, however, is not straightforward. As was the case in the allocation of many Brazil nut concessions in Peru (Chapter 8), the initial allocation of catch shares in fisheries has often been carried out on the basis of historical harvest efforts and the estimated harvest capacity of different fishermen or companies (McCay et al., 2011; Chu, 2009). The fact that quotas are often transferable is based on the assumption that, in this way, the harvest effort will be efficiently allocated among the fishers.

FIGURE 16.7 Many coastal fisheries, like in this area off the coast of Peru, were long unregulated, but nowadays most nations intend to regulate fishing in their coastal waters.

GOVERNANCE OF COMMON-POOL RESOURCES

As noted in this chapter, many wild species resources are, by nature, common-pool resources. If a common-pool resource is freely available to anyone without being subject to rules restricting and coordinating subtraction, the situation is called **open access**. When many harvesters act in a self-interested manner in this kind of situation, the result is often collective ruin as overexploitation and consequent depletion of the resource in a phenomenon called the **tragedy of the commons** (Hardin, 1968). However, in many other cases, common-pool resources are harvested and managed in a coordinated way among the collective of users. We refer to this kind of situation as **common-pool resource governance** (this is often also called common-pool resource management, but we prefer 'governance' to distinguish it from management as defined in Chapter 15).

In order to carry out such common-pool resource governance, it becomes necessary to resolve such questions as who is allowed to appropriate the resource; what are the timing, quantity, location, and technology used in resource subtraction; who is obliged to contribute labour to maintain the resource; how harvest and obligation to management activities are monitored and enforced; how conflicts are resolved; and how rules affecting these issues are changed over time. All of this leads to several dilemmas. How common-pool resources can be used in a coordinated way is just the first-level dilemma. Then, there is also the second-level dilemma of how people can get organised to create rules that specify rights and duties related to resource use and management. The third-level dilemma, finally, is about how to invest in monitoring and sanctioning in order to increase the likelihood that participants follow the agreements they have made. Solving each of these dilemmas requires that individuals make sacrifices for the common benefit. Empirical evidence, indeed, shows that people often are able to improve their resource use situation in such common-pool resource settings.

The exceptional lifework of the 2009 Nobel laureate in economics, Elinor Ostrom (1933–2012), included groundbreaking research on common-pool resource situations. She listed attributes of resources and their users (appropriators) that enhance the likelihood that **self-organisation** emerges with the aim to govern common-pool resources in a coordinated way (Ostrom, 1999). Some **attributes of the resources** themselves are conducive to self-organisation aiming at coordinated use.

The first rule is that improvement of the state of the resource has to be feasible. This means that the resource is neither so depleted that it seems useless to put effort into improving its use nor in such a good state that not much would be won by investing in organising. Second, there must be indicators available that can be reliably used to assess the condition of the resource at a reasonable cost. Third, the resource has to be such that its availability and abundance are quite predictable from the users' viewpoint. Fourth, the spatial extent of the resource has to be such that it is feasible to govern. In practice, this means that the resource can be delimited within external boundaries that extend over an area that can be reached with reasonable effort by relying on the transportation and communication technology in use. This, in turn, provides the resource users with a realistic means to develop accurate knowledge of the state of the resource and its distribution within the area's microenvironments.

It should be noted, however, that these attributes of the resource can also be seen as attributes of the *relation* between the characteristics of harvesters and of the resource. The first resource attribute, feasible improvement, hardly increases the likelihood of self-organisation unless the appropriators are **aware** of the improvement feasibility. Such awareness is therefore crucial, as shown in Chapter 6 describing river turtle management in Peru. Likewise, the second resource attribute, the availability of indicators at low cost, is an attribute not only of the resource but also of the technical, financial, and intellectual capacity of the harvesters or other actors supporting them. The same could be largely said about the third resource attribute, the predictability of abundance and availability, and, similarly, the fourth resource attribute, a spatial extent that is practical to govern, both of which largely depend on having the technical and financial means available for transport, observation, and communication (see Chapter 14).

Ostrom (1999) also listed a number of **attributes of the users** of the resource that seem to increase the likelihood that they engage in self-organisation for coordinated resource use. The first user attribute is related to the salience ('essentiality') of the resource in the harvesters' lives; the resource users have to depend on the resource in question to such a degree that a major part of their livelihood is based on its subtraction, or they have to depend on it in some other important way. The second user attribute is related to the perception that the harvesters have about the resource; they must share a common understanding or image of the resource, most of all regarding the four characteristics discussed here. This understanding also has to include a shared idea of how individual harvesters' actions affect others and the resource. The third user attribute describes the ways in which the users value the resource over time. The users should have sufficiently low discount rates in relation to future benefits to be achieved from the resource. This means that the users should think that saving the resource for future use is not a bad decision considering the expectations about the value of an immediate harvest in relation to that of a future harvest. Again, this attribute is also, in fact, an attribute of the *relation* between the characteristics of the harvesters and of the resource. As noted in Chapter 14, resources that have a growth rate that is lower than the discount rate of the harvesters are particularly susceptible to overharvest and depletion.

The fourth user attribute is related to the distribution of interests among the resource users. What increases the likelihood of efficient self-organisation to emerge is that the current resource use pattern affects all users in a similar way independent of their economically and politically defined power relations. If the wealthy and powerful are affected in a similar way as the poor, they may make use of their power to the benefit of all. The fifth appropriator attribute is trust. What seems to be needed for self-organisation to be successful is that there is sufficient trust among the resource users. In other words, the users must be able to expect that agreements are respected and promises are kept among the group of individuals involved. In this also reciprocity is essential. The sixth user attribute is related to decision making. For self-organisation to take place, it is important that users have sufficient autonomy regarding the rules of access and use. They must be able to define the rules of the game without fear of acting in contradiction with external rules and the authorities enforcing them. And, finally, as the seventh attribute, there must be at least a minimal set of skills related to organisation among the resource users (Figure 16.8). They must have some organisational experience based on participation in other associations or previous learning from neighbouring groups of people.

DEVISING EFFICIENT INCENTIVE STRUCTURES

It is important to note that the existence of functioning institutions for wild species harvest does not, as such, guarantee optimal harvest and resource management from an economic point of view, and this may be even less the case from an ecological perspective. Generally, the more complex the natural environments from which wild species are harvested and the more complex the social settings, the more difficult it is to formulate efficient ways to regulate harvest and

FIGURE 16.8 Community assembly in Sarayaku, Ecuador. Organisational skills and experience such as those shown here increase the likelihood that resource users self-organise around the governance of common-pool resources. *Photo: Eriberto Gualinga.*

management. This problem is often further exacerbated by the fact that in such contexts, the harvested species' characteristics are insufficiently known; the same rules are not appropriate for all kinds of different resources. For example, because of different resource characteristics, distinct norms are needed to regulate the use of sessile, slow-growing trees and mobile, fast-growing, animals. The same is true for the socio-economic roles of different wild species goods: the regulation of systems that are based on highly local subsistence resources may need other kinds of rules compared to those stemming from commodities that are globally traded.

A market economy requires adequate participation in an information framework that records ownership of property and other economic information, and those who lack legal ownership of their property have difficulties with obtaining credit, selling the business, or expanding. However, although clear legal rights and legal instruments can be the gateway to prosperity for some, it is not so for everybody. Access to the benefits from wild species resources is also about other factors, including technology, capital, markets, labour, knowledge, identities, and social relations. All of these factors are regulated or affected by governments and other bodies of governance from the global to the local levels.

Many institutions serve primarily to ensure a determined type of **distribution** of the resource among the different harvesters, but not to guarantee sustainability in the long term. Efficient governance, harvest and management for the purposes of distribution, maximised net benefits, and sustainability require awareness and understanding of the impacts of harvest interventions and the potential for improvement. Beyond the context of common-pool resource situations and property rights by large, there is much insight that can be taken by exploring Elinor Ostrom's work in general, particularly in relation to how the diversity of institutional arrangements worldwide shows that offering public or private property regimes as one-size-fits-all solutions for harvest problems is likely to fail (Ostrom and Cox, 2010).

It is important to keep in mind that rules creating rights and obligations are a reflection of larger societal and economic structures as well as political power relations. It is common that a promising harvestable wild species product rapidly attracts the interest of groups keen on harnessing part of the benefits it is expected to create (Laird et al., 2011). Often, powerful economic elites also control the political bodies that hold decision-making power over these questions, while the harvester-level actors commonly represent social groups that have little political weight in policy formulation and choice over the policy instruments used.

In practice, incentive structures are often formed by a combination of command-and-control and market-based instruments, and often the incentive structure is complex and difficult to unveil, with many competing motivations that function through different policy instruments that many times create incentives that work in contradiction to one another. Because of their direct implications on power relations among resource user groups, decisions about what mechanisms to apply when allocating resource rights are highly political, leading to struggles over resource policies and the policy instruments to be applied. Different kinds of resource use situations require different approaches, and these should take into account both the spatial and temporal scales at which the resource systems function. Large-scale systems with long-term dynamics need distinct approaches compared to local-scale situations with short-term dynamics. Societal responses to these needs depend on power relations that often create asymmetries of participation and mandate to govern and regulate resource use (see e.g. Cumming et al., 2006).

There are currently two major trends in resource governance that have particular importance for wild species harvest and associated power relations in the tropics: first, economic and political globalisation and international trade enhance connections, interactions, and transnational agreements between actors across the planet; and, second, political decentralisation processes underway in many tropical countries transfer decision-making power towards smaller administrative units, thus promoting local governance and increasing the role of the civil society (e.g. Hiedanpää et al., 2011; Agrawal et al., 2008; Larson and Soto, 2008). For example, in Peru, voices claiming local participation have been increasingly heard in policy formulation, and local consultation has been written into the legislation (Evans et al., 2013).

Although the existence of legislation requiring local consultation does not guarantee participation, it can be considered a victory for those emphasising the agency of local populations. The principle of **free prior informed consent** is implicitly and explicitly written into many international and national rules that particularly concern local and indigenous populations. In practice, this means that any regulatory or other reforms, or development projects that affect indigenous resource and land-use rights, have to be negotiated with such populations, and not only negotiated but also negotiated in such a way that all parties have equal access to information. For example, International Labour Organisation Convention 169 emphasises such an approach. **Cooperative incentives** that function through dialogue (Figure 16.9), roundtables, and stakeholder participation are closely related to **information-based incentives** that function through technical assistance and spatial processes such as use zoning and land-use planning.

Decisions about the use of particular forms of incentives and policy instruments to regulate wild species harvest and associated practical management decisions and activities should be based on repeated reassessment of information on critical aspects of use, such as the species themselves, their environments, the harvesters, and the markets. This, however, is seldom the case, and decisions are often made on an *ad hoc* basis instead, as reactions to changes in the state of resources or expectations pertaining to the markets for wild species products; there are few examples of countries that have developed comprehensive policies and regulatory structures concerning wild species harvest (Laird et al., 2011). While it is important that policies and policy instruments are designed in an adaptive manner following changes in factors such as resource availability and market demand, it is equally important that these decisions are made using the best available information, particularly when wild species harvest implies risks of severe or irreversible harm. Knowledge, communication and awareness are centrally important for improved harvest situations. These are the topics of the next chapter.

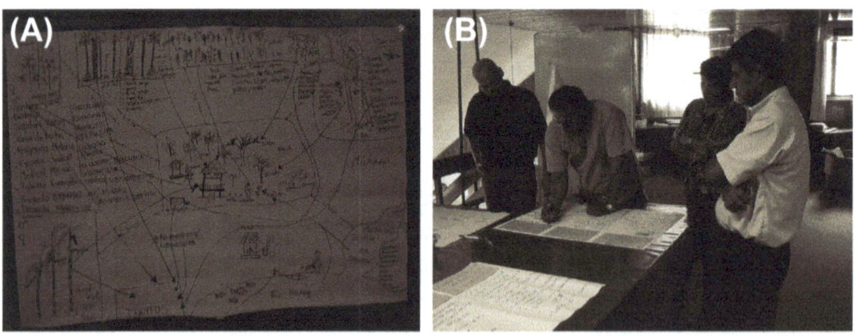

FIGURE 16.9 Capacity building can form an incentive to act if the action has been constrained by lack of know-how or other knowledge-based resources. (A) Community members have, jointly with the personnel of a cooperation project, produced a map that describes the important resources of the community. (B) The project personnel are generating a summary of the workshop findings to be discussed with all of the participants.

Chapter 17

Knowledge for Action and Interaction

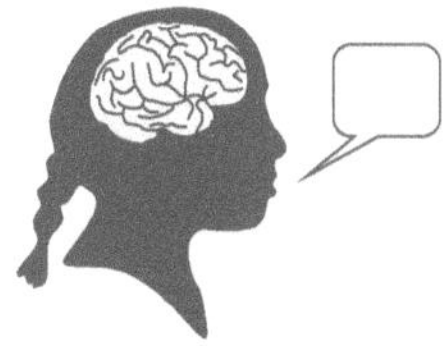

FROM DATA TO WISDOM

Which wild species are harvested in each particular place depends on the conditions of the natural environment as well as on the needs and preferences of human societies. These tend to change from time to time and from place to place. A species that has been present in some area for a long time may become a resource only after people discover that it can be useful. In the same way, something that has been always used may become ignored after specific societal changes occur. For example, new industrial products have often led to the abandonment of some previously appreciated natural products as has been seen in several of the previous chapters.

Resource dynamics, costs and benefits, management, and governance are all important topics to explore when setting the goal to improve the resource system. The more that is known about these thematic perspectives to wild species harvest, the better the possibilities are to comprehend what factors influence the success or failure in a harvest system and, if needed, how to improve the present situation. Any decision about harvest and management must therefore involve the application of **information** and **knowledge**. These concepts are often used interchangeably, but actually they mean distinct things related to the ways in which we interpret the world and **communicate** or transfer our interpretations to others. Whilst these themes about the human mind are cross-cutting throughout the process of wild species harvest (Figure 17.1), they also have some properties that are inherent to them.

For the purpose of analysing knowledge and communication, we first briefly define some of the key concepts that we use. Sometimes, the nature of interpretations of reality is represented hierarchically as a pyramid or triangle in which data become information that, in turn, becomes knowledge that, finally, can give rise to wisdom, as schematically shown in Figure 17.2.

In order to make a rough sketch of this conceptualisation, **data** are things that directly describe details of the physical world; they are the 'hard facts'. Data could be, for example, counts of the Brazil nut trees (*Bertholletia excelsa*) in a forest. The trees exist whether or not someone is there to witness them, but it takes someone to record the trees so that their existence becomes data. To transform the data into **information** would then take us one step further by

Diagnosing Wild Species Harvest. http://dx.doi.org/10.1016/B978-0-12-397204-0.00017-6

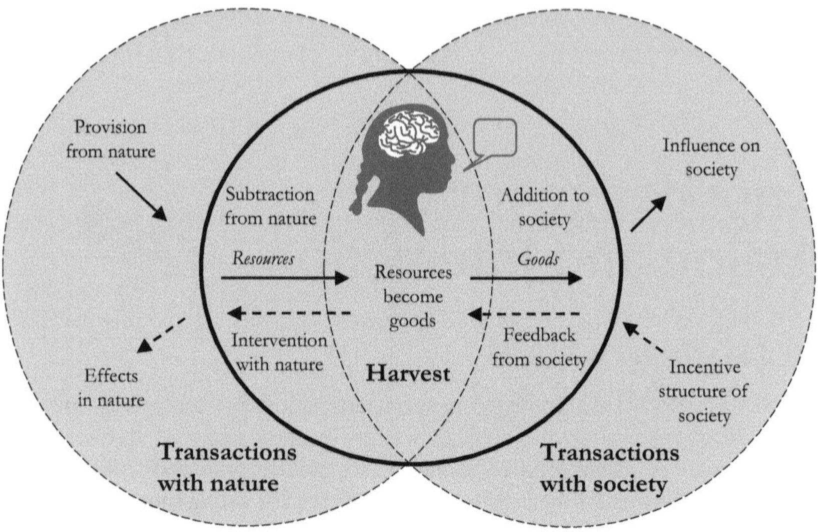

FIGURE 17.1 Knowledge and communication are vital in all aspects of resource harvest – they are crucial to the harvest act in itself. Knowledge is essential for the harvester's transactions with nature and society and for the management of resource systems and harvest as they are embedded in the socio-ecological system of wild species harvest.

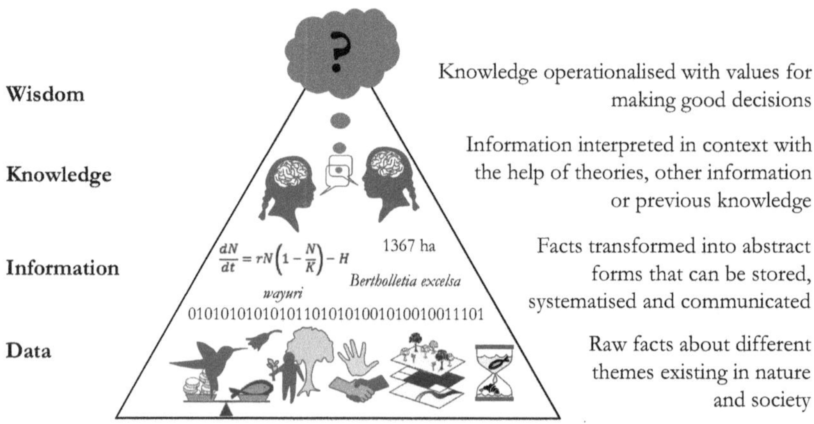

FIGURE 17.2 Knowledge triangle showing the hierarchical levels of knowledge from raw data, through information and knowledge, to wisdom.

changing the physical fact into something that can be communicated to others, be it through speech, text or numbers on paper, or images. So, when a harvester tells her neighbour a story about the more than 1000 Brazil nut trees she possesses, the data become information, an abstract representation of the hard facts. The concrete trees were not handed as evidence, but abstract information about them was. Thus, while data should, in theory, be something objective and concrete, information is abstract and can already be biased or erroneous. In science, information is usually derived from raw data through steps of data analysis. We continue with the example of a Brazil nut concession: 'My concession has a density of eight Brazil nut trees per hectare, whilst the concession of my neighbour has 12 trees per hectare'.

The further transition from information into **knowledge** is a more ambiguous process, but as a general rule it takes place in the human head when information is interpreted and put into a specific cultural context. It results from a cognitive process within the human brain and involves the processing of information in such a way that it helps interpretation, being essentially **culturally constructed**. It may represent the traditional practical knowledge existing in a distant community as well as information converted into the form of theories through a scientific process. Knowledge is always **context specific**; it cannot be taken out of someone's head and poured into someone else's head as such. For the transmission of a message that aims to generate knowledge, there is information; for example, when the harvester's neighbour receives the information about her numerous Brazil nut trees, say, in the form of a story, and then integrates the information into his previous perceptions about what the trees are like, how far from one another they grow, and how distant the forest is, he generates knowledge about her Brazil nut stand. The knowledge can further give rise to **understanding** and **wisdom** if used to achieve goals. However, although they occur at the top of the information pyramid, they are not entirely constructed atop the lower levels of it. Inputs from experience and values also have a role to play: 'Despite the fact that my neighbour's stand is better than mine, I choose not to rob Brazil nuts from her lands because I appreciate friendly relations'.

This conceptualisation can be criticised as an oversimplification, but it is not the purpose of this chapter to engage in a discussion about the hierarchical theories of knowledge. Rather, the level of detail in this chapter requires simplified use of the concepts of information and knowledge. In short, we consider knowledge to be in people's heads, whereas information is outside (e.g. Alavi and Leidner, 2001). This means that knowledge cannot be directly transferred from one person to another, but it can be transmitted in the form of information that bears a message. The more the message sent resembles the one received, the better the communication is. Knowledge is often different in different subgroups of people within the same society or even community. The conditions for a common understanding are usually best reached when the parties engaged in communication share a common knowledge base, when there is a suitable level of **cognitive proximity** among the interacting individuals and groups of people (Hautala, 2011). This means that people share sufficient knowledge and experiences so that the same things have similar meanings for them and lead to similar interpretations. However, even local knowledge may vary much between generations or genders within the same community (Pfeiffer and Butz, 2005).

OPERATIVE KNOWLEDGE FOR WILD SPECIES HARVEST

The most basic knowledge regarding wild species harvest has to do with the **utility** of some wild organisms or their parts, which transforms them into a potential resource. For example, the Neotropical vine cat's claw (*Uncaria tomentosa*) would be just another plant for humans if not for its medicinal properties that make its bark and roots useful (Figure 17.3). And similarly knowledge about the edibility of the seeds of the Brazil nut tree leads to the assignment of a **utility value** to the seeds, consequently making them a resource.

If the harvester keeps the Brazil nuts collected, the nuts would not pass from being a **subsistence resource** in the forest and a **nonmarket good** in the household. But as the information about the edibility of Brazil nuts is known by or communicated to other people, Brazil nuts will potentially be bought, markets may further emerge and the harvester exchange the nuts for money or some other reward. This turns the Brazil nuts into **market goods**. The generation of knowledge, and the associated flows of information may seem oversimplified and trivial but actually these kinds of interactions form the underlying reality in

FIGURE 17.3 The cat's claw, a vine that grows in several Central and South American countries. The plant is named after the claw-like curved thorns of its leaf axils.

all harvest systems from very simple subsistence economies to complex commodity chains.

Thus, there are yet other kinds of information flows that matter. For example, the harvester needs to expect a positive return resulting from planned work. If such a reward is not to be expected, it is most likely that no harvest takes place. But, when there is a signal from markets that indeed the supply of wild species goods will be reasonably remunerated, the harvester can then evaluate the level of the expected revenue, make decisions about future harvests, and maybe also make commitments that are reasonably realistic for both the producer and the buyer. The feedback that the harvester ultimately gets in the form of a price when selling the harvested material then hopefully is similar to or even greater than the expected price.

In order to explore the role of such information in wild species harvest, we first limit the scope to the level at which direct harvest decisions are made. This refers to decisions made by the direct extractors physically working in the harvest activities as well as those actors that hold formal or informal rights to the harvest of wild species in a specific area, pay salaries for labour, or informally exercise powers to make decisions about harvest. In short, the focus is on such knowledge that is central when choosing among different options of direct action. These decisions concern, for example, which organisms and individuals to target, what the harvest volumes are, what are the timing, places, and methods used. Yet other such concerns include, for example, what to do with the harvested materials, how to transport them, whether to engage in some form of processing or not, and, in the case of a market, to whom to sell and for what kind of price or other reward, what the quality and volume requirements are, and so on. In order to generate this kind of knowledge, the harvest decision makers have to use different kinds of information and also engage in communication using that information.

First of all, the harvester must have knowledge about the resource; where it occurs, when, and how abundant it is. The **timing of harvest** is an important practical issue, crucial in the assessment of the costs and benefits of harvest. It is a vital component of many forms of extractive activities that are bound to annual or longer cycles, either related to the phenology or regeneration of the harvested species or, depending on seasonal changes in the environment, facilitating or hindering harvest using particular methods. For example, fishing in cold regions can be practised throughout the year, but the methods used vary greatly depending not only on fish reproductive ecology, including migrations, but also on the air and water temperatures and the associated freezing and melting of superficial waters. On a yet shorter time scale, the daylight-related rhythms of both humans and some of the extracted species are important. For example in Sarayaku (Chapter 4), the nocturnal curassow (*Nothocrax urumutum*) is essentially hunted in the early morning hours, starting before the dawn. In those areas where hunting and fishing are important activities, it may even be that the whole rhythm of human communities is adjusted to the times when these activities are most productive.

Without information and subsequent knowledge about the provision from nature it is also impossible to estimate how much can be extracted or how much is harvested in relation to the existing resource base. In its simplest form, estimates about the levels of provision of harvestable resources in nature are achieved by entering the place in which harvest is to be carried out and finding out whether there is anything to extract there. This is common, for example, in subsistence hunting and fishing. In other situations, there might be information about the harvestable individuals, even to the level of their exact volumes and locations, acquired by using modern technologies. This is the case, for instance, in some advanced forms of commercial logging (Chapter 9) that apply remote-sensing, GIS (Geographical Information System) or GPS (Global Positioning System) devices, or in fisheries applying echo sounding. It is also important to know about the finer scale details such as population structures.

In remote areas, where informal economies often flourish, information about the levels of harvest may not be collected at all, at least as understood as **formal registers** of harvested material. When palm leaves are harvested from the wild and sold in town, it is not common that there is bookkeeping to record the extracted quantities (Chapter 7). In the case of illegal harvest, such statistics are even less likely to exist. However, harvesters engaged in the same type of activity year after year often accumulate knowledge about harvested volumes and thereby learn about resource dynamics. This is typical, for example, among many fishermen and hunters. Such implicit understanding, often called **tacit knowledge**, is valuable for harvesters, but it is difficult to transform into information that can be shared with people who do not themselves have similar knowledge (cognitive proximity). However, information about costs and selling prices can be more easily shared with other actors of the same community or a cooperative formed by several extractors. Well-organised hunters can report not only the number of different game animals killed but also information about the sex ratio and age structure of the animals caught as well as about the places and times of extraction. They may also write down and share notes about some hunting scheme that yielded good results. In the chapter about river turtles in Peru (Chapter 6), the local community members reported the numbers of eggs collected and hatchlings released as well as the numbers of eggs and hatchlings sold.

In large-volume branches of natural resource use, information about extraction levels is collected systematically, but it is not necessarily shared with others. Yet, in some cases, clients or authorities demand maximal transparency. For example, industrial logging is progressively becoming an activity in which the buyer should be able to track the timber back very accurately, sometimes up to the very point of felling an individual tree. Informality and illegal logging can sometimes dilute the veracity of such information, as seen in Chapter 9. Commercial fishermen commonly have to report their landings, but information regarding any side-catch is ignored (Figure 17.4). Similar situations also occur with many other extracted resources. For example, the selective logging of some target species may also affect many other trees in the same forest, but no records are made on it. Some

FIGURE 17.4 Side-catch is discarded from a fishing boat in Peruvian Amazonia, with some fish alive and some dead. It is common worldwide that no data are collected for side-catch statistics.

felled individuals of the target species may also be abandoned in the forest due to logistics problems, poor quality of wood, or some other reason. Thus, even if we have statistics about the sold timber, this only tells one part of the story.

Beyond directly describing the resource base and the flow of matter from nature, harvesters also need to have concern about the effects of their activities in the ecological system. People working in harvest-related activities inevitably get information and generate knowledge about the state of the harvested resources, and to some degree also about the surrounding ecosystem that provides the resource base. Maybe, as a consequence, they also get to know something about the impacts of their harvest on this system. This kind of feedback from the ecological system is essential in any long-term wild species harvest. The direct effects of extraction are often easy to perceive by the harvesters as overharvest decreases the availability of the harvested species, like in most of the cases described in Part 2 of this book. Thus, although it can often be easy to notice that harvest levels exceed regeneration, at other times it can be much more complicated. For example, skewed age structures or sex ratios are difficult to detect, and their significances are hard to understand. Also, indirect effects of harvest based on species interactions may pass unnoticed for a long time. Collecting and reporting information about the effects of harvest are complicated tasks, and they are commonly neglected by the harvesters. Sustainability, however, can be pursued only if the harvester develops a realistic picture of the state of the harvested resource and takes it into account in the planning of future harvest.

ADJUSTING PROFITABILITY

Based on available information and knowledge, harvesters try to adjust their efforts so that their work is profitable. In well-balanced conditions, the **expected feedback** and the **obtained feedback** eventually can be of similar magnitudes. This can be the case, for example, when a targeted wild species is foreseeably common, the costs of its harvest are predictable, and the product has a stable value. Under these circumstances, the harvest effort can be relatively easily adjusted to the existing needs according to sufficient information on factors that may change the balance. Often, however, factors related to uncertainty make harvest decisions more difficult to make. For example, when the resource availabilities or qualities vary or the selling prices fluctuate unpredictably, the risks can be high (Figure 17.5). Actually in most cases where a market mechanism is involved there is a degree of uncertainty, which will inevitably influence the actual harvest decision. In order to make informed decisions about the value of future harvests and thus avoid making disadvantageous contracts, the producers need to know about the development of

FIGURE 17.5 The Belén market in Iquitos, Peru, is the main spot of exchange for a diversity of wild species goods from live animals to fish, game meat, medicinal plants, and fruits. Information about daily price fluctuations is of prime importance for harvesters who want to sell their harvested matter.

resource availability, harvest costs, and prices. Sometimes, these uncertainties are shown in the value of the resource use rights. For example, Rocha et al. (2006) found that existing uncertainty over current timber volumes within Brazilian forest concessions tended to decrease the concession value.

Information about feedback and **incentives from society** can include volumes and other details about what has previously been consumed or traded. Whether or not extracted materials are formally reported, harvesters normally have a good knowledge about the extracted materials that they deal forward. When the activity is illegal or informal, the information about monetary transactions tends to stay in the shadows for obvious reasons.

In market situations, information about what was traded and what was obtained in exchange has vital importance for further activity decisions by harvesters: in the end, it is the net benefit resulting from the harvester's **transactions with society** that counts on this side of the equation. Harvesters thus often register, either in their minds or by using some way to keep the books, what they have sold and how much they got as a reward. The direct feedback can also be influenced by costs linked to transport, planning, and fees. Harvesters therefore need to be able to handle information about such things as prices of fuel, professional assistance, taxes, subsidies, and any other changeable fees. This can be demanding for some harvesters living in distant areas and make them vulnerable to fraud by intermediaries – although this kind of **information asymmetry** may be changing as wireless and mobile communications technologies are becoming increasingly cheap and widespread. Likewise, the necessity to fill in forms and submit plans and declarations to feed official procedures can be a burden and lead to neglect of such formalities.

The more complex the commodity chain of wild species-based goods is, the more difficult it is for the harvester to obtain reliable information about the overall socio-economic contribution of harvest (Figure 17.6). Only when harvested organisms are provided directly to end users, with little processing and further trade involved, can the harvesters have good chances to see the value of their work in society. Every step further along in the commodity chain adds value to the wild species-based material and also increases the network of actors involved. When one compares, for example, the price paid by an end consumer of Brazil nuts in Europe and the price paid to the collector, it is difficult for both to gain understanding about how the price difference has come about. Certified fair-trading mechanisms try to fill in this gap by securing the interest of the primary producers.

Direct feedback can be obtained from buyers or from registers of price fluctuations and subsidy levels, as well as legal rulings about fees and sanctions. Such information is often collected but seldom easily accessible, systematically organised, and made available to support the work of wild species harvesters. A **middleman's** place is central as he or she has access to information concerning both the provision from nature and the market situation. Similarly, documentation about laws and policies is often available from public sources,

FIGURE 17.6 (A) Small-scale timber harvesting may be important at a local level. For example, roundwood harvested in the rural communities near the city of Iquitos, Peru, forms an important sales article in the city. (B) These goods, however, are used almost as they are when they leave the forest, and there is very little further processing; the price of the final product sold increases only because of the work and profits of the merchants. (C) Commercial logging of larger timber species, however, gives rise to longer and more complex commodity chains employing directly and indirectly more people.

but again, the information is frequently not systematically organised, or it is hard to comprehend by many harvesters.

In situations in which wild species harvest initiates long and complex commodity chains involving a high level of informality, it is often difficult to quantify the contributions of different economic actors in the overall harvest-based economies. However, when the influences in society are apparent and well understood, the situation can be different. This is the case, in particular, when harvest forms part of formal economies in which its contributions to the local, regional, and national socio-economies can be traced and quantified (e.g. price formation or taxes paid and reinvested in public services). Under these circumstances, the roles of the different actors involved in the commodity chain can be better evaluated against the overall cumulative effects of harvest in society. Examples of these kinds of harvest economies at the global level include some cases of commercial fisheries (Upneja et al., 2001; Virtanen et al., 2001) and forestry (Palo et al., 1999), both supporting a variety of associated activities from the industries that produce the needed vessels and machinery to the post-processing that makes food from fish or furniture from timber.

The more important the raw material is, the higher is its demand, with stricter standards related to quality and timing. At some point when these aspects are highlighted, pure harvest activities will get accompanied with efforts for resource management and later on even plantations or other ways of industrial-scale production. Information about the influences of harvest in society actually makes a difference even at the level on which the actual harvest takes place. Such a situation is distinct in those circumstances where market mechanisms are underdeveloped, making indirect effects difficult to unveil. Here, the risk taken by the extractor in person may be high when, for example, a middleman sets the price level according to his or her own convenience.

There are also other kinds of knowledge that the harvesters may have to take into account when planning their activities in relation to the incentive structures that prevail. Knowing about the rules of the harvest is in many cases of vital importance; important issues in this respect include property rights, permitted methods, and extraction seasons and quotas, as well as the surveillance and sanctions used to enforce rules. It has been shown that knowledge about the capacities of authorities to monitor compliance with rules is a factor that influences how people follow the rules (Ramcilovic-Suominen and Epstein, 2012). Information about surveillance practices in general and controlling operations in particular can also easily leak from the authorities to other actors. In practice, when people know that the likelihood of being caught is low, they are often more willing to break the rules.

Awareness of other kinds of incentives also counts, for example when there are societal arrangements that promote and regulate wild species harvest through subsidies, certification, fair-trade buyers, or some other way. There might also be knowledge that helps improve the results of harvest, be it through improved harvest methods or management of the ecosystem. In addition, know-how concerning the product processing needed so that harvested goods of the required quality can be sold is vitally important. All of the types of knowledge described in this and the previous section can be called **operative knowledge** of wild species harvest. It refers to any knowledge that harvesters manage in order to make the most of their wild species economy on a daily basis. Optimising the use of wild species resources in the long run by various actors, however, also requires awareness and understanding of the impacts of harvest.

AWARENESS OF IMPACTS

Awareness that resource harvest causes some kind of negative impacts is a prerequisite for those actors attempting to do something about it. Governmental authorities, local communities, or individual actors will not take action to solve a problem unless they know that the problem exists. To start, people may not be aware that the abundance of a resource is decreasing. This may happen, for example, in situations in which the harvested goods are traded on wider markets, and awareness of the state of the natural populations fades away along the commodity chain. Locally collapsed species stocks may not be recognised as a problem as long as the wanted raw materials keep coming from other areas. This is how, for example, selective logging has emptied Amazonian river margins of the big individuals of valuable timber species over recent decades (e.g. Grogan et al., 2010). As long as sawmills got the timber they needed, this was not perceived as a problem by national governments or by the timber industry, because the short-term interests of the industry were fulfilled. It was only when all easily accessible timber sources began to be exhausted that problems of obtaining sufficient volumes of good-quality raw materials followed.

Another problem occurs when people note that resources are dwindling but do not recognise the causal mechanisms. People in general tend to have difficulties in understanding the dynamic nature of biological resources (Moxnes, 2000). In addition, people in rural communities sometimes **believe** in magic or have fatalistic beliefs that do not recognise causal links between the withdrawal of resources by humans and the depletion of the resource stock (Baland and Platteau, 1996; Burke, 2001).

The 'tragedy of the commons' (Chapter 16) occurs when resource harvesters are aware that they are overharvesting a resource but are unable to collectively take action in order to halt this. Alternatively, when they are unaware of the impacts of harvest on resource status, they see no dilemma between pursuing individual benefits and avoiding collective ruin, and therefore have no incentive to even attempt to coordinate harvest behaviour among the collective of resource users (Burke, 2001).

Conservationist thinking and practice tend to emerge only after experiences of resource scarcity or negative environmental changes. Such experiences are, however, not sufficient alone to inspire action. It is also required that people believe that there is a potential to improve the situation and that the existing social and political institutions have the ability to achieve this (Alvard, 1993; Holt, 2005). Sudden 'booms' of increased market demand may lead to over-harvest and resource depletion at such a pace that there is not enough time for societies to adapt. Once they have first become aware of the impacts of over-harvest and then have managed to organise in order to halt it, the resource may already be nearly gone. On the other hand, if change is very slow, it may go unnoticed and therefore spur no awareness and no action. For example, extinction processes in preindustrial times may have taken hundreds, if not thousands, of years, and may not even have been perceivable within the span of one human lifetime. Similarly, it may today be difficult to detect such slow changes as those that may occur in Brazil nut stands as a consequence of almost all of the seeds being removed from the ecosystem.

Finally, even if local harvesters are aware that their actions are causing resource depletion, and therefore agree on internal rules restricting harvest, this does not guarantee that the process is halted. Such rules must also be effective from an ecological point of view, and this requires that they are designed based on an adequate understanding of the resource dynamics (Chapter 13). The case of palm leaf harvest (Chapter 7) shows an example where local rules are in place, but these rules are based on an insufficient understanding about the dynamics of the resource. In that case, although there is a rule restricting the number of leaves that can be harvested at any single occasion, it does not put any limit on the frequency of harvest. Although the rule saves individual trees from immediate death at the first harvest, it does not ensure sufficient production of seeds, and it is therefore ineffective in preventing resource depletion in the long run.

In Nordic countries, the forest sector provides an interesting case from which many lessons can be drawn. Forestry and related industries are today important parts of the economy in Finland and Sweden. The situation was quite different a century and a half ago, when much of the forests in these countries had 'deteriorated' due to slash-and-burn agriculture, selective logging, and lack of management. The change that followed was strong and successful in achieving the goals that were set. The know-how on scientific forest management was developed using external assistance, including the establishment of national research bodies and faculties for forest research. Moreover, people were taught to appreciate and prioritise efficiently managed forests at all educational levels, from primary schools to the universities. New forestry practices were gradually adopted, and today almost all forests in these countries are subject to intensive and high-yielding management based on a cycle of clear-cutting, planting, thinning, and then clear-cutting again. These practices gained widespread acceptance, as forestry provided employment and wealth both directly and because it supplied high-quality raw material to sawmills and the pulp and paper industry. Although a few critical voices were heard all along this process, only recently has it become more commonly recognised, as a consequence of accumulated research evidence, that something also went wrong as this conversion of natural forests into intensively managed wood production areas also implied a loss of biodiversity and recreational values.

SCIENCE AND UNCERTAINTY

Science plays a very important role in the generation of reliable information and undesrstanding. Scientific research covers all levels of the information pyramid (Figure 17.2), giving it a special significance in human development. This also applies to the context of managing wild species resources. Although science is not a solution to the challenge of sustainability in itself, it forms links in a chain that can make sustainable uses realistic as alternatives to short-term opportunistic resource use. Science offers superior methodologies to generate knowledge and communicate information in a systematised manner and in such a way that it is exposed to criticism that contributes to a continuous increase in this knowledge. But it does not provide eternal truths carved in stone.

The **state of the art** is continuously challenged when competing scientific ideas are evaluated and compared with each other. Concurrent scientific interpretations and theories, then, are the best explanations available so far, but they only persist until they are overcome by better alternatives. This means that change is inherent to science since results can and should be challenged when new data or theories are available and interpretations can be revisited. Figure 17.7 shows an example of how the same data may give rise to very different conclusions depending upon the interpretation method that is used. The lesson that can be learned from this example is that an interpretation (in this example, interpolation) that is carried out mechanistically and is not coherent with other existing information can be alarmingly misleading. Wild species harvest represents such a complex and multidimensional activity that makes it particularly susceptible to biased interpretations. Any conclusion that is made using data and information from a specific place and time needs to be critically assessed as time runs on and new information becomes available.

No matter how much progress science may make, wild species harvest is such a complex phenomenon that a certain degree of **uncertainty** is inevitable. Such uncertainty may be of three distinct types: model uncertainty, parameter uncertainty, and inherent uncertainty. **Model uncertainty** refers to the level of understanding we have of the world and whether this is sufficient for the world in order to be logically or mathematically represented in order to use what concrete data we have to estimate the unknown. In other words, model uncertainty leads to the question, for example, of whether the logistic equation described in Chapter 13 really is an appropriate way to represent the dynamics of, for example, wildlife, fish, or timber. **Parameter uncertainty**, on the other hand, is about whether there are enough data of the quality needed. Given the considerable costs involved in, for example, the inventory of rare species such as the tapir

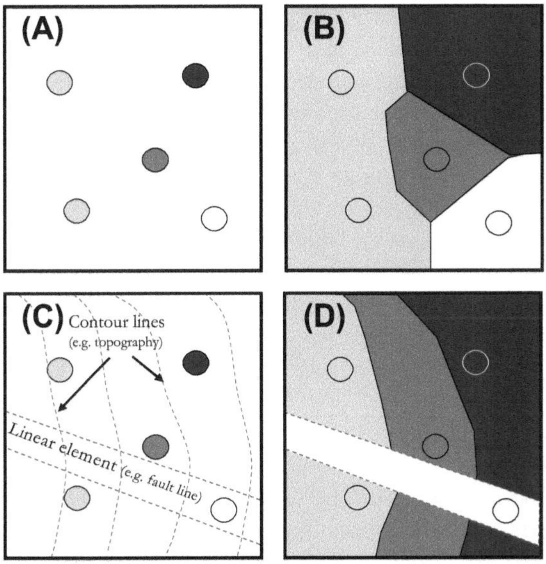

FIGURE 17.7 (A) Comparing spatial interpolation of point data using two different approaches. Circles represent different field study sites, and their shading indicates the kind of registered condition, for example soil type. To assess the likely values for the areas that lie in between, (B) applies a mechanistic distance-based interpolation approach where all locations are assumed to have the same character as the nearest field study point. In (C), some additional information about the research area's properties is shown, which is used to quide the interpretations presented in (D). The resulting maps (B) and (D) are highly dissimilar, and they may lead to very different conclusions regarding, for example, the distributions of habitats of the harvested species.

(*Tapirus* spp.) or big-leaf mahogany (*Swietenia macrophylla*) reliable data on their abundance will very rarely be available. Similarly, determining the rates of, for example, the growth, reproduction, or mortality of wild species can also be very challenging. Finally, **inherent uncertainty** simply means that the world is, in itself, stochastic and unpredictable. This is nothing we can avoid by building better models and collecting more data. We must simply accept it. An important challenge is therefore to identify such resource use schemes that work even in the absence of comprehensive data and are not too vulnerable to unexpected events (Johannes, 1998). When applying mathematical models, such as those presented in Chapters 13 and 14, to real-world management questions, one must also bear in mind that they are only approximations, and their validity for any specific real-world case should be carefully scrutinised.

In order to make decision making evidence-based and enable it to cope with inevitable changes, it is useful to constantly and systematically **monitor** the **status of the resource stock**, as well as **harvest levels**, and sometimes other socio-economic and ecological conditions too. Many harvest models require input data on **resource abundance**. Whereas this, in some cases, may be quite straightforward to obtain, it may in other cases be very difficult. Acquiring enough observations to make reasonably precise estimates of abundance of, for example, some hunted large Neotropical mammal species typically requires walking several thousands of kilometres doing a line transect inventory (Figure 17.8). Because of the high cost this implies, it is feasible to do only when considerable amounts of external funding are available. The same is probably the case for many other types of harvested species that are mobile, inconspicuous, and rare. Instead of direct measures of resource abundance, one may therefore use some proxy. One common proxy is the **catch per unit of effort** (CPUE), which is a measure of the amount harvested during some unit of time of harvesting labour using some defined type of technology (e.g. Rist et al., 2010). For many resources, such as game and fish, CPUE increases with resource abundance, although the relation is not necessarily linear (Harley et al., 2001).

There are also other alternative indicators of the status of the resource, which can be used particularly when it is difficult to obtain estimates of resource abundance itself. **Age structure** can be used as an indicator, as the proportion of young individuals often becomes higher in a heavily exploited population than in an unexploited or lightly exploited population, as the longevity decreases due to higher mortality. For example, for South American tapirs (*Tapirus terrestris*), collared peccaries (*Pecari tajacu*), and white-lipped peccaries (*Tayassu pecari*), the degree of tooth wear has been used as an indicator of age, allowing assessment of age structure (Bodmer and Robinson, 2004). Somewhat related to this, size structure can also be used as an indicator, as intense exploitation tends to lead to a size distribution skewed towards smaller sizes (e.g. Shin et al., 2005). This is both because the size distribution changes within each species

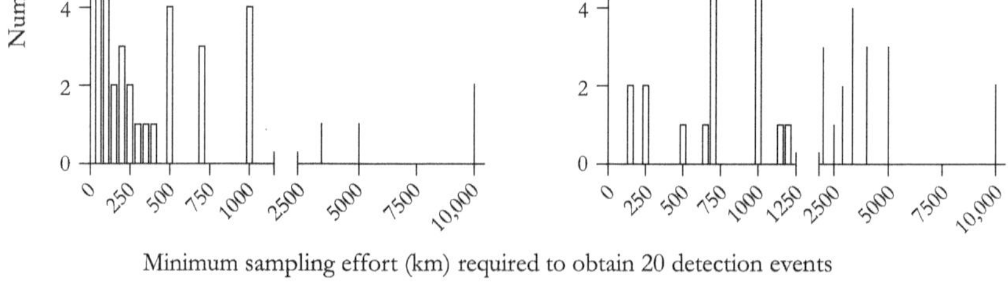

FIGURE 17.8 Minimum line transect sampling effort required in order to obtain 20 sightings (which is considered the minimum number in order to make robust density estimates) for 43 Neotropical mammal genera. Based on sighting rates from 11 tropical rainforest sites. *From Carrillo et al. (2000).*

and because, in multispecies harvest systems, larger species tend to disappear first (e.g. Welcomme, 1999).

In addition, the everyday **observations by harvesters** themselves can be useful. Wild species harvesters often note changes in resource abundance, even when they do not express this in absolute numbers. Relative measures are often enough for them to identify problems of resource depletion or to assess the successfulness of management measures, as noted for example in the chapters on hunting (Chapter 4), fishing (Chapter 5), turtle management (Chapter 6), and palm leaf harvest (Chapter 7). Sometimes, however, such everyday observations are insufficient, and scientifically designed observations can provide new insights. In the case of palm leaf harvest, for example, local people noted a general decrease in the abundance of the *wayuri* palm (*Pholidostachys synanthera*) near settlements, but only through a systematic inventory was it discovered that this decrease was much more severe for seedlings and saplings than for adult trees, such that recruitment was almost nonexistent in the most heavily exploited stands, and a total crash could be foreseen in the near future unless the harvest intensity was reduced (Sirén et al., 2013b).

For multispecies resources, the **species composition** can also be used as an indicator of the effects of management. Intensively exploited communities tend to have a higher proportion of small, inconspicuous, fast-breeding, and low-value species in comparison with unexploited or lightly exploited communities (e.g. Rowcliffe et al., 2003). Such comparisons can be based on direct observation in the natural environment, but they can also be based on observations of harvested individuals or even the supply in marketplaces, as has been done for example for hunted animal species in marketplaces in central Africa (Cowlishaw et al., 2005).

When it comes to monitoring the **harvest levels**, direct observations are rarely feasible, so often one may have to invent creative ways for indirect observations. For example, the number of remaining stumps of cut petioles on the *wayuri* palm trees (Chapter 7) provided a proxy for the past harvest level, although it is not well known how long these stumps would remain before they have rotted and fallen off. Logging intensity can, very roughly, be assessed by advanced remote-sensing methods used to detect the gaps created as trees are felled (Asner et al., 2002).

Often, however, the most feasible way to acquire information about harvest levels is to involve the harvesters themselves in reporting their harvests. Harvester reporting on a voluntary basis has been used, for example, in several studies of hunting in Amazonia. This, however, requires that trust is established among the harvesters and any other actors involved, such that they share the mutual interest and motivation to contribute to the collection of accurate data. For many commercially harvested resources such as timber and fish, such reporting is often compulsory, but then controlling the veracity of the reported volumes is difficult. Therefore, monitoring based on harvesters' reports is much less feasible in cases where unauthorised harvest is widespread and uncertainty can actually be a value to those harvesters and other actors involved in such activities.

Unreliable information sometimes may be useful as it can create 'shadows' for an illegal harvest. Without reliable inventory data on the volumes of valuable timber species – such as big-leaf mahogany within the concessions – it is easier to overestimate the stocks or 'invent' trees on paper in order to fabricate false documents for laundering timber that has been logged elsewhere (Smith et al., 2006). According to a study made by Amacher et al. (2009), smallholders selling timber in Brazilian Amazonia to intermediary merchants are responsible for a considerable part of the actual timber extraction, but it is not well known where this timber actually comes from. In Peru, a large part of logging also takes place outside of the forest concessions or other authorised areas and is carried out by individual teams of loggers supplying timber to forestry industrialists through middlemen, as seen in Chapter 9.

Although precise knowledge about the level of these activities would be useful for monitoring purposes, the problems of informality and illegality discussed here hinder data collection. Efficient monitoring would also need to take place at the level of individual concessions or even within annual felling areas, whilst in reality control often takes place at scattered control posts along the rivers or roads, or in the ports of cities. The needed spatial resolution of the available data thus cannot be cost-effective or obtained at a small enough scale, revealing a gap between the resolutions at which the small and the local actors operate, on one hand, and that at which authorities manage the official data on resource stocks, on the other (Salo, 2009).

Management planning should be based on analyses using information that is carefully systematised with all the original data preserved as far as possible and also clearly reporting the analytical methods that were used (Figure 17.9). Presumably, this point is evident for most of our readers, but it is worth making since it is not uncommon to see situations where a 'black box' approach is used, such that it lacks good information describing how the conclusions were obtained. A couple of decades ago, this problem was particularly severe in Amazonia. A number of unique natural resource inventory reports just showed their conclusive maps on subjects like forest resource distribution or land suitability for agriculture. As almost no supporting evidence was provided, however, the use value of these reports was limited. With such information feeding political decision making on resource use, problems cannot be avoided (see Chapter 12).

Contemporary planning schemes in Amazonia are nowadays fortunately based on a better working culture. In addition to improved professionalism in all respects, the data sources at hand have also improved a lot. For example, all-inclusive spatial data from remote sensing is widely available and can be used to support inventories made in the field (Chapter 12). In addition to such methodological improvements, national and international development projects have started increasingly to support **evidence-based** planning. For instance, the Amazon Cooperation Treaty Organization (ACTO/OTCA) established by the

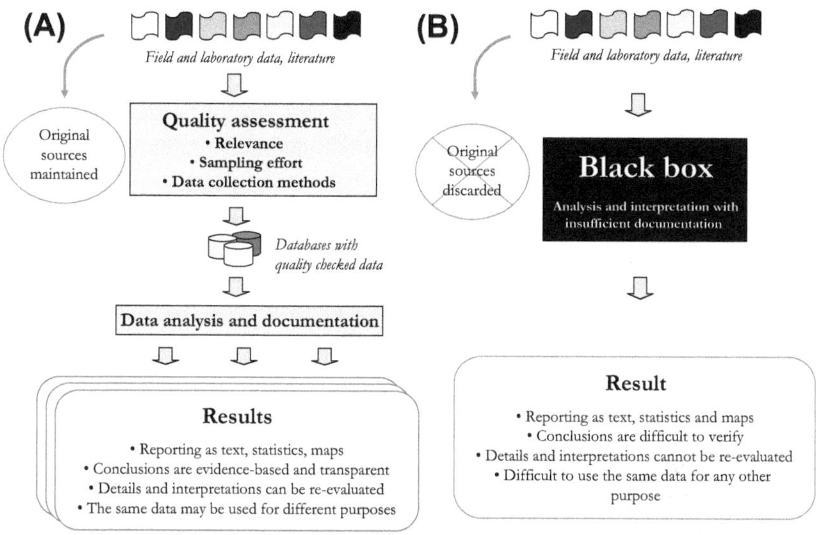

FIGURE 17.9 Comparison of rigorous and inaccurate research methodologies as used in, for example, natural resource inventories. Saving the raw data and appropriately describing the used methods facilitate further reconsideration of the conclusions obtained (A); it also facilitates the use of collected data for other purposes. This does not occur in the alternative case (B), where the raw data are discarded and the methods are poorly described.

Amazonian countries has put forward Ecological-Economic Zoning (Zonificación Ecológica Económica, or ZEE) to support sustainable development in the region. It aims to contribute to such viable economic, social, and environmental solutions that allow the use of Amazonia's natural resources without destroying its ecosystems (Rodríguez Achung, 2006).

These laudable goals, however, are not easy to reach, as many observers have noted (e.g. Dourojeanni, 2012). It would be ideal if the zoning practitioners had good access to any relevant existing information in a form that is easy to use and understand so that further conflicts arising from overlaid land use designations can be avoided. There are, however, reasons to doubt how much the ZEE actually affects what happens on the ground. Actual decision makers at the level of national governments, provinces, or municipalities do not necessarily take the ZEE into account when making decisions affecting the environment, and thus far, zoning has been mostly used to distinguish between coarse categories of land use such as agriculture, forestry, and conservation. Its utility as a planning tool for finer scale phenomena such as wild species harvest activities others than timber extraction remains unclear.

In contrast to many industrialised countries in the global north, where human activities destroyed natural ecosystems long before planning regimes were introduced, many if not most parts of today's Amazonia have been assigned science-based use categories that often protect them against outright land-conversion. The legacy of past land-use in these areas is thus less binding in terms of what has happened to the environment. However, areas designated for many seemingly antagonistic purposes, including logging, agricultural expansion, conservation, indigenous territories, mining and insfrastructure development already saturate a major part of the landscape. Furthermore, deficient level of knowledge often still hinders evidence-base decision-making.

FACILITATING LEARNING AT THE GRASSROOTS LEVEL

Information concerning wild species harvest originates from a myriad of sources and is distributed in society in an infinite variety of ways. Meanwhile, the larger society is engaged in associated dynamics, including decision-making processes in which a wide array of information sources is used along with a diversity of fora and media. As evidence of the importance of such diversity, most countries have organised their legislation, administrative bodies, research forces, and educational systems to support knowledge-based exploitation of natural resources. It is not guaranteed, however, that these arrangements succeed to link all relevant actors with relevant and mutually accepted information. There may even be a risk of elitism, when the world of researchers, planners, and other formal actors is too far away from the practitioners.

Good practices of resource use and management develop at the ground level, here and there, every day. The problem is that even if they exist and have been adopted in some places, they simply fail to diffuse and adapt to take on new forms sufficiently widely or at least fast enough. No matter how valuable these innovations are, their failure to spread to other areas may lead to their disappearance. Information about good practices is also hard to find even when one tries to do so. Although the proximate actors responsible for harvest decisions operate on the basis of operative knowledge, their **communication** with other actors is essential in the search of solutions to sustainability problems. The gradual spread of ideas occurs continually when harvester groups working within the same areas meet, as was seen in the case of turtle management in Chapter 6. However, this type of transversal diffusion among peers is quite slow in the context of large areas. Speed of diffusion may be limited by low population density, lack of transportation and communications infrastructure, national borders, language barriers, or other hindrances. That Amazonian natural ecosystems have by far remained rather intact is largely due to its low population density and relatively poor transportation infrastructure, but these very characteristics of the region also constitute major impediments for communication and the diffusion of ideas.

Overcoming the hindrances that limit the spread of information is a challenge. For example, exhausting resources have in some cases led communities to establish functional solutions for more sustainable harvest practices. Hunting regulations in Sarayaku, although still limited in their scope, are an example of this (Chapter 4). Such community-level solutions demand strong vision and collectively determined actions. They have a great potential for application in those cases in which wild species harvest is linked to the cultural meanings that are appreciated within the community. As they work well in real-world conditions in some areas, it would be worth trying them elsewhere, too. However, such locally emerging solutions are evidently often far too weak to spread and be adapted to new conditions to solve major problems in Amazonia or elsewhere. Their impact is often limited to the local scale unless their spread to other areas is facilitated by efficient communication. Internet and mobile communication devices have revolutionised communication in much of the world, and Amazonia is no exception (Dourojeanni, 2012) (Figure 17.10); it remains to be seen, however, whether the increased access to such media even in remote communities will have any effect on the spread of ideas about the harvest of wild species.

In an ideal case, there would be a continuous flow of ideas among innovators involved in wild species harvest and those whose work could benefit from the new ideas. The flow of information should naturally be **multidirectional**, involving the people engaged in wild species harvest and others learning from them. In the real world, there are barriers to communication between different groups of people. One limiting factor is the fact that many scientists involved in research in the tropics publish their results in scientific journals in English (this book, too, is published in English). Even though this is common knowledge in the scientific community, in practice it may take a long time before academically published findings start to have an influence at the ground level. **Facilitating mechanisms** would again be useful if they could speed up this process. This requires, however, that all relevant parties will activate themselves and participate, which is easier said than done.

Access to scientific results could, to a degree, be facilitated by publishing results in **open access journals** with multiple language versions and by applying **open data policies**. These developments are becoming stronger in science, but they often require additional work or funding, and many important scientific contributions are still difficult or expensive to access by people and institutions with restricted economic possibilities. Many conservation scientists actually think that participation in capacity building and local initiatives are more efficient ways to contribute to conservation and development than publishing articles in peer-reviewed scientific journals; few actually take part in such activities, and they find it frustrating that they do not have possibilities to diffuse their research findings to those who could make best use of them. In the highly competitive world of academics, articles in peer-reviewed journals are what counts when one applies for a job or for research funds, and dedicating precious time to other kinds of activities implies putting one's whole academic career at risk (Shanley and López, 2009).

The need for a stronger interaction between science and practice was already embedded in the Biodiversity Clearing-House Mechanism that was defined in the Convention on Biological Diversity (Laihonen et al., 2004), part of the Earth Summit outcomes from Rio de Janeiro in 1992. Clearing houses aim to work like marketplaces by providing channels to share exper-

FIGURE 17.10 The Sarayaku community (Chapter 4) lies in the midst of the rainforest 30 km from the closest road connection, but it has an Internet centre (A) functioning through an antenna capturing the signal from satellites (B).

tise, data, and information. Such mechanisms have been established on digital platforms in many countries and areas, with some of them designated specifically for species-rich areas like Amazonia (e.g. SIAMAZONIA in Peru; Kalliola et al., 2008). However, the success of such Biodiversity Clearing-Houses in facilitating information sharing among different parties has thus far been lower than expected. From the perspective of a scientist, dedicated participation requires additional motivation and resources. Although researchers might in principle be interested, in practice they often are too occupied with other activities. The same applies to the people working with wild species harvest in the field. Even if they knew how things could be done better, it is not easy to engage in changes beyond one's own immediate surroundings. There are also crucial differences in the ways in which things are communicated in science and elsewhere in society. Harvesters who seek clear answers and need to work for their daily income may easily become frustrated and continue their work as before if information is not available for useful purposes and in adequate formats.

Summing up, it is important to believe in, and focus on, **education** and **capacity building** (Figure 17.11). Conventional educational activities can be effective, but many wild species harvesters have little formal schooling, which restricts their skills to adopt new information. This is, however, more complicated than to simply say that the local resource harvesters are unable to understand scientific information and to use it to complement their own knowledge. Researchers,

FIGURE 17.11 Schools participating in management activities have been an important arena for the recovery of river turtle populations in Peruvian Amazonia. The participatory approach to environmental education has involved conservation professionals, teachers, pupils, and parents.

in practice, also have a limited understanding of local people's perceptions of the problems related to wild species harvest. One way to overcome this limitation is to use research approaches that aim to generate information that is relevant to local resource users and that promote collaborative learning (Shanley and Rodrigues Gaia, 2002). Local people can be involved in formulating research questions as well as designing methods, collecting data, and interpreting results; such involvement can provide efficient learning and often seems to stimulate some action. Some examples of this are shown in the stories in Part 2 of this book, such as those about hunting (Chapter 4) and palm leaf harvest (Chapter 7), and another example is the successful management of the arapaima fish (*Arapaima* spp.) in the Mamirauá Reserve in Brazil (Box 17.1).

Such **participatory research and planning** often benefit from the use of maps and other geographical data that help people to relate their own activities to the surrounding circumstances (participatory geographic information systems). This is useful since local cultures have significant accumulated information about their living environment with specific spatial associations (Fagerholm et al., 2012). When such local knowledge, or geographical aspects of the area's **social capital**, is incorporated into development actions through spatially oriented participatory mechanisms, it can make a significant contribution to the people's empowerment and at the same time promote good governance of the local environment (McCall and Dunn, 2012). Thus, knowledge and spatiality play crucial roles in wild species harvest from both the natural scientific and social viewpoints. Natural boundary conditions for sustainability, costs, and benefits, as well as management and governance, are also often about how to make actions fit the spatiality of the resource base.

Box 17.1 Arapaima Management in the Mamirauá Reserve, Brazil (based on Castello et al., 2009).

The arapaima, in Brazil known as *pirarucu* (*Arapaima* spp.) (Figure 17.12), is very important in the economy and culture of the traditional inhabitants of the Amazonian lowlands. The arapaima is a very special fish, distinguished in the first place by its rapid growth and impressive size, typically reaching about 1.7 m in just 3–5 years, when they become sexually mature. The fish can acquire a size of about 3 m and 200 kg of weight. In addition, arapaima tastes good and has fillets without bones, making it a very attractive prey.

Unlike most other large fish species in the Amazon River, the arapaima do not perform any long-distance migrations but only short seasonal migrations from the main river channel to floodplain lakes where they spawn during the high-water season. Moreover, it has the very unusual characteristic for a fish that it breathes air, therefore coming up to the surface every 5–15 min. This implies that, in addition to fishing arapaima with gillnets, it is also possible to harpoon them when they come to the surface to breathe, although this requires special skills that only some fishers possess.

Box 17.1 Arapaima Management in the Mamirauá Reserve (based on Castello et al., 2009)—cont'd

The arapaima was the most important species in Amazonian fisheries in the nineteenth century, but the catches have been declining since then. The populations have in some regions been extirpated, and it is even considered vulnerable to extinction. Government regulations have involved a minimum allowed length of 1.5 m, a closed season, and, in the State of Amazonas, even a total ban on arapaima fishing from 1996, but enforcement has been almost absent, as poorly staffed and equipped agencies have been in charge of the surveillance of huge areas.

The situation in the Mamirauá Reserve in the State of Amazonas in western Brazilian Amazonia was not very different from this general picture. Research during 1993–95 showed that fishing took place all around the year, irrespective of the closed season, and at least 70% of the harvested arapaima were shorter than the minimum allowed length, so it was predicted that the population soon would collapse if the current fishing practices continued. In 1996, a management plan for the reserve was implemented, which divided the reserve into two zones, a strictly protected 'preservation zone', and a 'zone for sustainable use'. However, attempts to enforce restrictions on fishing, as well as management regulations in general, were largely ineffective. Monitoring large areas was very costly, and local people were unwilling to participate in the surveillance, as it often implied having to sanction their own friends or relatives.

By 1998, however, fishers in four communities in an area called Jarauá had started to show interest in conservation of fish resources and had themselves banned the use of gillnets during one month a year, the time when the arapaima migrates into the lakes. Then, the Mamirauá Institute of Sustainable Development took this opportunity to begin a fishery management program with an entirely new approach. A first step was to develop a method for counting the arapaima, based on known scientific inventorying methods and also relying on the skills of specialised arapaima fishers, who could even distinguish individual fish from each other. Based on this, technicians of the Mamirauá Institute calculated the harvest quota, which would be 30% of the adult population of arapaima in the area, which the fishers accepted on the condition that they would be able to sell the fish legally and thus achieve a higher price than they had done before while selling the fish illegally. Then, the Mamirauá Institute negotiated with the governmental authorities and managed to arrange an exception from the ban on arapaima fishing in 1999.

Based on this total quota, the fishers themselves developed a system of individual quotas, where each fisher was given a standard quota, which, however, could be given away or sold to others. In addition, the elected leaders of the fishers' association had the right to change these quotas based on an individual fishers' behaviour. Fishers who broke the rules were punished by reducing their quotas, and fishers who participated in the arapaima counting instead got higher quotas as a payment for their work.

After only 1 year, the arapaima counts indicated that the population had increased, and a higher harvest quota was requested, but this was rejected by the authorities. After only 3 years of management, the arapaima population had

Continued

Box 17.1 Arapaima Management in the Mamirauá Reserve (based on Castello et al., 2009)—cont'd

increased fivefold, but, in spite of this, the government authorities decided to abolish the harvest quota altogether. After paying a visit to the area, 600 km away from the state capital, however, government officials were convinced about the effectiveness of the scheme. After 8 years of management, the population of adult arapaima had increased 23-fold, the harvest quotas had increased 10-fold, and the per capita income of involved fishers had increased eightfold. Moreover, the number of communities involved in the management scheme increased from four to 108.

To analyse the reasons behind this success, a team of researchers used the framework of the eight **design principles** for robust common-pool resource management institutions developed by Ostrom (1990), which is similar to, but somewhat less detailed than, the list of attributes conducive to self-organisation developed by the same author (Chapter 16). They found that the change from failure to success could be explained, initially, by the fact that four out of the eight design principles were absent before the management program started, but the support provided by the Mamirauá Institute to the local fishers led to improvements on all of these four points, such that afterwards, all of the design principles were present (Table 17.1).

FIGURE 17.12 Fishermen with their catch of *arapaima* in the Mamirauá Reserve, State of Amazonas, Brazil. *Photo: Leandro Castello.*

TABLE 17.1 Assessment of the Presence or Absence of the Institutional Design Principles for Enduring Common-Pool Resource (CPR) Institutions for the Case of the arapaima Fishery in the Jarauá Communities before and after the Participatory Management Scheme was Implemented (from Castello et al., 2009)

Principle	Before	After
1. Clearly defined boundaries that define who has access to the CPR	Present: zoning system	Present: as before
2. Appropriation and provision rules that are tailored to local conditions	Absent: size limit and closed season violated	Present: fishers follow management rules
3. Collective-choice arrangements that allow most resource appropriators to participate in the decision-making process	Present: opportunities existed for individuals to participate	Present: as before with increased participation
4. Effective monitoring by monitors who are part of or accountable to the appropriators	Absent: underreporting of illegal catch	Present: fishers themselves count arapaima
5. Graduated sanctions for resource appropriators who violate operational community rules	Absent: illegal fishing widespread	Present but weak: sanctioning done by reducing fishing quota
6. Mechanisms of conflict resolution that are cheap and easily accessed	Present: regular meetings were held	Present: as before
7. Rights to organise recognised by higher level authorities	Absent: fishing of arapaima banned	Present: scheme provided the legal permit to harvest arapaima
8. Larger CPRs are organised in the form of multiple layers of nested enterprises, with small local CPRs at the base level	Present: communities, the Mamiraua Institute, and IBAMA	Present: as before

Chapter 18

Spatiality in Nature and Society

PATTERNS IN SPACE

Things are not equal everywhere. This is something of which anybody involved in wild species harvest is perfectly aware. To harvest wild species, you choose to go to some place where you think it will be worthwhile to go, and you avoid going to certain other places. Most scientific theories and models about wild species harvest, however, ignore this fact. This does not mean that they are wrong. On the contrary, a characteristic feature of all scientific theories is that they disregard some aspects of reality in order to be able to highlight others. However, it is useful to be alert when applying non-spatial models to real-world problems, as well as to assess carefully to what extent any form of spatial heterogeneity may influence the interactions between harvesters and the species they harvest.

Any act of wild species harvest occurs in a specific location and is linked to particular geographical realities. First of all, there is a location for every individual plant or animal of the harvester's interest. Consequently, in order to work, the harvester needs to move in relation to the places of harvest – and many times the resource also moves. Furthermore, consumers, government officials, merchants, and other actors making decisions or taking action in relation to wild species resources or harvested goods are located somewhere and can also move in space. Usually those people who work in the harvest system at the grassroots level are aware of such issues of practical spatiality and routinely use place-related knowledge when making decisions about harvest. These decisions are based on knowledge about, for instance, search and logistics costs related to distance from home or from processing facilities and markets (Chapter 14) and the limits and extent of resource use rights (Chapter 16), among other spatially meaningful variables.

It is thus clear that wild species harvest cannot be assessed properly without addressing the spatial dimensions of the harvester's transactions with nature and society (Figure 18.1). Spatial patterns in nature form many important boundary conditions for harvest by affecting species and population distributions, but it is not only harvesters as individuals who must adapt their practices to these basic patterns in nature. Humans acting collectively also should seek to make their harvest-related institutions better fit these boundary conditions.

Diagnosing Wild Species Harvest. http://dx.doi.org/10.1016/B978-0-12-397204-0.00018-8

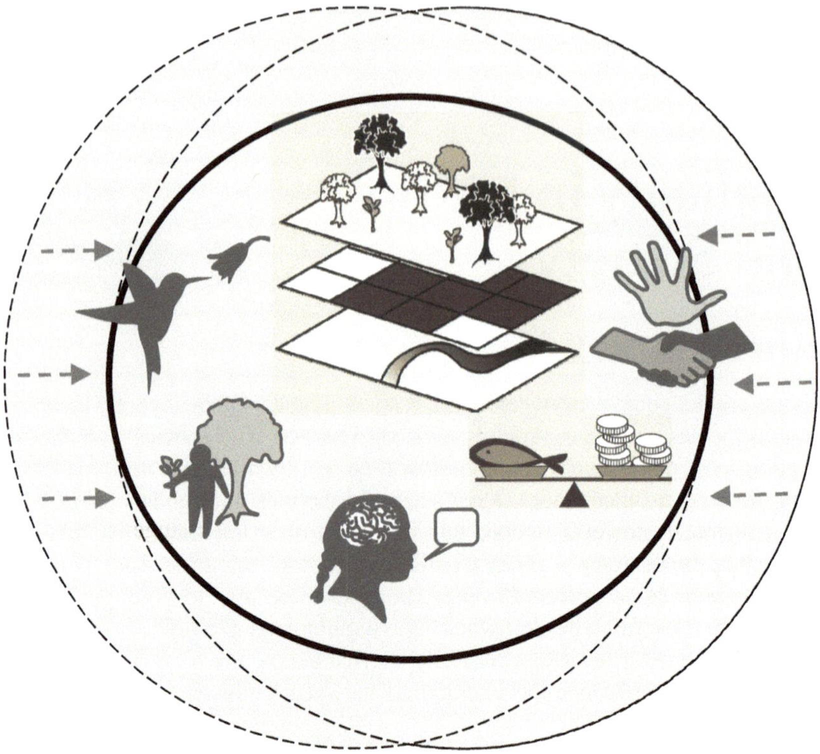

FIGURE 18.1 Spatiality is related to all aspects of the resource system; in a way, it forces them to shrink into a single whole that is thematically multidimensional. Any particular theme can be seen through its location-specific nuances as well as assessed within the geographical reality that is made up from all these different themes together.

Resource abundance and productivity often vary from place to place, either due to natural factors or as a consequence of human activities. For example, local productivity variations resulting from natural factors such as microclimate are familiar to any berry picker who has to search for the best patches and abundant areas of berries year after year. Combined with the variability in physical accessibility, this natural variability creates spatial patterns of harvest activity and resource abundance. Evaluation of the impacts of wild species subtraction from tropical forests, in turn, is usually made over relatively large areas, thus missing out on the variability over small spatial scales; however, field inventories of even large organisms, such as timber trees or game mammals, in tropical forests are time-consuming and costly (Chapter 17). This makes it difficult to obtain numbers of observations large enough to allow reasonable error margins in the resulting abundance estimates.

In particular, if one wants to estimate the abundance of harvested species at a fine spatial resolution the workload easily becomes far too great for the assessment to be realistically carried out. Things can become even more complicated when the resources are highly mobile. This applies not only to migratory animals but also to many other wild species that are mobile at least to some extent and at some state of their life. Actually, this concerns most, if not all, species (even mahogany moves when it is a seed).

NATURAL ENVIRONMENTAL VARIABILITY

From a **natural history** perspective, different regions on Earth develop and can be identified on the basis of environmental characteristics that vary according to latitude, distance from coastlines, elevation or water depth, topography, currents, salinity, and many other far-reaching geographical attributes that affect climate, soil, water, and other critical natural conditions. Any specific harvest location is embedded in larger environmental contexts with heterogeneity and dynamics of their own, thus giving rise to distinctive habitats and contributing to their distribution. Sometimes, geographical characteristics change rather gradually from one kind of environment to another, such as the change from humid tropical rainforests to increasingly seasonal rainforest types. However, there are also many clear discontinuities in nature related to such features as coastlines, steep slopes, river margins, or shifts in soil properties on a border between rocky ground and sedimentary deposits.

In terrestrial ecosystems, similar conditions often support structurally similar vegetation types characterised by analogous sets of plants presenting the same kinds of functional plant types or life forms (e.g. trees, bushes, lianas, epiphytes, herbs), leaf types, and spacing between individuals, to list a few characteristics. A series of other more detailed environmental variables then contribute further, creating environs at a more refined level, characterised by their respective plant and animal communities. In terrestrial habitats, these locally recognisable finer-scale environment types are usually closely linked to such underlying natural variables as local moisture and microclimate - circumstances that together give rise to **edaphic factors** (soil properties) and other specific growth site conditions. In turn, in the aquatic realm (ranging from rivers and lakes to seas and oceans), many additional factors such as water temperature, acidity and salinity, sediment and oxygen content, currents, and river flow lead to the formation of distinct habitats. Moreover, between the land and the water there are countless types of wetlands with their own typical characteristics.

Wild species harvest must essentially be adjusted to the spatial framework provided by edaphically and hydrologically distinctive growth sites and habitats. Habitat types are the most obvious predictors of potential places of harvest and knowledge about the spatial distribution of those habitats that potentially support vigorous populations of targeted wild species or those places where the individuals of these species concentrate. This knowledge is crucially important to any harvester and often relatively easy to obtain just by getting one's boots dirty, at least concerning relatively small areas. When information needs expand from small areas to larger extensions of land or water, detailed knowledge about the role of spatial variability behind the provision from nature becomes increasingly important.

These challenges are particularly prominent in the species-rich tropical rainforests. Fortunately there are increasingly efficient tools which help to 'dissect' the extraordinarily high biodiversity of species-rich natural areas into spatially distinctive units (Tuomisto et al., 1995), making **species distributions** look less random and at least to some degree understandable and predictable. This also helps one to assess the impacts of resource harvest with enhanced precision. When a given species of interest is absent in some place in a region where it is generally found, it should not be automatically interpreted to be a consequence of overharvest because its absence may equally well depend just on the fact that the habitat in that particular place is not suitable for the species. To comprehend these aspects of resource spatiality in more detail, one needs to understand the environment through its spatial patterns and the processes from which these patterns result. The better the knowledge base is at this level, the better are the possibilities to sustain harvest (see also Chapter 17).

One alternative is to rely on the knowledge of the people who know best. As far as local-scale environmental characteristics and resource-availability questions are concerned, the best possible experts usually live in the area under focus. Many harvesters and other local dwellers are highly skilled in distinguishing between different habitats, their species assemblages, and harvest possibilities. They are also well aware of the spatial distributions and extensions of the different habitats. **Local ecological knowledge** accumulates over time and, for example, ethnobotanical studies have revealed that the local inhabitants of Amazonia have rich nomenclatures for the different kinds of vegetation and other environment types of the region where they live. These knowledge systems can be in many ways comparable to the resolution of scientific inventories made in the same areas (Halme and Bodmer, 2007).

Some harvested species are specialists of stages of succession whereas some other prefer waterlogged places, and some perhaps require high levels of soil fertility. For example, balsa trees (*Ochroma* spp.), whose lightweight wood is exported from Amazonia to over the rest of the world, or the *sangre de drago* or *sangre de grado* tree (*Croton lechleri*), whose latex is used for medicinal purposes (Chapter 10), are pioneer species that often grow at disturbed river margins or where humans have felled the forest, commonly on sediment soils recently deposited by migrating rivers. Some animals, such as the large rodents pacas (*Cuniculus paca*) and agoutis (*Dasyprocta* spp.), also thrive in mosaic landscapes of agricultural land and secondary forest, and often even become agricultural pests. In pure natural conditions, these species may abound in treefall areas or other places of vegetation damage by disturbance. Still other factors affect the distribution of aquatic life, including flooding patterns, the content of sediments and organic matter in river waters, and water acidity.

Although spatial environmental variations can tell much about resource distribution based on edaphic variability and the succession resulting from

FIGURE 18.2 A flowering tree (probably *Cassia* sp.) stands out from the rest of the rainforest canopy of Peruvian lowland Amazonia, illustrating the diversity of tropical rainforests, where rarely two trees standing side by side belong to the same species.

disturbance, these factors only explain a degree of the real distributions of species and individuals. After sieving out the contribution of such local habitat mosaics in potential resource availability, highly diverse forests still contain lots of species that co-exist in a more or less random-looking manner in areas where no major habitat variability can be distinguished (see Chapter 2 about measures of diversity). Such local-level diversity is challenging for resource harvesters from a spatial perspective because it takes much effort to inventory and map the individuals of any particular species of interest. The flowering of one tree species during the time when the photograph in Figure 18.2 was taken illustrates, how scattered the individuals of the same species often grow in a diverse rainforest.

Let us now imagine that our interest is in that particular species. As harvesters, we need to find this individual even if it was not very conspicuous after the flowering period is over. We also need to find other individuals of the same tree species that are reasonably near each other in order to guarantee profitable harvest. As actors with long-term interest in sustainable harvest, we also need to determine ways to find the same trees again during the next year. Searches are costly, as are extraction, the logistics of transport, and many other phases of the harvest activity, not to mention the need to spatially delimit resource-use rights and enforce them to retain rights in order to harvest the annual production of the same trees year after year. All of these considerations, as highlighted in Figure 18.1, take us back to the previous chapters in Part 3 of this book, but now let us consider the situation from a spatial perspective.

The example of the flowering tree also illustrates some of the challenges met by, for example, Brazil nut (*Bertholletia excelsa)* harvesters (Chapter 8), or more generally the exertions faced by many other harvesters of forest tree resources: wild rubber tappers, palm leaf collectors, or extractors of barks for medicinal purposes, as seen in Chapters 3, 7, and 10, respectively. Selective loggers also face situations of the same kind but with different outcomes (Chapter 9). When the harvest is lethal, harvesters often have to move to new areas before they can return to a place already logged and perhaps find new, sufficiently big individuals to harvest. Such space-consuming overharvest is a particular threat for highly valued timber tree species. When all large trees are removed by loggers, the number of mature individuals producing new offspring is reduced, thus threatening the entire population's viability over large areas.

ACCESSIBILITY AND COST-DISTANCE

Natural environmental variability and, of course, variability caused by human impacts on habitats can give rise to significant spatial variability in the abundances of those wild species that constitute resources of interest to harvesters. In addition to this, however, the mobility of harvesters and of the individuals of the harvested species interact to produce distinctive spatial or spatio-temporal patterns.

As a general rule, people prefer to harvest resources close to where they live or where there are markets for the harvested goods. Human settlements have often been established in places with good accessibility to natural resources and transport routes. Nevertheless, harvesting wild species almost always implies a need to **travel to the place** of the resource, possibly with some equipment, and then transporting the harvested matter back home or to the markets or processing facilities. This inevitably involves a cost in terms of time and, often, payments for vehicles and fuel, as well as many other costs. For example, in the cases of the fishermen and the loggers in Chapters 5 and 9, people often have to live in remote places where the harvest is carried out, and they engage in arduous work transporting the extracted material or make an effort to store it in ways that help to keep it from perishing and prevent any deterioration of quality.

Distances and physical **accessibility** therefore are important factors in practical decisions about wild species harvest. Accessibility, which has to do with the physical and geographical constraints to harvest, should not be confused with **access**, which concerns institutional aspects of property rights and their enforcement (Chapter 16). In analysing the relative distances and costs spatially, we find the term **time-distance** is often useful, meaning that the accessibility of any particular place is measured as the time of travelling there from a defined origin, taking into account how the travel speed differs among different means of transport (e.g. travelling on foot in the forest, or on different types of roads and waterways). The time-distance can then be converted to a **cost-distance**, by assigning logistics costs and an opportunity cost to time. For example, Figure 18.3 shows time-distances from the city of Iquitos. It

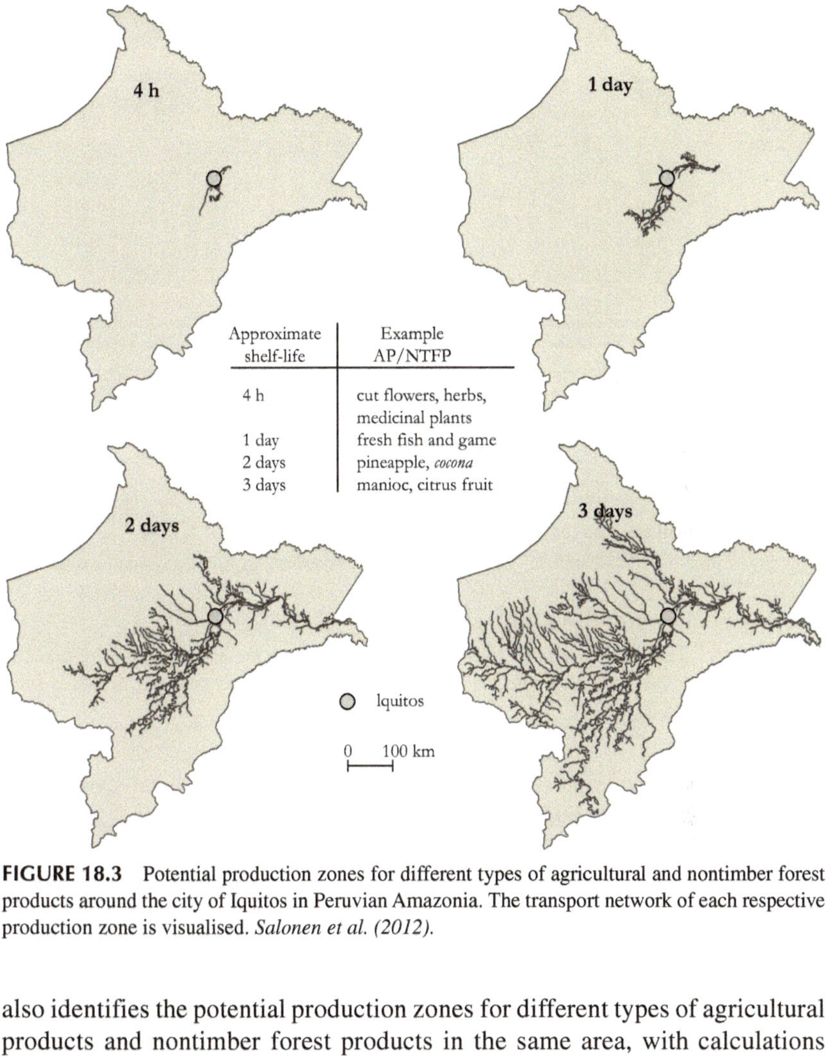

FIGURE 18.3 Potential production zones for different types of agricultural and nontimber forest products around the city of Iquitos in Peruvian Amazonia. The transport network of each respective production zone is visualised. *Salonen et al. (2012).*

also identifies the potential production zones for different types of agricultural products and nontimber forest products in the same area, with calculations based on the time-distance and the perishability of the different goods. Even with the same means of transport, the cost-distance is not necessarily linearly proportional to distance itself, but may increase in a convex manner because the cost per unit of distance increases as the harvester moves further away from home (Sirén et al., 2013a).

Particularly in open access situations and when access and harvest rights are not efficiently enforced, the physical and geographical barriers to accessibility become critical. In fact, although Amazonia commonly is imagined as a remote wilderness, the extensive river network reaches almost any place within the basin; it has been calculated that only a fifth of the Amazonian lowlands are further than 10 km away from roads or navigable rivers (Peres and Lake, 2003; Toivonen et al., 2007). Because travelling by road and navigable rivers is

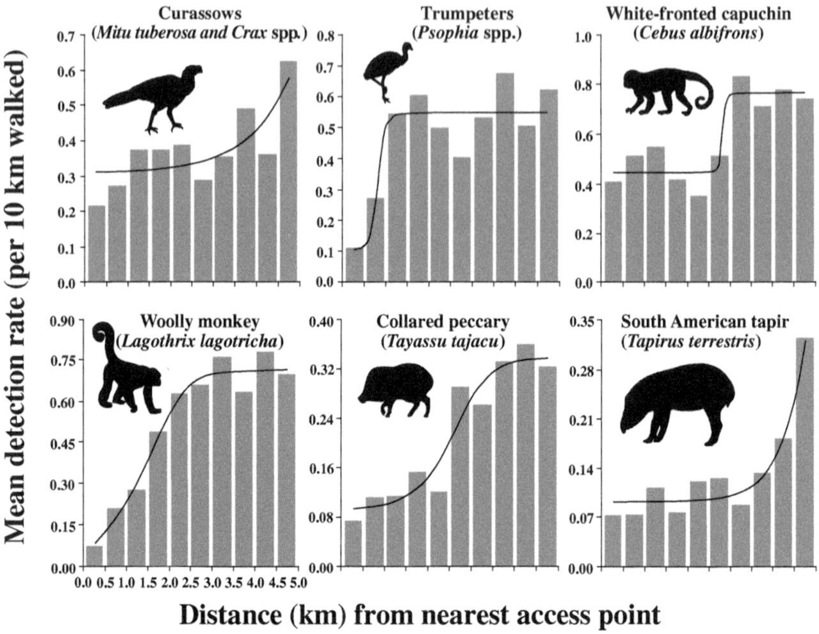

FIGURE 18.4 The abundance of large vertebrate species, which are preferred prey for hunters, shows strong positive correlation with the distance from villages, roads, or rivers. *Modified from Peres and Lake (2003).*

much less costly than transport using trails or creeks that only small canoes can enter, resources are typically more intensively harvested near rivers or roads. For example, hunting tends to be intense near settlements, roads, and rivers, whereas areas at distances over some 9–15 km from transportation routes tend to be very little affected by hunting, which results in marked gradients in the abundance of various species of game animals (Figure 18.4).

Distinctive spatial patterns of resource abundance and harvest often result for different types of resources (Box 18.1). For the case of tropical forest hunting, the levels of harvest, in terms of total biomass hunted per square kilometre and year, are often highest near human settlements because people prefer not to travel very long distances in order to hunt. This, however, leads to the situation where the populations of the prey species are reduced, and therefore the amount of harvested matter per unit of search labour, commonly called **catch per unit of effort**, and used as a proxy for overall wildlife abundance and also a highly relevant indicator of the economic return of hunting, is much lower near settlements than in more remote places (Figure 18.7), as also noted in Chapter 4.

The spatial pattern of *wayuri* palm *(Pholidostachys synanthera)* leaf harvest in the communities along the Bobonaza River (Chapter 7), however, is very different from that of hunting in the same region. Also for the *wayuri* palm, the abundance of the resource increases with distance from the village. However, a

Box 18.1 Using Spatial Patterns to Evaluate Hunting Impacts

When conducting field research on wild species harvest, it is important to use methods that take into account any existing spatial structure – for example, that harvesters may mostly operate near villages, roads, and rivers, and much less in more remote areas. This is the case, in particular, for tropical forest hunting, and this was explicitly taken into account in a study of hunting in Sarayaku, Ecuador (Chapter 4), based on a rotating sample of hunters voluntarily reporting their hunting activities, including their trajectories walked as well as the location of each hunting kill. It turned out that the aggregated search labour of hunters showed a very clear spatial pattern, being highly concentrated around the village and gradually decreasing with distance from the village (Figure 18.5).

There were more than 60 species that were hunted, but it turned out that most of these were almost only hunted very near the settlements where more preferred species were uncommon. When going further away, hunters did not bother to hunt such species but concentrated mostly on 14 species that, due to their size and good taste, were hunted almost wherever and whenever the opportunity appeared. The spatial pattern of harvest for these 14 species was therefore examined more closely. Based on these results, the respective species were grouped into four categories of different degrees of impact, for which category 1 is the one that is most impacted by hunting and category 4 is the one that is least impacted. The results for one species from each one of these categories are shown in Figure 18.6. The left column shows the number of harvested animals per unit of area and time, and the right column shows the catch per unit of effort (CPUE), which can be used as a proxy for resource density. Although this should ideally be calculated as the number of animals harvested per unit of time, it was in this case more practical to calculate it as the number of harvested animals per unit of distance walked by hunters.

In order to interpret these results, it is useful to compare the graphs in the right column with Figure 14.5(B) and the left column with Figure 14.5(C). For the species in categories 1–3, there is a clear decline in CPUE (i.e. in resource density) with increasing effort (i.e. decreasing distance from village). However, for woolly monkeys (*Lagothrix* spp.) (Category 1), this decrease is steeper, indicating that this species is almost extirpated from all but the most remote parts of the study area, where the aggregated hunting effort, or search labour, is very low. For woolly monkeys the harvest rate also declined with increasing effort (decreasing distance), indicating that the range of hunting effort (i.e. search labour) for this species corresponded to the right-hand side of Figure 14.5(C); that is, all through the study area, the search labour was greater than what would lead to maximum sustained yield. On the other hand, category 2 species, such as nocturnal curassow *(Nothocrax urumutum)*, show a harvest peak at some intermediate level of search labour, suggesting that the variability in search labour in the study area for this species spans the left as well as the right side of Figure 14.5(C). For category 3 and 4 species, finally, harvest rates increase with increasing search labour, suggesting that the variability in search labour in the study area span only the left side of Figure 14.5C meaning that not even in the vicinity of the village, where the levels of search labour per unit of area and time are highest has the hunting brought the density of these species below the level providing maximum sustained yield.

Continued

Box 18.1 Using Spatial Patterns to Evaluate Hunting Impacts—cont'd

Things are, however, a little bit more complicated than this because animals actually do move around. For example, the harvest peak of nocturnal curassow occurs in the hunting effort zone 6 (Figure 18.5), but this may depend on a surplus biological production occurring further away, such as in zone 5 or 4. To understand such source-sink dynamics in detail, it is necessary to study also the movements of individual animals. There are now techniques available using collars with global positioning system receivers that are feasible at least for studying large mammals (e.g. Tobler, 2008), but still very little is known about the movements of game animals in lowland tropical forests.

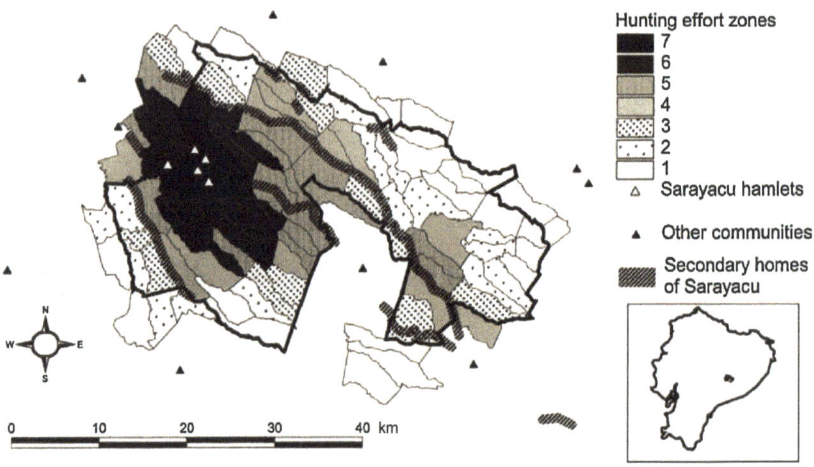

FIGURE 18.5 The spatial distribution of search labour or "hunting effort" in the hunting grounds of the community of Sarayaku, described in Chapter 4. The darker the shading, the greater is the hunting effort, which ranges from 8 km/km^2 per year, in the zone least frequently visited by hunters, to 285 km/km^2 per year in the zone that is closest to the settlements and subject to the most intense hunting. *From Sirén et al. (2004).*

count of the stumps of cut petioles that remain attached to the trunk after harvest revealed that harvest levels near settlements are low and, contrary to the case of hunting, gradually increase with distance (Sirén et al., 2013b). This continues, however, only up to a certain point. Then the level of harvest suddenly drops to zero, and beyond that point the *wayuri* stands are in their natural state, not having ever been subject to harvest within living memory. This difference in the spatial pattern of harvest seems to depend on the fact that different types of costs (Chapter 14) are involved in the harvest of these two distinct types of resources. For hunting, the costs of searching for conspicuous prey species are very significant, whereas palm leaf harvesting, on the contrary, involves very little search costs. This can theoretically be explained with the help of a model

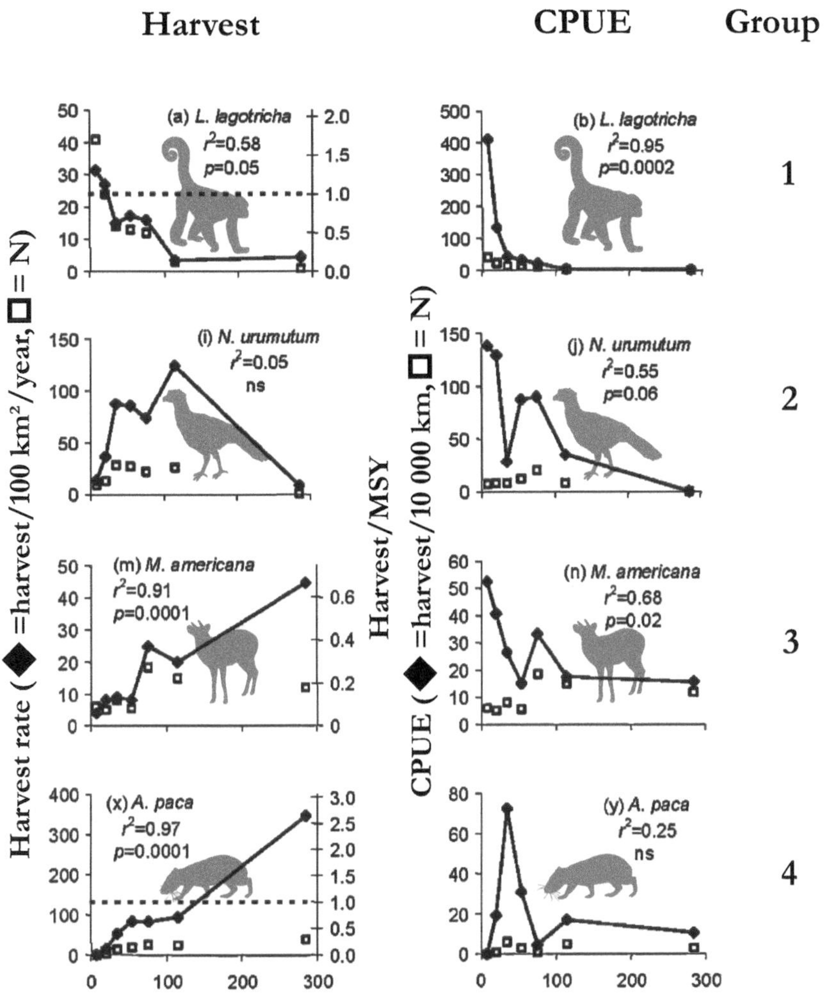

FIGURE 18.6 Harvest rate (left) and catch per unit of effort (CPUE), which is a proxy of resource density, (right), for a few representative species. The horizontal axis represents hunting effort, and the seven data points on each graph represent the respective average value of the respective seven hunting effort zones shown in Figure 18.5. Based on these spatial patterns of harvest and CPUE, conclusions can be drawn regarding the impact of hunting on the respective species. It is useful to compare the graphs on the left side with Figure 18.5C, and the graphs on the right side with Figure 18.5B. *Modified from Sirén et al. (2004).*

that is similar to the bioeconomic model presented in Chapter 14, but it also incorporates the notion of different kinds of costs, as well as a spatial dimension (Box 18.2). The phenomenon that harvest mostly takes place in a narrow band between areas that already have been depleted of a resource and areas that are still untouched has also been observed for timber trees, particularly big-leaf

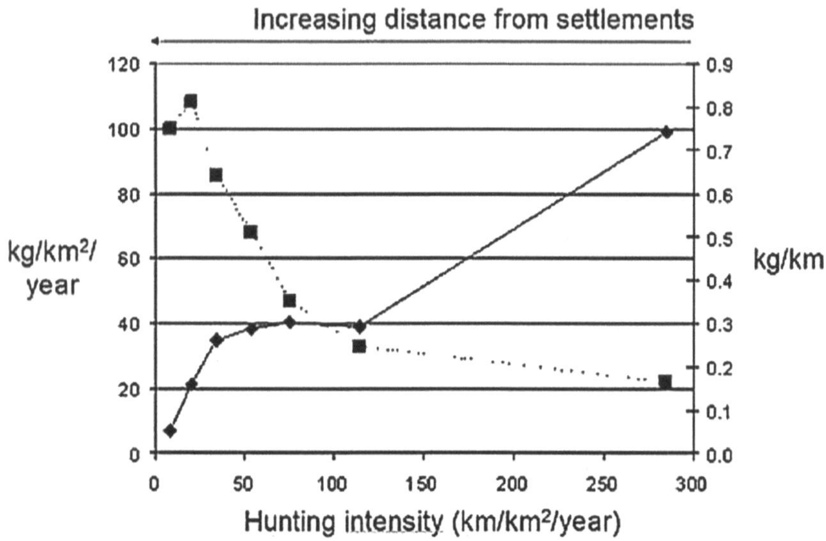

FIGURE 18.7 Harvest level per unit of area and time (solid line, left y-axis) and catch per unit of effort (CPUE) (hatched line, right y-axis) of wildlife hunting in the comunity of Sarayaku, described in Chapter 4. As the distance from settlements increases, the harvest level decreases, but CPUE increases. *From Sirén (2004).*

Box 18.2 A Spatial Bioeconomic Model

The following model explains how different spatial patterns of harvest can appear. The model has a one-dimensional spatial structure, consisting of a large (infinite) number of linearly spaced infinitesimally small patches, thus mimicking the situation where people live along a relatively straight river or road, surrounded by large forests, and travel into the forest in perpendicular angle from the river or road in order to harvest. It is assumed that there are no other costs than the opportunity cost of time, such that the total cost of harvest equals the opportunity cost of time multiplied by the amount of time devoted to harvest labour. This labour, in turn, is considered to consist of three different components: searching, handling, and transporting. Harvest is assumed to be the product of resource density, search time, and the catchability coefficient, as in the non-spatial model in Chapter 14. Total handling time (e.g. felling trees, cutting leaves, pursuing or eviscerating animals) is assumed to be linearly proportional to the total amount harvested. Total transport time, finally, is assumed to be linearly proportional both to harvested amount and to the distance from the origin of harvesters, such as their home village, to the point where the resource is harvested. This latter implies that harvesters are assumed to always return carrying a full load, which is not always the case in reality. However, for the current purpose of the model, it is justified to make this assumption in order to keep the model relatively simple. The price is assumed to be exogenously determined (i.e. fixed), as is the case on a perfect market, and resource growth is assumed to follow the logistic model presented in Chapter 13. Finally, open access

Box 18.2 A Spatial Bioeconomic Model—cont'd

harvest is assumed, leading to rent dissipation and a bioeconomic equilibrium where resource growth as well as profit is zero.

The total cost of labour in each patch is

$$C = c_{\text{time}}\left(S + \frac{dH}{v_t} + \frac{H}{v_h}\right) \tag{18.1}$$

where c_{time} is the opportunity cost of time, S is the time spent searching in the patch, d is the distance to that same patch, H is amount of resource units harvested in the patch, v_t is the speed of transport, and v_h is the speed of handling. Solving the economic equilibrium $pH - C = 0$ (i.e. setting that income equals costs) yields the equilibrium harvest rate:

$$H = \frac{c_{\text{time}}S}{p - \frac{c_{\text{time}}d}{v_t} - \frac{c_{\text{time}}}{v_h}} \tag{18.2}$$

where p is the output price. Given that $H = qSN$, where q is the catchability coefficient and N is the resource population density, the population density at economic equilibrium is

$$N = \frac{c_{\text{time}}}{q\left(p - \frac{c_{\text{time}}d}{v_t} - \frac{c_{\text{time}}}{v_h}\right)} \tag{18.3}$$

The biological equilibrium (i.e. the state where net resource growth is zero) is achieved by combining the logistic equation (Eqn 13.2) with the harvest function $H = qSN$, and solving for N, which results in the equation

$$N = \frac{K(r - qS)}{r} \tag{18.4}$$

By combining Eqns (18.3) and (18.4), the equilibrium local search time, S, can be found, and then easily also the equilibrium local harvest rates, H, and local resource population densities, N. The equation that is of most interest to us now is that of the harvest rate in each patch, which is

$$H = \frac{c_{time} r v_t v_h \left(Kpqv_t v_h - c_{time}(v_t v_h + Kq(v_t + dv_h))\right)}{Kq^2\left(pv_t v_h - c_{time}(v_t + dv_h)\right)^2} \tag{18.5}$$

This equation is messy and there is no need to try to understand it in detail. The important thing is that this equation shows how the harvest rate in each patch depends on its distance from the origin or hunters (Figure 18.8). In addition, the factors determining the spatial distribution of harvest can be explored by first, solving Eqn. 18.5 for distance, d, in order to identify the distance d_{Hzero} at which harvest drops to zero:

$$d_{\text{Hzero}} = v\left(\frac{p}{c_{\text{time}}} - \frac{1}{Kq} - \frac{1}{v_h}\right) \tag{18.6}$$

Continued

Box 18.2 A Spatial Bioeconomic Model—cont'd

and then, taking the derivative of Eqn. 18.5 over distance, *d*, in order to identify the distance d_{Hpeak} at which harvest reaches its maximum value:

$$d_{\mathrm{Hpeak}} = v\left(\frac{p}{c_{\mathrm{time}}} - \frac{2}{Kq} - \frac{1}{v_{\mathrm{h}}}\right) \tag{18.7}$$

From this, it can further be shown that the higher are the catchability coefficient q and the carrying capacity K, the closer d_{Hpeak} gets to d_{Hzero}, meaning that the highest harvest rate is found at a distance very close to the distance at which harvest drops to zero, occurring in a narrow band between areas already depleted of the resource and areas still untouched by harvesters. In addition, if the price p increases, d_{Hpeak} as well as d_{Hzero} increase, meaning that the harvest front moves further away. The same happens if the opportunity cost of time, c_{time}, decreases, and also if the handling speed v_{h} increases.

This spatial pattern of a narrow harvest front is very characteristic for the harvesting of palm leaves described in Chapter 7. According to this model, then, this can be explained by the very high natural abundance of this plant (high K) and that the species is very easy to spot and does not hide or run away (high q).

In the case of hunting, on the other hand, as described in Chapter 4, the spatial pattern is very different. Although the abundance of game animals increases with distance, there is some game also near the village, and this is where most game harvest takes place. The harvest rate then gradually decreases with increasing distance. According to the model, this can be explained by the fact that wild game animals typically occur at low densities even when not subject to harvest (low K) and are difficult to hunt because they hide and run away (low q). It should be noted, however, that in contrast to the model assumptions, hunting is a multispecies harvest activity, which implies that the relationship between the amount of search labour or harvest effort, S, and the equilibrium harvest rate, H, is somewhat more complex than what this single-species model predicts (e.g. Rowcliffe et al., 2003).

Source: A. Sirén and K. Parvinen (in preparation).

mahogany *(Swietenia macrophylla)* in Brazil, until a moratorium on harvest of this species was issued in 2001. In that case, however, it was not a question of a stable equilibrium, as in the model in Box 18.2, but of a harvest frontier that was rapidly moving forward (J. Grogan, personal communication).

The spatial pattern of logging is essentially linked to road and river accessibility (e.g. Grogan et al., 2010; Veríssimo et al., 1998) and the effects of varying cost-distances accumulate over time. For example, the Amazonian river network has offered a historically somewhat stable accessibility to the forested landscape, enabling logging along many of the rivers for decades or even centuries (Barros and Uhl, 1995). The cumulative effect on the riverside forests has thus most likely built up over a long period of time as a combination of accessibility and demand for different timber species. Road building instead

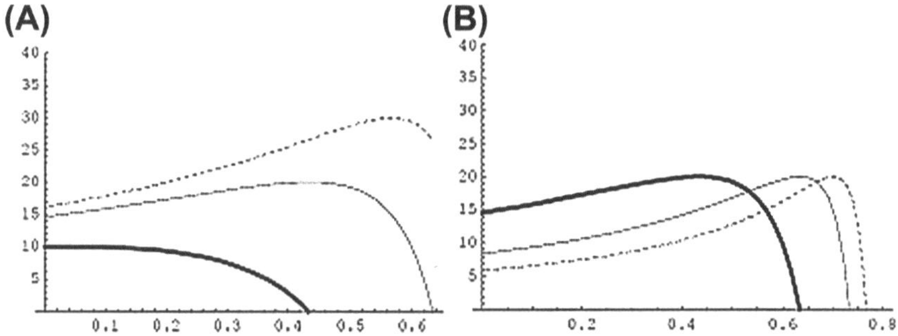

FIGURE 18.8 Numerical simulations of the spatial bioeconomic model described in Box 18.3. (A) Harvest as a function of distance at three different levels of carrying capacity (K). Bold = low K, thin = medium K, dashed = high K. (B) Harvest as a function of distance at three different levels of catchability coefficient (q). Bold = low q, thin = medium q, dashed = high q.

opens the logging frontier in a very short time; once good accessibility is in place by roads, it enables the extraction of all kinds of timber (see Chapter 14). Thus, time and space are intricately connected as they together determine many harvest patterns and associated effects in the natural system.

Compared to Figure 18.3, which depicts time-distances from the city of Iquitos in Peruvian Amazonia, Figure 18.9 shows the same region of Loreto to illustrate the distance of different forest areas from the closest river (Salo and Toivonen, 2009). This kind of distance measure illustrates the cost-distance for terrestrial transport of logs that must be hauled over skid trails and forest roads, after which the timber can be transported to the sawmills by river. Trees are felled only in those locations in which the combined cost of terrestrial and river transport of all the logs to the sawmills is not prohibitively high. For these kinds of calculations, it is important to take into account the general topography of the landscape and the soil characteristics in addition to the river system's flooding pattern, river types, and flow directions. Moreover, other species-specific variables also count, including the wood density (floating characteristics) and market price.

MOVEMENTS AND MISFITS

In the case of sessile organisms such as timber trees, it may for many purposes be enough to take into account the mobility of the harvesters and the harvested matter. The picture becomes more complicated, however, when one also takes into account the movements of the resources themselves. Such **mobility patterns** of harvested species have profound effects on how they are affected by extraction. There are basically three different types of mobility: home ranging migrations, and dispersal.

Home ranging consists of relatively frequent movements carried out by an animal within a fairly well delimited area that varies in size according to the species, the individual, and the environment in question. Considering Amazonian species of birds and mammals that are commonly hunted for food, their home

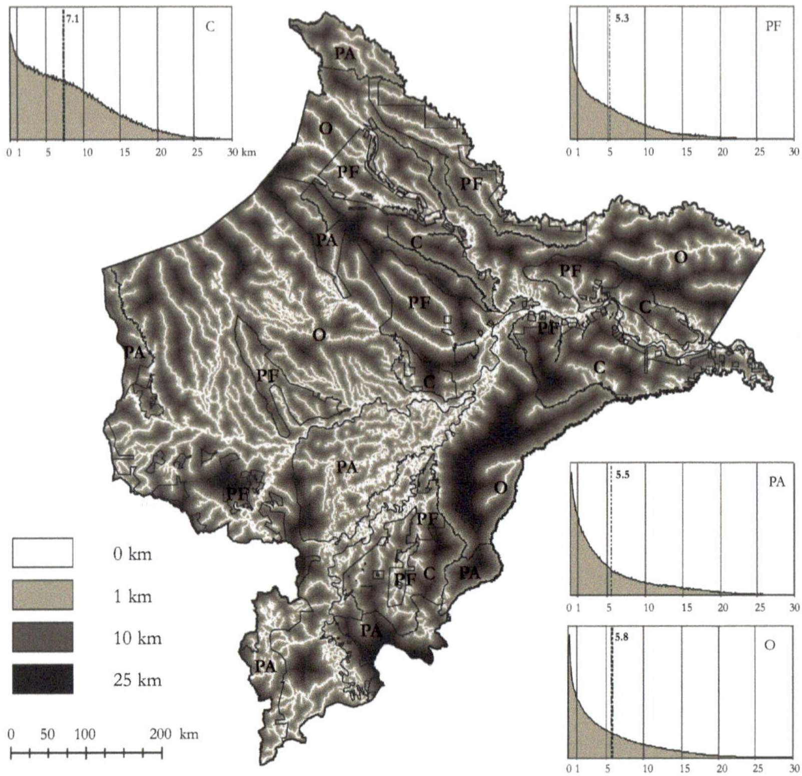

FIGURE 18.9 Euclidean distances (km) from rivers in Loreto (Peru), overlaid with concessions (C), other production forests (PF), protected areas (PA), and other areas (O). The graphs show the land area at different distances (km) from rivers in each land use assignment. Dotted lines show the average distances (km). *From Salo and Toivonen (2009).*

ranges are relatively small in comparison with the distance a human hunter can walk during a day (Sirén et al., 2004). For example, the rodent species paca (*Cuniculus paca*) has a home range of just a couple of hectares, whereas collared peccaries (*Pecari tajacu*) and large primates have home ranges extending over a few square kilometres. The largest terrestrial animal of them all in Amazonia, the South American tapir (*Tapirus terrestris*) has been shown to have larger home ranges, varying between 6 and 39 square kilometres, but this is still much less than what a hunter can pass through during a one day hunting excursion. Only one species, the white-lipped peccary, (*Tayassu pecari*) roams over much larger areas, having home ranges of from 22 up to 110 square kilometres or more, such that the species often is considered not to be home-ranging but instead migratory.

Migrations constitute movements back and forth over medium or long distances, often between specific types of habitats fulfilling different roles in the life cycle of the migrating species. Migration occurs among many different kinds of animals, including mammals, birds, reptiles, and fish. In the context of Amazonian wild species subject to harvest, it is particularly important among fish because most of the species that are important for subsistence and commercial fisheries are migratory (Chapter 5), with the arapaima (*Arapaima* spp.) being the most important exception (see Box 17.1).

Such movements of the harvested species can have profound consequences for resource abundance and harvest, as well as their spatial patterns. If the harvested species are migratory, this may reduce the need for harvesters to move themselves because they may just wait for migrating individuals to pass by. On the other hand, the harvest that is carried out in one certain place may have impacts that are felt also very far away, over the whole migratory range of the species in question, as in the case of Amazonian fisheries (Chapter 5).

Migratory birds are also often subject to important harvest pressure taking place in breeding and wintering areas as well as along the routes of migration. Bird hunting can be particularly intensive in migration bottlenecks or other areas of concentration (Kirby et al., 2008). Media reports in 2012 and 2013 from India, for instance, described how migratory birds are trapped using nets in order to sell them for human consumption. According to the non-governmental organisation Conservation India, local hunters using nets may annually capture and kill more than 100,000 Amur falcons *(Falco amurensis)*, migrating through northern India on their way from the species' breeding areas in Siberia to the wintering areas in eastern and southern Africa (Dalvi et al., 2013). The falcon's habit of roosting in immense flocks in the same places year after year makes it particularly vulnerable to this kind of hunting. Similarly, migratory birds are trapped with nets in many areas along the coasts and on some islands of the Mediterranean (Magnin, 1991), a topic that has attracted surprisingly little scientific attention during recent decades.

Dispersal refers to the one-way movement of an individual from the place of birth to its place of breeding, but it can also refer to subsequent movements that are of a permanent or semipermanent character. Dispersal occurs among plants as well as animals. Plant propagules such as seeds or spores may disperse with the help of, for example, wind or animals (see Chapter 13), including fish, as is the case for many tree species in the seasonally flooded forests of the Amazonian lowlands (see Chapter 5). Among animals, it is common that young individuals disperse considerable distances, particularly among social species such as monkeys, where the young leave their original group and may spend long times alone before they finally join another group (Glander, 1992). Dispersal may be density-dependent, such that individual animals move away from areas where they face a lot of competition for food or other resources to areas with a lower density of competitors.

Dispersal may also lead to so-called **source-sink dynamics** characterised by a net flux from an area with a birth surplus to an area with a death surplus, as shown in Figure 18.10. Such source-sink dynamics may result from natural environmental variability, such that there is a continuous flow of individuals from a 'good' habitat with a birth surplus to a 'bad' habitat with a death surplus.

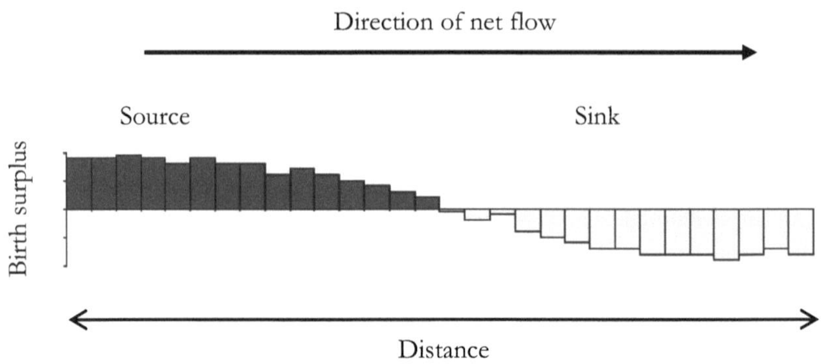

FIGURE 18.10 Schematic view of how source-sink dynamics can arise as a consequence of spatial variability in birth surplus.

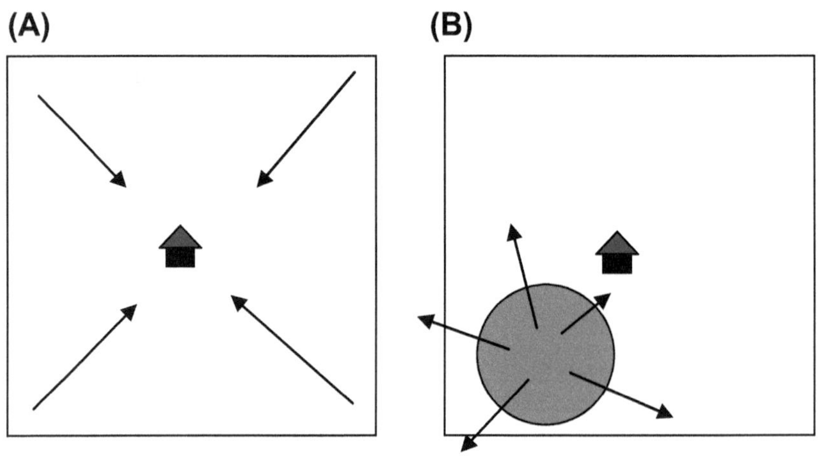

FIGURE 18.11 Two possible cases of source-sink dynamics as the result of human activity. (A) When hunting is conducted near the village, this results in a sink area to which new individuals come from the source areas compensating the loss. (B) Alternatively, a wildlife refuge, or 'no-take area' may serve as a source area from which individuals disperse out to areas where they can be hunted.

Sources and sinks may, however, also be the result of human activity. Commonly, high harvest in a limited area, such as the surroundings of a village, is sustained by an influx from the surroundings (Figure 18.11A). In this way, for instance, hunting levels that may seem unsustainable can in fact be sustained thanks to this continuous influx of new animals. However, there is also the risk that such harvest finally leads to depletion of the species in question, even in the unharvested areas, if dispersal rates as well as mortality rates due to harvest are too high. When populations in source areas deteriorate for any reason, their potential to supply surplus individuals weakens. Collapsing harvest levels may then occur for reasons that are not due to any changes in the hunting intensity in the main hunting area itself. When the reproducing areas are different from the areas where most of the harvest occurs, changes in the former may influence the latter, and vice versa (Jonzén et al., 2001). There are, however, also examples in which humans have created reserves where harvest of wild species, such as hunting, is prohibited, such that the reserves become source areas in which birth surplus disperses to the surroundings where harvest is allowed (Figure 18.11B).

The fact that many wild species move over large distances often contributes to a **spatial misfit** between ecological and social systems (Cumming et al., 2006). An example of this is presented by the difficulties that the people of the Sarayaku community encounter when they attempt to manage their fish resources in a sustainable way, abandoning dynamite fishing and reducing fishing with natural fish poisons (Chapter 5). In spite of these efforts, they do not see much result because many of the fish species migrate over long distances and their stocks are also affected by the fishing carried out perhaps hundreds of kilometres downriver, as well as in a number of communities upriver. Here, the spatial scale of the ecological system is much greater than the spatial scale of the social system. The spatial scale of the ecological system is that of the range of migrations of the fish, whereas the spatial scale of the social system is that of the part of the river system that is effectively under control of the about a thousand inhabitants of Sarayaku who regularly gather to discuss community affairs, including fishery management, but they rarely have the opportunity to discuss such issues with other inhabitants along the river. The opposite may also occur, namely that the spatial scale of the social systems is larger than the spatial scale of the ecological systems. An example of this is the cod fisheries in the Baltic Sea, where harvest quotas are set for large expanses of the sea, although cod stocks actually may be made up of several genetically distinct local substocks. The consequence of this can be that some substocks are exterminated, and nobody notices it as long as the other substocks are in good shape (Svedäng et al., 2010).

DISCOVERING ENVIRONMENTAL PATTERNS IN AMAZONIA

Amazonia, the region that was the scene for the stories from the forest floor in Part 2 of this book, is a good example of an area where knowledge about the relationships between the ecological communities and environmental parameters has taken huge steps during recent decades. However, many decisions about human activities have been made with little spatial fit with the boundary conditions formed by the natural environmental variability. For example, as described in Chapter 12, until just a few decades ago detailed thematic maps from the lowland forests of Peruvian Amazonia were virtually lacking when many important land-use decisions were made in the region. As a consequence, landscape features such as straight penetration roads through the green carpet of forest started to appear, but they were not based on documented processes of data collection and interpretation.

For decades, or even for centuries, there has been a general agreement among scientists about the extreme levels of species-richness that characterise Amazonian rainforests. However, well into to the 1980s, there were no tools available for efficiently mapping local-scale environmental heterogeneity embedded in the overall diversity and, as a consequence, information on Amazonia's biogeographical patterns remained scarce. For this reason some road lines and land-use delimitations were drawn using insufficient spatial information with the assumption that the forest was almost as homogeneous in its characteristics as could be suspected by observing the vegetation from a distance. Perhaps tools for detailed analysis were not even missed because it can be easier to rely on simple assumptions compared to getting engaged in complex spatial considerations in species-rich tropical forests where numerous more or less similar-looking trees co-exist, without clearly dominant species indicating any specific types of environment.

Increased availability of remotely sensed images has significantly contributed to the understanding of geographical settings in the Amazonian lowlands. **Aerial photographs** have been available from some selected areas from the

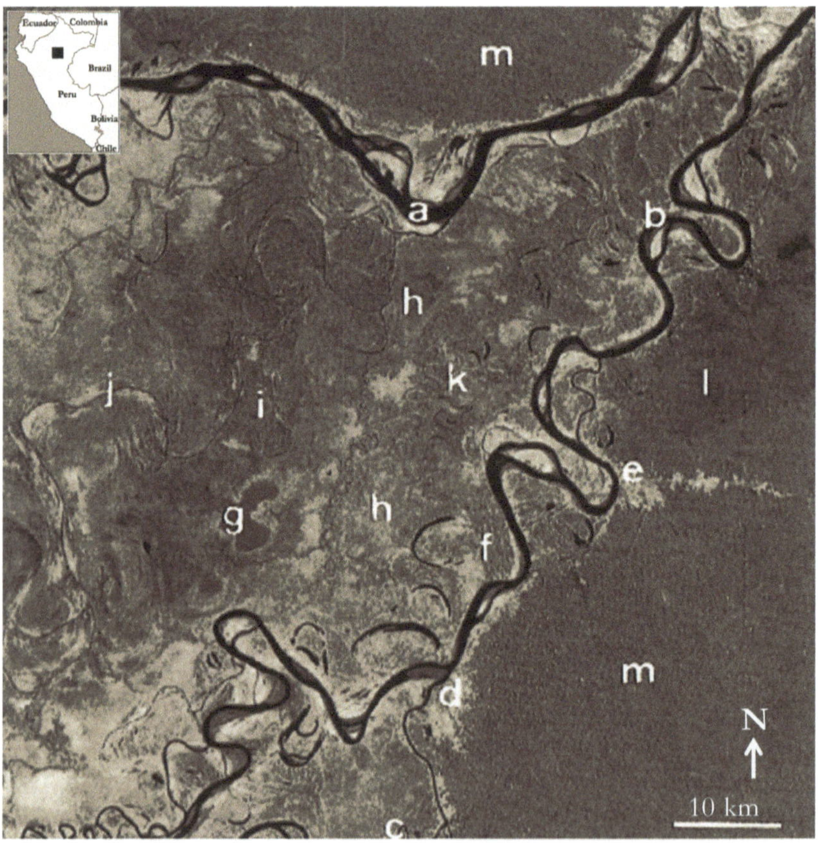

FIGURE 18.12 Satellite image (Landsat MSS) from Peruvian lowland Amazonia near the confluence of River Marañón (from west to east) with River Ucayali (from southwest to northeast). This picture clearly reveals mosaic-like landscape structure in the floodplain areas between the two rivers as well as the non-inundated lands that surround this area. Small letters on the image indicate particular vegetation types discussed in the paper by Tuomisto et al. (1993).

mid-1900s, but because individual photographs only cover small areas and mosaics of several images are difficult to use, many important habitat patterns still remained undiscovered. The real change in scientific understanding began in the 1970s when side-looking **airborne radar** and optical **satellite imagery** (Landsat satellites and their successors) become widely available and provided uniform data over very large expanses of forest. These images showed plentiful eye-catching patterns in the tropical lowland areas that were earlier considered to be more or less featureless. For example, in Peruvian Amazonia, such imagery revealed highly unstable river channels, waterlogged swamps, palm-rich forests, and non-inundated lands with a particular geomorphology and vegetation structure (Salo et al., 1986; Kalliola et al., 1991). Figure 18.12 shows an example of one such image. Finally, it was only after remotely sensed imageries became widely available from the Amazonian lowlands during the late twentieth century that mapping tools and evidence-based scientific interpretations about the spatial variation in the region's environments started to develop. This also contributed to an improved understanding of the dynamic factors that determine environmental patterns and species distributions. Box 18.3 illustrates how scientific research now has access to a number of efficient tools and working procedures to help verify spatial patterns in highly diverse tropical forests.

Box 18.3 Scientific Tools to Understand Spatial Variability in the Amazonian Lowlands

Thanks to technological advancement and generously open data policies, incredibly abundant and accurate remotely sensed spatial data are now available for most parts of Earth at little or no cost. For mapping of vegetation and landscape features, **multispectral satellite imagery** is useful. Such imagery records how the vegetation canopy, or the surface of Earth in general, reflects radiation of different wavelengths, including visible light of different colours as well as near- and mid-infrared radiation. **Digital elevation models** can also be very useful, such as those produced by the Shuttle Radar Topography Mission, which has almost global coverage, including the entire Amazonia. Using such data, one can clearly observe and study features such as geomorphology, river channels, or patterns of structurally distinctive vegetation. The interpretation of such characteristics is often a stimulating research phase for scientists. They also aid in mapping natural environmental variability in the rainforests of Amazonia and help to find places where the harvestable species thrive.

To confirm the veracity of the interpretations that are based on the use of space-born data, also more detailed survey methods need to be used. Oblique viewing and photographing from low-flying aircraft window is useful because many forest characteristics can be clearly distinguished when they are seen from the air. Even some canopy-forming tree species, particularly palms, can be identified. Good advance planning helps to make such flights cost-efficient so that as good a selection as possible of different kinds of vegetation patches or other places of interest can be checked during one mission. Also, more advanced airborne surveying technologies are increasingly available. For example, **LIDAR**

Continued

Box 18.3 Scientific Tools to Understand Spatial Variability in the Amazonian Lowlands—cont'd

(Light Detection and Ranging) technology can be used to measure forest canopy height which, in turn, can be as input data to compute above-ground biomass. Such information offers valuable opportunities for enhanced forest monitoring, management, and planning (e.g. Lim et al., 2003).

Although the Amazonian rainforest at first sight may appear to be uniform, **field research** has shown that there is, in fact, considerable edaphic and floristic variability. Information about these patterns can be reached through geographically oriented inventories, which are designed in such way that they efficiently reveal spatial variability in both plant species distributions and different environmental variables, such as topographical variations, soil types, soil moisture, and so on. Such edaphic conditions may vary between different sites, even between those that are relatively nearby, and this reality reflects to the distribution patterns of plant species. Congruent floristic patterns have been found in taxonomically unrelated plant groups, such as ferns, palms, trees, and the shrubs and small trees of the botanical family Melastomataceae. For example, in a study including 23 inventoried sites, soil differences explained 50% of the variation in floristic differences, whereas pure geographical distances explained 16% (Ruokolainen et al., 2007).

Botanical inventory of a tropical rainforest is time consuming and challenging because there are so many species present. Identifying most of the species found in a plot of just a couple of hectares may even take years. When the purpose is to map environmental heterogeneity in space, it is therefore practical to inventory only some particular groups of species that can be used as indicators of edaphic conditions. For practical reasons, understory plants are much easier and faster to study than canopy trees; in particular, ferns (Figure 18.13) have been found to be a suitable indicator group because they are abundant and relatively easy to identify (e.g. Tuomisto and Ruokolainen, 2005).

There is also a very high correlation between the assemblage of fern species and the spectral reflectances recorded by satellite sensors (Sirén et al., 2013c). This implies that environmental variability can be mapped over large extensions of rainforest by inventorying the fern flora in a limited number of sites (e.g. line transects of 2 × 500 m), comparing the relationship among fern species assemblages and satellite data in these sites, and then extrapolating the result to the whole area of study.

Spatial analyses are typically carried out with the help of special software for digital image processing and geographical information systems. Further steps in data analysis may involve various kinds of statistical assessments that aim to specify details of some particular patterns or to generalise the ground truth data by the means of spatial interpolation. The details of such analyses are each time specific to the actual information needs. For example, researchers may compare the patterns and magnitudes of wild species harvest with the distributions of the different types of nature occurring in the region, thus getting indispensable information to help management planning.

FIGURE 18.13 Ferns are very useful as an indicator of edaphic factors in neotropical forests because there are very many species of them but they are nevertheless relatively easy to identify.

The Amazon Basin has developed as a result of a complex geological history with many dynamic agents of major landscape changes over time (Chapter 3). As a result, the present-day environments are diverse at different spatial scales from very local heterogeneity to wide-ranging regional patterns. Figure 18.14 shows examples of some distinctive landscape structures and phenomena in Peruvian Amazonia. Much of this spatial variation can be attributed to the geological evolution of the region in the face of such forces as the uplift of the Andes Mountains, their erosion, and the transportation of sediments onto the Amazonian lowlands. Additionally, the leaching of nutrients from soils together with other soil forming and hydrological processes continue shaping the local environmental conditions and contribute to the overall diversity of the region. Awareness about the presence of large-scale landscape-structuring units and their significance for the vegetation patterns is steadily increasing as a consequence of accumulating scientific evidence (Irion and Kalliola, 2009; Hoorn et al., 2010; Higgins et al., 2011).

Examination of satellite images reveals indications of major landscape changes near the unstable river channels, while the non-flooded uplands seem to be less affected by such drastic changes. A meandering river induces forest regeneration in floodplains through the destruction of old forests at the erosional banks and the contrary process occurs on the newly deposited fluvial bars through primary succession (Figure 18.15). Despite these changes being particularly characteristic for many Peruvian Amazonian floodplains, this example also serves to make a more general point. Natural factors inducing events of ecological disturbance and succession processes create distinctive spatial patterns that then become inseparable constituents of the

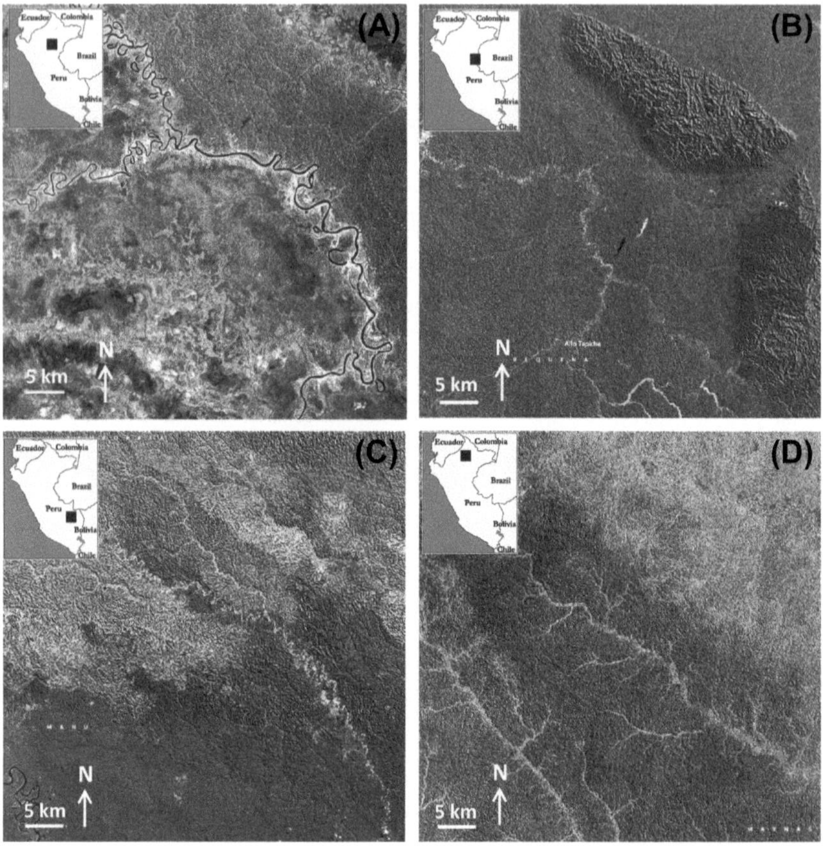

FIGURE 18.14 Landsat imagery from the Peruvian lowland Amazonia revealing details of landscape variations within the region. (A) Proximity between non-flooded (top right) lands and partially waterlogged floodplain areas. (B) Uplifting Serra do Moa mountains at the border between Peru and Brazil. (C) Bamboo-rich forests (bright) in contact with other types of non-flooded forests near the Manu National Park. (D) Two types of broad-leaved non-flooded forests with different characteristics (dark and light tonalities in the image) west from the city of Iquitos. *Source of imagery: BING Maps.*

region's landscape structure, including suitable habitats for some harvested species. Likewise, in the non-flooded areas of Amazonia there are other characteristic processes shaping the landscape with indisputable biological consequences; as an example the slow but unavoidable leaching and weathering impoverishes recent superficial soil layers and those deeper-lying soil layers that have become locally exposed. When soil layers of different geological strata become exposed, they often have different mineral constituents. This can result in a mosaic of different kinds of soils within the same region that, in turn, determines on a smaller scale much of the available soil nutrients, ecosystem productivity, forest structure, and species composition of flora and fauna. Understanding such patterns can have direct value for the harvest and management of wild species, but perhaps this kind of knowledge most often

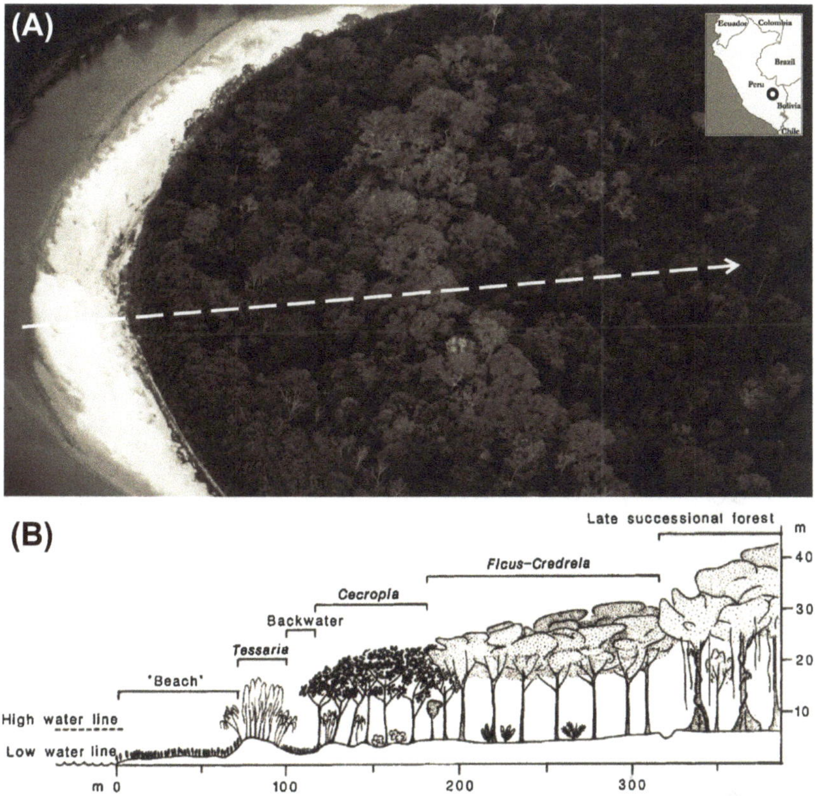

FIGURE 18.15 Example of a dynamic landscape forming process inducing age-structured pattern of forest. (A) Oblique view of the meandering River Manu in Peruvian lowland Amazonia. (B) Riparian successional zones in a meander tip along a transect line. *From Salo et al. (1986).*

affects harvest indirectly, as it can form the basis for other, often more intensive, human activities shaping natural landscapes.

SCALING SPACE AND TIME

Like Amazonia, all other places on Earth also have their unique characteristics resulting from complex combinations of co-occurring factors. This implies that the many of the points that were made above are of a universal nature on a level that surpasses local details. Climate, geology, geomorphology, soils, hydrology, and vegetation all have their say, as do the various kinds of human activities that also shape particular characteristics of a specific place. Despite local-scale heterogeneity, in Amazonia as elsewhere those places that are close to each other tend to have similar characteristics whereas those areas that are further away tend to be increasingly unlike. This reality was referred to as the first law of geography by Tobler (1970): 'Everything is related to everything else, but near things are more related than distant things.' This means spatial dependence and co-variation of a set of specific variables are associated with different

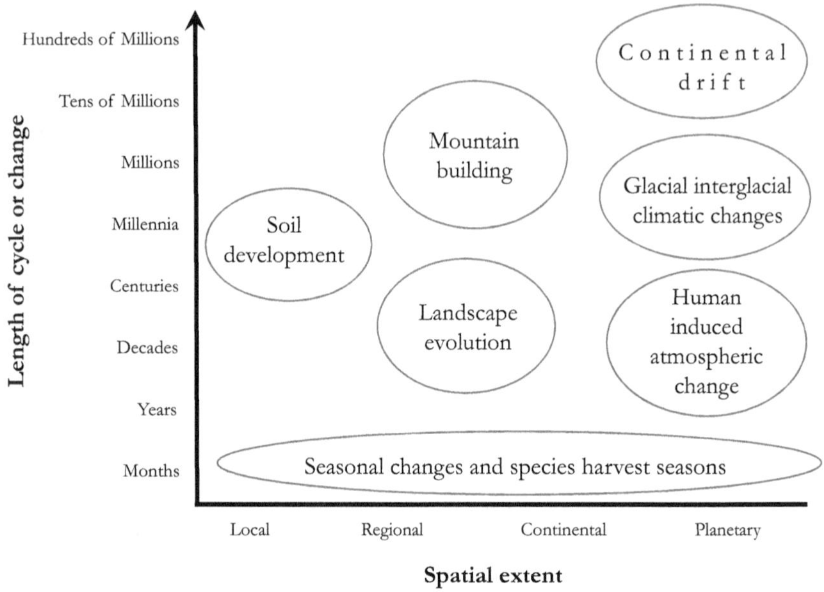

FIGURE 18.16 A schematic diagram illustrating the scales of space and time involved in a set of physical and biological processes important in nature and for humanity. Together with an infinite number of other ongoing processes, their co-occurrences and interactions shape the Earth.

geographical locations. For example, many nearby harvest situations resemble each other in terms of their provision from nature and the local knowledge and technologies that are in use. This theme is also critical in the statistical analysis of geographical phenomena: occurrences in adjacent places are not fully independent from each other.

This takes us to the issue of scales. Many features on Earth look different when they are addressed in different spatial scales, from the global and continental to the local level. Every place is inevitably embedded into an innumerable number of inseparable space realities and, moreover, space is tied to time as different processes appear on different scales of time (Figure 18.16). No species distribution pattern, no habitat and no landscape structure remain the same over time. Understanding this reality helps one to realise how human actions also are occurrences among other mechanisms, all influencing the world concurrently yet to different degrees. Considering these interactions and the transactions between humans and the natural environment takes us close to the study framework called **time geography**, which is a conceptual framework for understanding human activities and constraints (e.g. Pred, 1977). This may sound self-evident, but the complex settings and constraints that space and time make up together, over all of their different scales, are most critical to almost any specific issue of wild species harvest. When the practical-level activities are as local as they usually are, then the wider picture may not become evident without a particular notion on the roles of different scales of space and time.

Chapter 19

Legacies from the Past and for the Future

THE MOMENT OF HARVEST IN THE FLOW OF TIME

The past can help us to understand the present in many important ways. As a consequence of events in the past, we inherit resource systems with different kinds of imprints from the natural and human processes that have influenced them. For example, past harvest or other human interventions with nature may have altered the resource stocks affecting the abundances and distributions of species. Humans may also have changed the resource base through purposeful activities, such as management. Similarly, if people have left their imprint on nature, past developments obviously have left their traces on our present society. Harvest systems at various levels of human organisation, from subsistence livelihoods to more stratified socioeconomic structures, have evolved through time, finally giving rise to complex markets, institutions, and specialised incentive structures, governance, and power relations. Moreover, what has been learned in the past has profoundly influenced today's knowledge base and the culture at large. In total, everything that happened in the past has left its legacy in today's world – as subtle it may often be – and today's decisions and actions similarly shape tomorrow's world.

From this perspective, the ways in which societies have evolved in relation to wild species resources form an accumulated body of **legacies**, seen both in the state in which we inherit the provision from nature and in the shared heritage of sociocultural practices and values related to wild species harvest. Sometimes it is possible to identify specific conditions, processes, and events in the past that have contributed to the current state of the **provision from nature** or that have shaped the **sociocultural heritage** of wild species harvest. Legacies are found in relation to the harvester's transactions with nature and society – and for that matter, in relation to everything that has been dealt with in the previous chapters in Part 3 (Figure 19.1). Because any case of wild species harvest therefore requires an analysis of its temporal dimension, the present chapter aims to systematise the exploration of the legacies relevant for diagnostic purposes.

As described, harvest possibilities at the present are conditional on a series of past natural circumstances, processes, and events, which are often related to consecutive human decisions and actions. This approach considers harvest as a momentary episode within a continuum of time (Figure 19.2), in

Diagnosing Wild Species Harvest. http://dx.doi.org/10.1016/B978-0-12-397204-0.00019-X

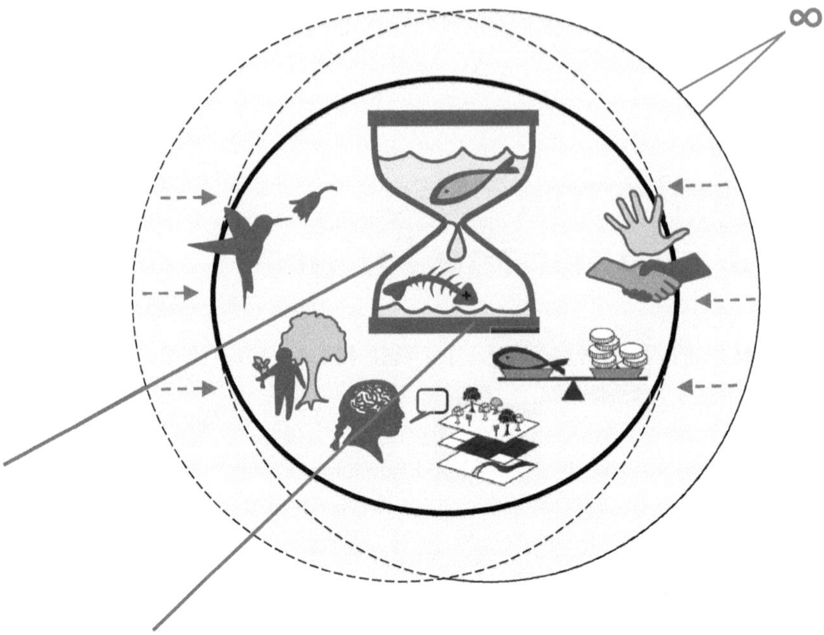

FIGURE 19.1 Time cuts through the complex multidimensional resource system. The present circumstances constitute legacies from a variety of past developments, and whatever comes to occur at the present will form part of tomorrow's legacies.

which there are three distinct time domains to consider: the **past** from which we receive our legacies, the **present** in which we make decisions and act, and the **future** that will inherit whatever the present produces. Landscapes, ecosystems, communities, habitats, and harvested populations have undergone changes over time, with some of these changes taking place over geological timescales, some on scales stretching from prehistory to historical times, and yet others being very recent. As shown in Chapter 18 (Figure 18.16), space and time are also tightly interlinked because slow and large-scale processes co-occur with local and short-term events. Likewise, harvest – even if very local and short in duration – will have at least some after-effects. The future is nebulous but it will nevertheless become the present as time passes.

When legacies are evaluated in any concrete situation of wild species harvest, then the focus must be in a clearly defined ecological system and flow of provision from nature on one hand and a corresponding social system with its dynamics on the other. The legacies relevant to the set of circumstantial factors that describe the present harvest refer to any such concrete events in the past that may have had influences on the options available today. As a very simple example, someone who caught a fish yesterday left a legacy that affects you as you go fishing today because there is one fish less out there that

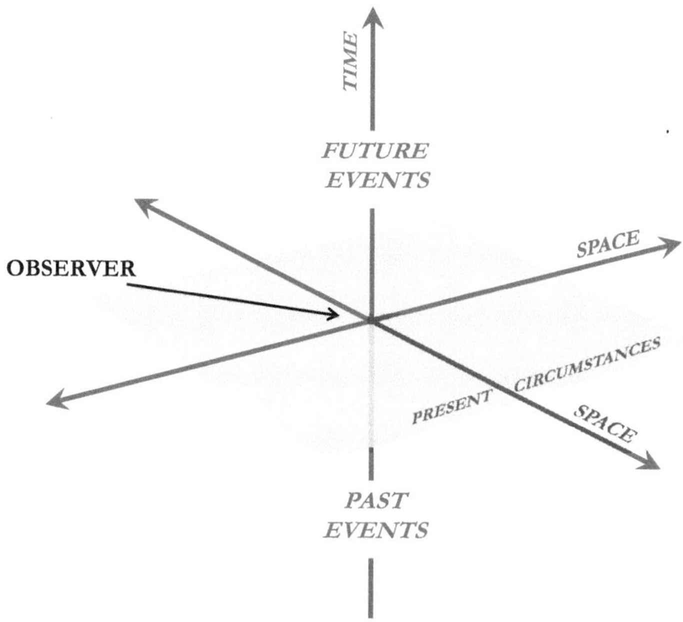

FIGURE 19.2 Present circumstances and their relationship to events in the past and in the future. Time in a way cuts through the present circumstances with all of its natural and social systems, of which extractive economies form part. As time passes, the present moment becomes part of the past and the nearest future will, in turn, be the present for a short moment.

you can potentially catch. The situation is however relevant only if you fish near the place where the other person was fishing yesterday or if you both try to sell the fish catch on the same limited market. Otherwise, yesterday's harvest by someone else does not affect you. If the time scale is extended by a few years, somebody else's past fishing may still affect you because it may have reduced the number of reproducers and thus the recruitment of new fish. In this case, the effect is not limited to the very near vicinity of the place where the fish was caught, particularly if the fish belonged to a migratory species. In that case, fishing occurring at any single place within the migratory range of the fish can have an impact that is distributed over the entire range.

Similarly, felling a tree has an immediate effect only at the place where the tree was standing and in the gap created when the falling tree crushes the understory. With time, however, the absence of the seeds that the tree would have produced if it had not been cut down will affect the species' regeneration also in the surroundings. These examples help to illustrate the point that relevant legacies are always specific to a particular situation and a location. In practice, there is thus a limited (yet still infinite in theory) set of past events that may have impacted the present circumstances in any perceivable way and, in the same way, the present actions can only influence a limited array of future settings and harvest possibilities in a discernible way.

To illustrate this point schematically, we have borrowed the concept of a **time cone** (or light cone) from general physics. It helps us to define the boundaries of the causal past and the causal future of a given point in time and space. The causal past means the things in the past that can be identified as having causally affected the present, whereas the causal future is that future that can potentially be shaped by present things that may affect the circumstances ahead in some specific way. Accordingly, only those past events are relevant as legacies for a specific location and situation that may have theoretically influenced its present state (Figure 19.3). Considering the closest time in the past that has been left behind, possible events influencing the present are most commonly events that took place nearby. But the further back we go in time, the broader the possible space becomes from which legacies may have left their traces in today's circumstances. For example, the centuries-long history of booms and busts in the use of wild species resources in Amazonia after its world market integration (Chapter 3) can be relevant for many current harvest situations in the area – if not through any direct causal mechanism, then through some existing yet incidental traces that nevertheless occur in the region's provision from nature and in its cultural landscape. In the same way, the **after-effects** of the present actions are **space-time limited**.

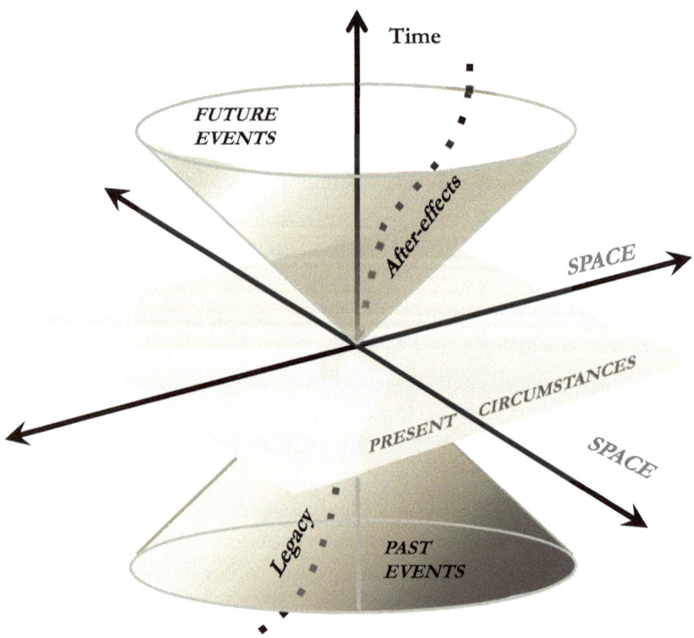

FIGURE 19.3 Schematic presentation of the time cone to illustrate how legacies are formed. Events from the near past may have influenced the present circumstances only if they occurred near the locality, while events from further away may have influenced if they occurred longer times ago. The cone that opens to the future can be interpreted using a similar logic. The figure also shows an imaginary timeline for a set of occurrences, such as the harvest path of a berry picker.

The legacies that directly affect wild species today can to some extent be understood thanks to the records left by people in the past (Reitz and Shackley, 2012), which can be classified into four broad categories; biological records, artefacts, ecofacts, and documentation. **Biological records** can consist of very old fossilised plants and animals, pollen, and phytoliths. They can, however, also consist of living organisms, such as, trees bearing scars in the bark as vestiges of human actions centuries ago (Östlund et al., 2009; Figure 19.4), or the *wayuri* palms (*Pholidostachys synanthera*) that bear the stumps of petioles cut up to a few months before (Chapter 7). **Artefacts** are things made by humans, such as arrowheads, hunting pits, fish traps, and alike. **Ecofacts** are the garbage produced during everyday life, such as fish or animal bones, shells of molluscs, seeds, pieces of wood, and a variety of other items that commonly have accumulated at sites of human habitation ever since prehistory (Kirch, 2005; Dincauze, 2000). Finally, there is purposeful **documentation** by humans, which can take many forms; the earliest forms are cave paintings and other prehistoric

FIGURE 19.4 The inner bark (i.e. the thin layer of tissue between the wood and the hard outer bark) of Pine (*Pinus* spp.) was long a staple food among Native Americans as well as the Sami people of the Nordic countries. The practice ceased in the nineteenth century, and, in some places, early twentieth century. Old pines (*Pinus sylvestris*) like this one in the Tjeggelvas forest reserve in northern Sweden still constitute biological records of this practice. *Photo: Lars Östlund.*

art, which often were inspired by hunting, fishing or other forms of wild species harvest (McDonald et al., 2012). For the last few centuries, there is often written documentation available, such as travel accounts and maps by missionaries and explorers, and, more recently, official trade records, newspaper or magazine articles, and research reports. Photographic records can also provide important information about, for example, landscape change. Similarly, oral history, such as accounts of self-lived experiences or narratives passed on from generation to generation, can be useful sources of information.

The further back one goes in time, the more problems arise regarding the availability as well as the quality of data. An ancient pile of fish bones may tell what species people ate, but it tells little about the quantities of fish harvested. Historical source documents and oral history also are often imprecise when it comes to quantitative matters. In addition, historical information, in the form of textual and other kinds of source documents, is often accurately dated and thus temporally relatively precise, but it can be spatially very imprecise (Szabó and Hédl 2011). When it comes to archaeological remains, on the other hand, it is often the other way around: typically they have a very specific location but, despite the advanced technologies available today for their dating, there may still be a degree of uncertainty about the age of such remains. For example, if two events took place at the same location, it may be nevertheless difficult to achieve certainty about their coincidence or chronological order if they are dated very close to one another. Of course, even if temporal coincidence could be reliably demonstrated, it does not necessarily imply causal links between the two co-occurring events or things.

These are some of the issues that have to be addressed in order to better understand the past and thus also be able to create reliable baselines against which to adjust harvest activities in the present, as well as to predict future developments and create reliable future scenarios (Zu Ermgassen, 2012). Luckily, though, the recent past is often what has had the most important effects on the present, and for the recent past typically more data are available – if nothing else, at least in the form of the accumulated experiences of the resource harvesters themselves.

THE INHERITED PROVISION FROM NATURE

Harvest-based economies are in constant change because the interface between the natural and human systems is dynamic. Simple chronological assessment often helps to reveal trends that can provide important clues about the status of the resource and its potential for sustainable harvest. For example, if harvest levels or the populations of the targeted species have gone through drastic changes in the recent past, there may be reasons for concern. An increase in the level of harvest often has the consequence that the resource decreases in abundance – a process that can continue until a new equilibrium is reached. In the best case, that new equilibrium is at just a slightly lower resource density than it was previously, but in the worst case it can be at a very low level or even result in the complete extirpation of the harvested resource.

A recent decrease in harvest, on the other hand, may have any of a wide variety of causes, including decreased demand or the implementation of conservation measures. However, it may also be a consequence of decreased resource abundance due to previous overharvest. Information on the past abundance of resources, the extent of habitats, and any changes in habitat quality can help to evaluate the reasons behind the decline. However, if resource abundance or harvest levels have shown great fluctuations in the past, this may have to do with natural causes, such as climate variability or interactions with other species. In such cases, one must be careful about extrapolating current trends into the future.

Possible reasons behind different levels of harvest at different moments of time can include, for instance, changes in harvest technology, markets, institutions, and human demography. In general, if the levels of harvest of a given wild species have remained constant (i.e. have been sustained) over very long time periods, it is often reasonable to conclude that the level of off-take does not significantly exceed the regeneration of the resource in the long term and that it will be possible to keep on harvesting the same amount also in the future. In other words, the harvest is likely to be sustainable. However, even if the quantity of wild species goods entering society had remained constant in time, the sustainability of harvest sometimes can be an illusion. This is the case, for instance, when the output level of stable wild species goods is based on harvest in a forward-moving frontier (as seen in Chapter 9) or when harvest moves from one set of species to another set of species with similar characteristics and uses (Chapter 5).

The legacy of intensive harvest may, for example, consist of shifts in terms of both the spatial distribution of harvest and the species composition of harvest (Figure 19.5). Often, the large, conspicuous, and high-value species are those that disappear first and get replaced in harvest by species that are smaller or shy or less profitable to harvest. This happens first in areas of high accessibility, whereas it happens only later in more remote and less accessible areas. As a consequence of such a process, the species that are harvested today in some areas may be those that were rejected by harvesters in the past but have increasingly come under extraction because the primary species of interest was depleted.

Examining existing records may also allow assessment of the causes of past changes in the intensity and character of harvest. **Technological changes** have often had profound impacts on harvest practices, as also have changes of cultural, demographic, and economic character. One prime example of how drastic the effects of technological and cultural development can be is the case of seafaring. There are abundant examples of how species have disappeared from island environments soon after the first arrival of humans in their boats. In addition, because species on isolated islands often are endemic (i.e. occur nowhere else

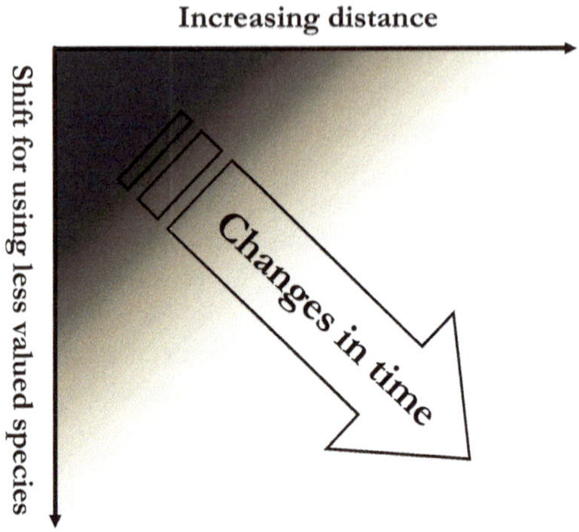

FIGURE 19.5 Harvest affects target species. With time, typically the most valued species are harvested at ever-increasing distances from settlements and transportation routes, whereas at shorter distances, ever less valued species are harvested.

on Earth), this has often implied not only local, but even global, **extinction**. Famous cases of overharvest include the sad stories of the huge flightless bird dodo (*Raphus cucullatus*) driven to extinction soon after the arrival of European sailors on the island of Mauritius in the 1600s (Quammen, 2012; Gooding, 2005) and the North-Pacific Ocean's Steller's sea cow (*Hydrodamalis gigas*) exterminated by 1768, only a quarter of a century after its scientific discovery (Turvey and Risley, 2006). In both cases, ruthless overharvest was most likely among the reasons, if not the main cause, of extinction. At least in the case of the dodo, however, other human activities, including land conversion and introduction of rats and various other new predator and competitor species, played a central role in the extinction of the species.

Another famous example of a human-induced extinction is the North American passenger pigeon (*Ectopistes migratorius*); it was once counted in the billions but was killed out in only a few decades by the beginning of the 1900s, with the last known individual dying in captivity in 1914. The species had been hunted by Native Americans and European settlers on a subsistence basis for a long time, but overharvest became a serious problem only after urban demand gave rise to commercial hunting along with increased hunter mobility due to railroad construction. All of this was reinforced by the development of more efficient hunting technology, particularly better guns (Soll, 2012). Contributing to the rapid decline and eventual extinction of this species was, however, also the massive habitat destruction through logging and clearing of deciduous forests.

Humans have, indeed, also favoured some species, whether consciously or unconsciously. The densities of Brazil nut trees (*Bertholletia excelsa*) in many places within the species' natural range are probably higher than would be just because of its natural regeneration. As was discussed in Chapters 3 and 8, this is because of anthropogenic enrichment by pre-Columbian inhabitants of those areas (Levis et al., 2012), either purposefully or coincidentally, or both. On the other hand, the virtual eradication of some species in the past can also be reflected in the present resource base. This is how, for example, the eradication of large predators in combination with favourable habitat changes and efficient game management measures have made elk (*Alces alces*) very abundant in the Finnish forests.

Continuous harvest during prolonged periods of time can also affect the gene pool of the harvested populations (Chapter 13). For example, harvest of phenotypically desirable animals from wild populations imposes selection pressures that may induce undesirable changes in exploited populations, such as earlier sexual maturity and reduced body size, the latter which is likely to negatively affect the potential for sustainable production and harvest of the affected species. This is not something that may happen in the future merely hypothetically. On the contrary, it has most likely already happened to many populations of harvested species (Allendorf and Hard, 2009).

Although the growth of individuals and the regeneration of populations can take any time from days to decades, the felling of a tree, the capture of a fish, or the picking of a fruit often takes only a very short while compared to the time the harvested biological matter subsequently spends in human hands during transport, processing, and further use. Similarly, the effects of harvest are longer lasting than the harvest itself. Because the effects of harvest are often seen only after a **time lag**, it may be difficult for the harvesters to perceive the long-term consequences of their harvest. This is true both for the ecological effects and the effects on the future benefit streams that can be expected from harvest.

THE SOCIOCULTURAL HERITAGE OF HARVEST

In addition to the resource base that today's harvesters inherit from the past, different kinds of sociocultural legacies have also developed through multiple human generations and contribute to the present resource-harvest-consumption systems. These legacies are the result of long-term adaptive development and **cultural evolution** taking place on the basis of a variety of human behavioural and societal processes, and they represent the sociocultural heritage of harvest. This involves not only the harvest activities in themselves but also the surrounding societal circumstances that may favour or discourage some particular forms and practices of harvest. This **heritage** in the form of a shared cultural capital may appear self-evident in such way that one may not come across to think about it. Even if there were strong current traditions of wild species harvest and use, it may also be difficult to reveal the specific causal paths from past practices and traditions to present-day harvest. Things just have acquired a certain form, some may say.

In the same way as a plant or an animal species becomes a resource only after its utility is discovered, its status as a resource may fade out as a result of **sociocultural change**, including shifts in technology. For example, many natural fibres are today used much less than before, not necessarily because of overharvest and resource exhaustion but rather because strings made of plastic or other artificial materials are both cheap and handy. The globally increasing opposition to whaling is a different kind of example of how sociocultural changes affect wild species harvest; the legacy of whaling traditions can be either dismissed as cruelty or transformed into a whale-watching tourism industry (Neves-Graça, 2006). Sometimes the sociocultural legacy of resource use can also reinforce and enhance the **demand** of specific wild species goods through, for example, urban-rural links (Cocks et al., 2011) or preferences linked to ethnic culture (van Vliet et al., 2011). The story of the açaí fruit (*Euterpe oleracea*) – a rural staple in the Amazon estuary that was first taken to the city by rural-urban migrants and then became a booming urban commodity – illustrates how deep-rooted practices of harvest and use can adapt and transform through time and through the **rural-urban interface** that is permeable for many legacies of wild species harvest (Chapter 11).

Changes in demand for wild species goods imply changes in the harvester's transactions with nature and society. How these transactions have developed over time can be explored by focusing on institutions. Particularly, as the demand for wild species goods has increased, different kinds of institutions have been devised – sometimes more, sometimes less. The emergence and subsequent change of wild species-related institutional arrangements can tell much about how the importance of harvest in society has developed over time. This can be seen in the changing policies and incentive structures including property rights, laws and other regulations, and markets. Whose voice has prevailed and whose interests have been written into policies and laws in the past forms an implicit legacy for the present regulatory apparatus. For example, as seen in Chapter 9, the Peruvian forest legislation bears an important imprint of scientific forestry (Dourojeanni, 2009) – sometimes seen as a valuable expert contribution and sometimes as an intervention by outsiders lacking local insight (Salo et al., 2013).

Commitment to rules and overall law compliance is related to the consistency and stability of institutions. Constantly changing or poorly enforced rules can make it difficult for people to follow the formal harvest regulations and, furthermore, decrease their willingness to even try to do so. In the long term, this kind of **institutional instability** can become the business as usual and as such it can constitute a powerful legacy that affects any future policy or regulatory reform (Smith et al., 2006). Similarly, if the harvest system is traditionally based on **informality**, it may be difficult to implement such reforms that essentially require formal actors to function, as has been witnessed in the Peruvian forest sector reform (Chapter 9; Salo et al., 2013).

Historical legacies can also be felt in the ways in which people become organised or self-organise in different places and in relation to different resources, and in the ways in which they depend on wild species resources. The roles of patronage and intermediary merchants are often based on robust forms of adaptation in local environments and societal settings (e.g. Laird et al., 2011; Sears and Pinedo-Vásquez, 2011). Many institutional arrangements have the objective of distributing benefits obtained from wild species harvest, often implying relevant historical legacies that can be seen in the present incentive structure, its efficiency and legitimacy, and governance at large. For example, in some places it is natural for people to form cooperatives, whereas in others they prefer private microenterprises. Different mechanisms for collective ownership and joint economic activity or different incentives may be feasible in different contexts depending on historical tradition. For example, the ways in which Brazil nut collectors are organised and the ways in which the property rights are established varies in different countries as well as across social settings (Chapter 8).

Historically developed **ownership structures** also count. In Europe, for example, it is typical that there are many private smallholder forest lots instead of most of the forest being centrally owned by the state or other large landowners. However, not all smallholders are equal. Some may be farmers, whereas others depend more on the use of timber and other forest products. Increasingly, forest owners are also urban dwellers; some inherited the forest patch and value it as a family forest, whereas others may have merely economic motives for forest ownership. All of these different kinds of forest owners and their distinct backgrounds may require different incentive structures to promote, for instance, sustainability (Wiersum et al., 2005).

Different kinds of institutional cultures and histories in general also create boundary conditions for the governance of wild species harvest. A historical **culture of top-down policy formation** and command-and-control implementation, for instance, can form a legacy that limits local participation. When this is combined with persistently deficient harvester organisation, the result can easily be the exclusion of harvesters from important decision-making processes (Laird et al., 2011). Even the **legacy of colonial rule** can still be seen in the particular forms of land and resource ownership in many parts of the world (e.g. Larson et al., 2010; Hersoug and Holm, 2000), and the centralised ownership of forests common throughout the tropics is often the result of the national elites' views about public interest and common good on one hand, intertwined with more earthly desires related to private interests on the other (Larson et al., 2010; Larson and Soto, 2008). Any reform related to property rights institutions thus builds upon people's perceptions about this kind of historically constituted resource ownership and tenure, as well as about the institutions and power relations emerging therein.

A lack of awareness or biased views about the **temporal perspective**, or legacies from the past and to the future, can be a problem when looking for solutions that would respond to the demands of sustainability. There are several important perspectives to this issue, including the disparity between the long timescales many organisms need to grow and regenerate compared to the short duration of the act of harvest, the long-term effects of harvest in evolutionary processes, the rates of discounting and considerations of intergenerational fairness that affect the allocation of benefits over time, and the commitment that people show over time in relation to collective agreements. Legacies from the past with implications to present-day wild species harvest are studied by scholars representing a number of fields, including ecological, environmental, geographical, economic, institutional, cultural, and political history.

However, if we have sustainability and responsible **environmental stewardship** as goals, we not only need to reach back into history but also need to understand the after-effects of any present actions and take them into account in decision-making (e.g. Driscoll et al., 2012). Knowing about the past events helps one to anticipate future developments more reliably. In general, different kinds of decisions about harvest, management, and governance are made with distinct time horizons in mind (Table 19.1). As was described above, it is common that inconsistent policies and unstable institutions hinder the search for solutions to problems that develop over long timespans (e.g. unsustainable forest management practices). Likewise, rigid institutional arrangements that resist change over time may be unable to respond to such situations that would require constant adjustment of rules (e.g. small game management). The institutional arrangements and organisational structures of governance thus need to fit the scales of time inherent in the harvest and resource use systems they are intended to regulate (e.g. Cumming et al., 2006).

Wild species harvest, in any given case, happens in a particular place and at a specific moment of time; the exact moment in which individual organisms are harvested from the wild may be just a blink of an eye. However, as all of the chapters of Part 3 have shown, the ephemeral instant of any harvest decision melts together countless conscious and unconscious considerations that are all limited by a series of legacies from the past – and that all form part of the legacy the harvester leaves for tomorrow.

TABLE 19.1 Time-horizons of different levels of decision-making related to wild species harvest, management and governance. Strategic decisions are made with long time-horizons in mind while tactical and operative decisions are related to increasingly short time-horizons, respectively. As concepts, the time-horizons of strategic, tactical, and operative decisions are typically related to forestry (Bjørndal *et al.*, 2012) where the strategic planning can span even to centuries and tactical decisions can have time-horizons of decades. However, these concepts can be applied to all kinds of wild species harvest – only the length of the time-horizons vary.

Time-horizon of decisions	Level of decision-making: Harvest	Management	Governance
"Strategic" decisions Long-term (years to centuries)	Investments in technology and infrastructure; learning from experience	Ecosystem manipulation; adaptive changes in methods; education	Value-based policies; legislation with incentive structures; information system development
"Tactical" decisions Medium-term (weeks to decades)	Timing of harvest and sales, selection of harvest areas, what species to harvest	Removal of competitors or predators, planting, stocking; place and methods of harvest	Setting quotas, monitoring; harvest bans; harvest incentives; planning requirements
"Operative" decisions Short-term (sub-seconds to months)	Harvest or not, place and methods of harvest, transport of harvested matter, (pre-) processing	Quantities and kinds of individuals to harvest	Patrolling, control, seizure, bookkeeping

Part IV

Diagnosis and Action

Any discussion about wild species harvest is relevant only where there are ecosystems left in such a state that they can continue providing harvestable resources, and when there are uses and markets for the wild species-based products. Whereas the value of harvestable wild species in some places certainly is a threat to the integrity of the natural ecosystems, in others the same value has been shown to provide incentives for maintaining natural land cover. Overharvest is not just an ecological problem of extracting more than what nature is able to provide, but rather a complex tangle of interrelated mechanisms originating both in the inherent characteristics of the natural system and in the properties of the social system.

This is relevant to anyone whose attitudes or actions affect wild species harvest, not only in tropical forests but also in a more general sense regardless of place. People's interests, incentives, trade-offs, and possibilities to take effective action vary greatly, and although actors are often clumped together as stakeholder groups, there is also a great internal diversity within any such group of people involved. We hope that this last part of the book will give the reader both a practical framework for systematical analysis of wild species harvest in any specific context, and a wider perspective to this phenomenon at a universal level.

Preview to the Chapters of Part 4

The three chapters of Part 4 draw on the stories from the forest floor as well as on the theoretical review to present the Diagnosing Wild Species Harvest Procedure (DWiSH Procedure), and discusses some implications of the lessons learned in the previous parts of the book.

Chapter 20 puts the core message from all the preceding chapters together into the form of a simplified conceptual model. The DWiSH Procedure is presented as an overarching approach to analyse any type of wild species harvest. Its application involves the use of various sources of information, their systematisation and assessment, and the use of a number of helpful methodologies. For example, the use of mathematical or other models in this context can consist simply of drawing graphs by freehand with pencil and paper in order to structure one's thoughts, and to communicate what one knows, believes, or knows that one does not know.

Chapter 21 asks the question 'Who?' As is common in human society, preferences and benefits, as well as obligations and costs, look dissimilar when considered from different people's perspectives. It is easier to say how things should be from a certain point of view than it is to find ways to fulfil everybody's interest. This implies that pinpointing visible and possible problems is not enough if attempts are not made to identify how these are related to particular actors and their roles in wild species harvest. Successfully unveiling the roles of different stakeholders with their respective interests and actions facilitates the endeavour to propose actions where wild species harvest could be better accommodated into the global quest for more sustainable use of natural resources, not only in the tropics but also worldwide.

Chapter 22 takes readers back to the fundamentals of renewable natural resources, in this case in the form of the diversity of species that can be useful to humans, as well as the big challenges of our times such as environmental degradation, living standards, and sustainable development. To what extent is the harvest of wild species related to these? Some may consider that this book is about a vanishing human profession, characteristic of "underdeveloped" areas and conditions, merely having significance as a sideline activity in the "developed" world. In this kind of view, anything from nature that is really important for human beings should finally be produced in plantations or by growing animals in captivity, or be replaced by synthetically produced substitutes. Others may argue that evolving cultures of wild species harvest present, in as many places and conditions as possible, are a way to safeguard the future of humankind in case our globalised and highly interlinked socio-technological systems were to collapse. There are also many other possible views about the relevance of wild species harvest in our world. We do not leave readers with a roadmap about how things should go but rather share our considerations in such a way that they can be digested, criticised, and developed further.

Chapter 20

Diagnosing Wild Species Harvest: The DWiSH Procedure

TOWARD A HOLISTIC VIEW OF WILD SPECIES HARVEST

Earlier parts of this book (particularly Part 2) have shown in concrete ways how short-term gains in the use of wild species easily take place in contradiction with the interest of long-term sustainability. Wild species harvest typically leads to an immediate gain but also to negative consequences – if not right here and now for yourself, then maybe elsewhere, later, or for somebody else. Yet, management decisions or recommendations based on only ecological considerations easily become unrealistic and impossible to apply in practice. Similarly, if ecology is ignored and the focus is set on economic and sociocultural aspects alone, any good intentions may soon become derailed if the resource base is exhausted.

Fortunately, there are prospects for even complex harvest situations to be comprehended and controlled to a degree by systematic analysis and planning, as evidenced by a growing body of literature presenting principles, criteria, and indicators used to form standards for the sustainable management of different kinds of wild species resources (Jennings, 2013; Clark and Kozar, 2011; Datta et al., 2010). Likewise, important contributions to the analysis of wild species harvest accumulate, such as the insightful review of the role of nontimber forest products (NTFPs) globally by Shackleton et al. (2011a). However, similarly to the aforementioned book, most of them are specific to a particular type of resources (e.g. fisheries, timber trees, or NTFPs), a particular social setting (often community-based management), or some particular components of the harvest system (e.g. economic, cultural, social). Even deeper insights that can help in finding solutions to sustainability-related problems can be reached by applying a holistic diagnostic view, drawing on a wide range of perspectives on the socioecological system of harvest.

It is the goal of this chapter to provide an approach including a framework and a set of tools for such a holistic analysis. We also aim to propose further insights for diagnosing harvest and management situations when the goal of the analysis is to form a basis for actions that seek not only to minimise negative ecological impacts but also to maximise long-term socioeconomic gains. All of these goals are often (but not always) widely shared by different actors across society, and an impressive body of research constantly addresses related topics

Diagnosing Wild Species Harvest. http://dx.doi.org/10.1016/B978-0-12-397204-0.00020-6

in the field of wild species harvest. Nevertheless, we believe that this field is still too fragmented and emphasise that efficient analysis and planning only arise from a balanced and systematic understanding of the phenomenon.

The panorama presented here aims to be universal and holistic yet simultaneously useful for focusing on wild species harvest separately from several specific viewing angles. Wild species harvest is based on a socioecological system formed by overlapping and interacting natural and human spheres. Through the harvest activity, people are involved in a series of transactions with the natural world; humans extract biological matter and their interventions affect those components of the ecological system that form the natural "counterparts" in these transactions. Likewise, on the social side of the system, a series of transactions between different actors and institutions of society keep harvest activities in movement. Often, these transactions take place among a multitude of people with varying social positions and backgrounds as well as through a diversity of societal mechanisms and forms of organisation that address divergent interests and concerns. Yet, the authentic conditions of harvest decision-making are still practical and always to some degree unique. Figure 20.1 shows how harvest decisions are culturally rooted but operationally depend on the ephemeral conditions present at the juxtaposition of nature and society.

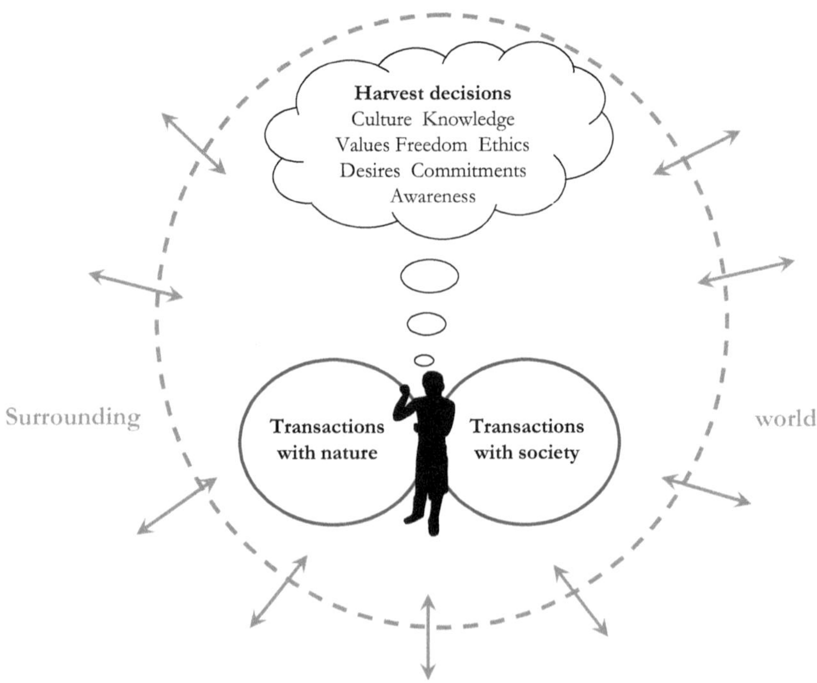

FIGURE 20.1 Engaged in a series of transactions with the natural world and society, the harvesters face unique circumstances of the surrounding world. All harvest decisions are also affected by the personal characteristics of the harvester.

The theory chapters in Part 3 of this book presented a series of different viewing angles to wild species harvest, constructed as seven separate thematic perspectives that help to structure the theoretical and practical assessment of harvest situations. Together, these chapters showed how, in any particular case of harvest, various types of thematic details must be considered if the aim is to acquire a balanced view of the case. Figure 20.2 schematically presents the entire socioecological system of wild species harvest. The harvester, as the key actor, is positioned in the middle. Around the harvester, there are a series of different spheres of interwoven and interacting circumstantial conditions that form the harvest situation embedded in the overall world socioecological system. Together, the spheres of the diagram create a representation of the complex and dynamic settings in which the natural and societal realities meet.

Because everything is connected to everything, there is a gradient from the immediate circumstances of the harvest towards the broader phenomena taking place in the overall ecological and human systems. On the left-hand side of the graph in Figure 20.2, there are factors that build up the provision from nature, ranging from those features that characterise the harvestable individuals, their populations, and the communities of which they form part, and finally up to those factors that contribute to complex ecosystem-level settings. A similar gradient of interconnectedness is also found from the harvester toward those mechanisms that tie different actors together in society. On the right, first, there are the direct incentives of the harvester that form the driving factors of his or her interests and decisions, followed by those factors that arise from the existing human organisational and societal arrangements forming the incentive structure for harvest and use, and, finally, there is the underlying sphere of the larger human society. Although few harvesters may be consciously aware of all these convolutions, this does not mean that the harvest act itself would be separate from them.

Figure 20.4 develops this idea further by also visualising the role of time – the legacies from the past and for the future. The present in any place and moment of interest – including its contextual complexities of an ecological, spatial, social and cultural character – is but a result of momentary circumstances influenced by a multitude of factors operating interactively and simultaneously. Figure 20.4 differs from Figure 19.3, in which the time cone was presented with a spatial dimension; the present situation now includes all of the spheres of the socioecological 'owl's face' of wild species harvest (Figures 20.2 and 20.3). The present state of affairs in these different spheres is determined by a specific set of legacies from the past. Those events that are close in time mainly determine the details of the present conditions at the local level; however, the more we go back in time, the wider is the variety of factors that may have influenced the present circumstances.

The upper side of the time cone, in turn, illustrates the potential future, including the after effects from this moment. Small-scale momentary consequences occur upon each of the harvester's decisions and actions. Sometimes, they at least seemingly remain so: small in scale and momentary in time. However, although the after effects of harvest may on the one hand be restricted to a

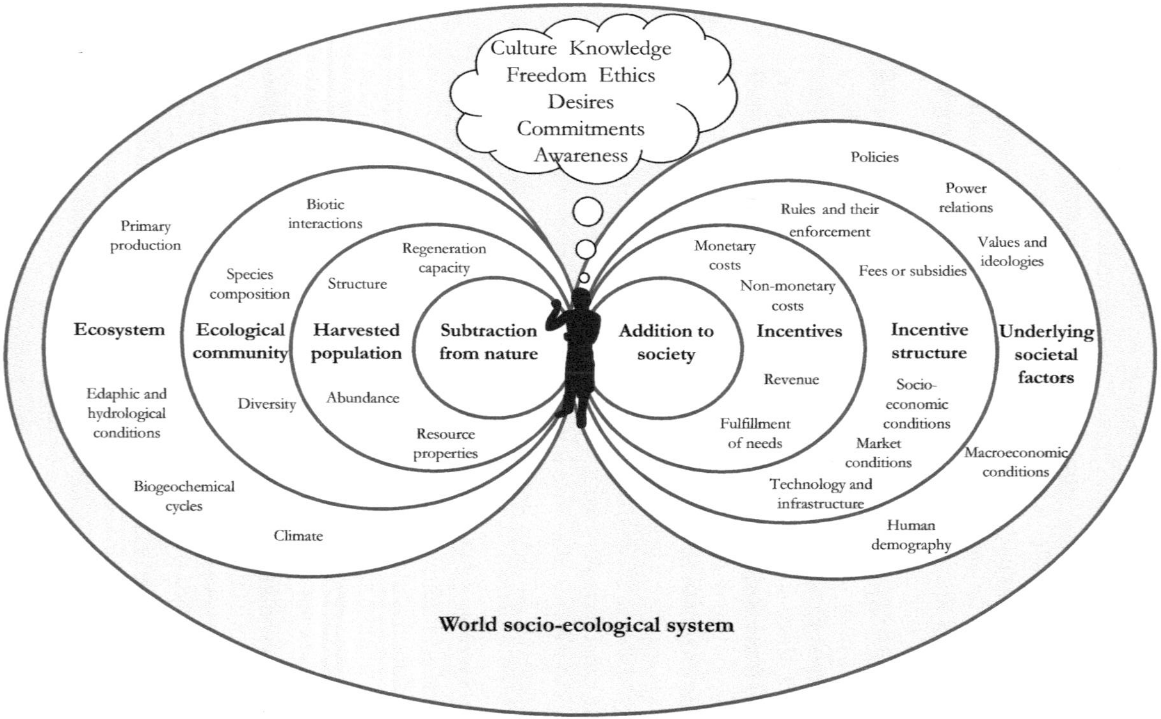

FIGURE 20.2 The 'owl's face diagram'. The (potential) harvester of wild species faces the conditions provided by nature as well as society, and forms the connection between the two. On the left hand side of the graph is the domain of nature, out of which what most immediately affects the harvester is the provision from nature – that is, the existence of wild species resources that could potentially be harvested and become goods. On the right side, on the other hand, is the domain of human society; what the harvester faces here is most of all the expected feedbacks from society in form of monetary payment or satisfaction of subsistence needs, for example. The provision from nature comes from the population of the harvested species, which, in turn, forms part of and is affected by the ecological community – which, in turn, forms part of and is affected by the entire ecosystem. Similarly, the feedback from society depends on the direct incentives, which depend on the incentive structure, which all are affected by and form part of the underlying society. The decisions the harvester makes, however, are not an outcome only of all of these factors, but also of his or her own particular way of thinking. In addition, all of these factors are embedded in the world socioecological system. Selected keywords from Part 3 of this book have been included to exemplify the different spheres.

FIGURE 20.3 The stare of the great grey owl (*Strix nebulosa*). When designing how to visualise the harvester in the challenging position between the two realms of nature and society, the outcome of which is in Figure 20.2, we realised that our sketches started to resemble the face of this bird. Therefore, we started to call it the 'owl's face' diagram. The coincidence was not a bad one; in many fables, owls represent wisdom. *Photo: Matti Sillanpää.*

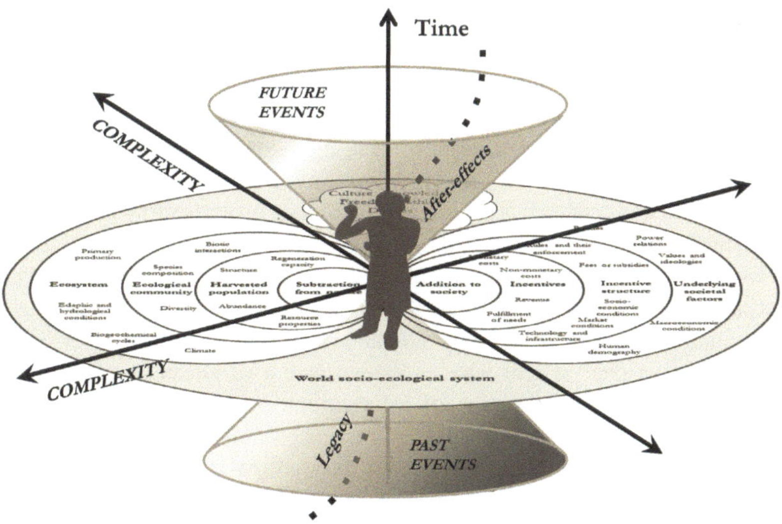

FIGURE 20.4 Harvest situations form part of the socioecological systems that are complex, both organisationally and spatially. All components of these systems change in time, with legacies from the past and after effects in the future.

very confined set of alternative outcomes, they also have potential to influence the outer spheres of the socioecological system. Whatever occurs today in the form of wild species harvest will form part of tomorrow's legacy, again bringing us close to the concerns of sustainability.

RATIONALE OF THE DWiSH PROCEDURE

As an outcome of all the earlier chapters of this book, we hereby present the **Diagnosing Wild Species Harvest (DWiSH) Procedure** as an analytical framework and systematic procedure. The purpose of this procedure is to provide a more holistic understanding of specific cases of wild species harvest than what conventional scientific research usually provides, thus also aiding the search for solutions to real-world problems. The approach of the DWiSH Procedure is such that it helps one to systematise the analysis of wild species harvest for qualitative and quantitative diagnostics that can be used as input and reference for academic study, management planning, institutional design, policy formulation, and the identification of sustainability standards based on case-specific principles, criteria, and indicators. The DWiSH Procedure synthesises information produced through research in various different disciplines, as well as knowledge that originates from sources outside science. Given the multifaceted nature of the topic and the variable reliability of different sources of information, dealing with uncertainty is an important part of the procedure.

In its most basic form, the DWiSH Procedure may be carried out based on any source of information, including personal observations, narratives, informal conversations, or newspaper articles. In a more advanced form, it can be carried out using ever more rigorously validated reliable source data. It can therefore be seen as an iterative process (Figure 20.5) that gradually takes the analyst towards an increasing understanding without the need to arrive at a point in which final answers can be given and the case closed. Ideally, it functions as a part of an **adaptive management or governance cycle** in such a way that experiences obtained from intertwined activities of harvest, management, and research feed into the procedure itself, leading to ever better understanding (Westgate et al., 2013).

Essentially, the procedure involves systematically analysing a case or cases of wild species harvest with particular emphasis on the perspectives offered by the various chapters in Part 3. The outcome of an initial round of 'DWiSHing' may well be an overview of what is known and what needs to be found out, leading to the further collection of information and consequently improved knowledge and understanding. Many different processes related to wild species harvest and management can be fed through the DWiSH Procedure. For instance, there are numerous schemes of principles, criteria, and indicators for sustainable forest management elaborated and proposed by the Food and Agriculture Organisation of the United Nations (FAO) (Castañeda, 2000), the Center for International Forestry Research (CIFOR), the International Tropical Timber

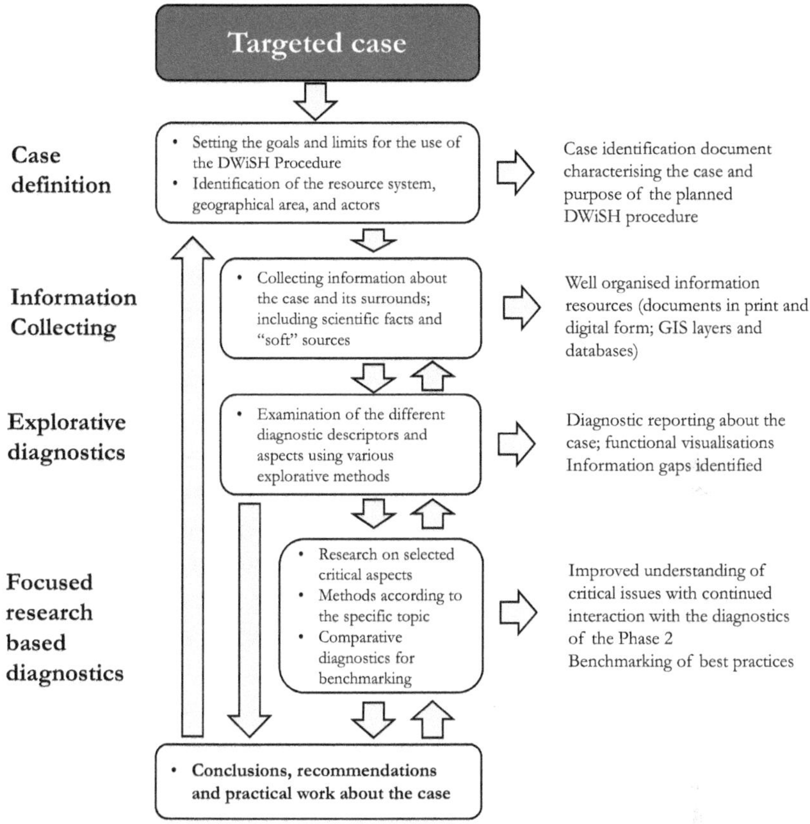

FIGURE 20.5 Schematic presentation of the proposed DWiSH Procedure.

Organisation (ITTO), the Forest Stewardship Council (FSC), and the Tarapoto Process, among others (Pokorny and Adams, 2003). Often directly related to this, there is a series of certification schemes emphasising sustainability, conservation, and fair and equitable distribution of benefits (McDermott, 2012), whose application can also benefit from the kind of broad and systematic approach the DWiSH Procedure represents.

The goal of improving the information basis through a series of formal approaches, deepening the level of understanding about a variety of themes, makes the application of the DWiSH Procedure well suited to dealing with management and planning problems in a way that can be connected with **multicriteria decision-making** (MCDM) methods. These are suited to deal with mixed goals and sets of data, both quantitative and qualitative. In resource management planning situations, the usefulness of MCDM methods has been discovered, for example in forestry (Mendoza and Martin, 2006; Ananda and Herath, 2009). As another practical example, the **driver-pressure-state-impact-response** framework that helps to structure information about interactions between society and the environment (e.g. Maxim et al., 2009) can be enriched through assessments using the DWiSH Procedure.

As a diagnostic process helping practical decision-making, it may be illustrative to compare the application of DWiSH to the examination of a patient by a medical doctor who is wise enough not to concentrate on one mere field of examination but takes into account a variety of factors within and outside the patient's body. When using the DWiSH Procedure, however, the goal is to contribute to the best possible use of the resources nature provides, taking into account the ecological impacts of wild species harvest as well as its contribution to the well-being of people. A series of natural and human-related characteristics of the selected harvest situation are observed, assessed, measured, and related to each other, and any final diagnostic judgments about the harvest situation should be made upon the sum of them all. Then, if needed, the most fitting remedies can be proposed and their impact assessed using an appropriate monitoring scheme. In the use of the DWiSH Procedure, the process outcomes may also include active participation and communication in resource management oriented expert panels and discussion groups by the means of the Delphi method (MacMillan and Marshall, 2006). Thus, the DWiSH Procedure is not an alternative to any of the above-mentioned approaches to resource governance and management planning, but rather it can enrich all of them.

WHAT IS THE DWiSH PROCEDURE MADE OF?

In principle, the application of the DWiSH Procedure is simple; it follows a structured approach for organising the seven thematic perspectives to the harvest of wild species and analysing them one after another. The procedure can accommodate both qualitative and quantitative criteria and their interplay is welcomed. On one hand, an example of an applicable qualitative measure is people's perceptions about how to distinguish resource-rich areas from those areas that are resource poor. Quantitative measures, on the other hand, can include such data as the distance from settlements or transportation routes to the harvested resource units (cf. Sirén et al., 2004; Salo and Toivonen, 2009; Salo et al., 2011) or the quantities and the size distribution of harvested units of wild species material.

Because the seven perspectives of the DWiSH Procedure are necessarily overlapping to a degree, comparing information and describing harvest from all of these different viewing angles is not always straightforward: they are not independent from each other. Because everything relates to everything, too much hesitance or an overly detailed approach may blur the overall picture. In practice, only a balanced level of detail among the different viewing angles allows a holistic assessment and comparisons between and among the seven perspectives. Inconsistencies or difficulties of drawing a line between the different perspectives, however, should not be seen as barriers to the application of the procedure. The iterative nature of the DWiSH Procedure also encourages one to re-evaluate any previous step and every single detail left behind whenever alternative interpretations emerge or new information becomes available.

Along with new understanding, new issues will emerge, calling for attention. After finishing one round of DWiSHing, the next round should be able to provide better answers to those questions that were identified as needing attention, helping to form an increasingly coherent and holistic view of the studied case. New rounds of DWiSH can then be carried out in the same way, each time with increasing detail (Figure 20.6). This means a growing level of sophistication and detail, possibly including extensive empirical or theoretical research. The quality of information often needs improvement; some of it may not be fully reliable or may be out of date, or information may be available only for similar cases in other areas. Sometimes, information about the studied species itself may be lacking, but there is information available for other similar species. Reporting such uncertainties and information gaps is a crucial constituent of the DWiSH Procedure.

This said, it is important to distinguish the above practical needs for improved information for the purposes of the DWiSH Procedure itself from the thematic perspective of knowledge in Chapter 17. The difference is that the practical needs in the information collection supporting the DWiSH Procedure concern and cover all of the thematic perspectives of the studied case ('what I know about the case and on what basis'), whereas the thematic perspective of knowledge

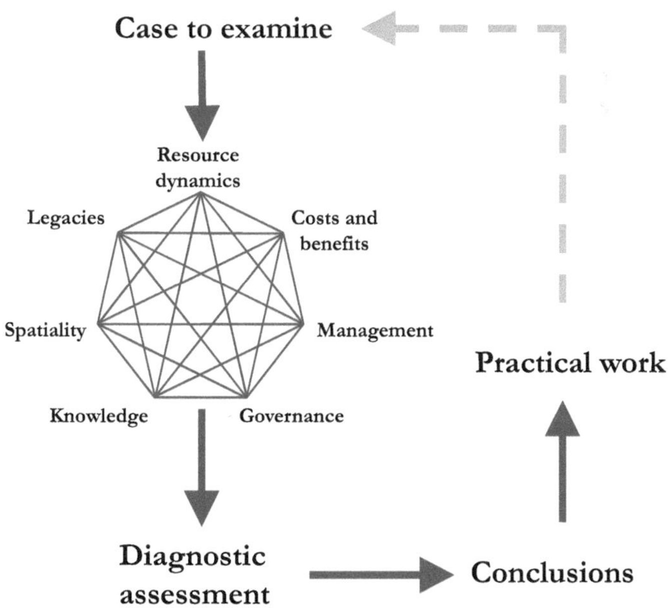

FIGURE 20.6 The DWiSH Procedure starts with a case, continuing with its analysis through the four steps of the procedure and final diagnostic assessment and conclusions. It can lead to further practical work and follow-up with the case.

focuses on the knowledge managed by the different actors involved in the studied wild species harvest system ('what do the actors involved in the harvest case know and on what basis'). Thus, in the analysis of this thematic perspective, the focus is on the knowledge, science, know-how, education, information, communication, and other related phenomena present and taking place within the social system of wild species harvest – independent from the person(s) carrying out the DWiSH Procedure. Of course, addressing these two viewpoints can be a mutually enhancing process.

When following the DWiSH Procedure and carrying out the thematic assessments, it is important to try to leave behind any preconceived opinions one may have and to make an effort to look at issues beyond the limits of one's own special field of competence. This facilitates finding new viewing angles and ultimately contributes to the capacity of the procedure to help encounter answers to questions of central importance for the rational and sustainable harvest, management, and use of wild species, such as the following:

- What is the potential to improve the economic and sociocultural benefits by changing the praxis of harvest and management, and what impact would that improvement have on the conservation of biodiversity and the state of the ecosystem?
- What is the potential to improve the conservation of biodiversity and the state of the ecosystem by changing the praxis of harvest and management, and what impact would that have on the economic and sociocultural benefits of harvest?
- To what extent can improvement of economic and sociocultural benefits of harvest, on one hand, and conservation of biodiversity and the state of the ecosystem, on the other, be made compatible with each other?
- What are the actions needed in order to realise any of these potential improvements?
- What are the most significant uncertainties and information gaps, and how can they be handled?

Below, we outline the practical application of the DWiSH Procedure, also suggesting some more detailed guidelines on how to carry out the analysis of the seven thematic perspectives and relate the evaluation results among the perspectives. However, the DWiSH Procedure should not be interpreted as a recipe that must be followed word for word, but rather as a source of reference and instructions for the general approach. Some parts of the guidelines below may appear irrelevant for certain cases, in which case it is perfectly fine to skip them. For some people, it may be useful to begin by following the procedure quite closely, writing out the results of each step; however, experienced researchers and practitioners may prefer to jump over something. Furthermore, the guidelines below are flexible and accommodate further contributions and refining. We hope that this first attempt to outline the DWiSH Procedure will stimulate others to continue contributing to its further development.

LET'S DWiSH: APPLICATION OF THE PROCEDURE

Case Definition

In practice, defining a case of wild species harvest often implies defining three main features: the harvest activity in question, the geographical area, and the main actors involved. These general denominators of the harvest are important in order to limit the scope of the analysis and also to communicate the case in brief. For a focus to be established and thematic assessments to be carried out, the case must have its limits. Ideally, the scope is wide enough to include a complete socioecological system of harvest, but in practice it may often be necessary to address some subsystem of a whole. Ultimately, the particular roles of the three denominators below in the case definition vary from case to case, and it is often difficult to include complete harvest systems.

First, it is important to define the **resource** that is the target of harvest. One possibility is to define the resource as a certain set of somehow related species or even one single species. Whenever feasible, the vernacular as well as the scientific names of the species should then be identified. If all of these names are not immediately available, however, that is not a major problem. Examples of definitions of the species in focus thus may be, for example, 'frogs' or '*aguaje* palm (*Mauritia flexuosa*)'. Another alternative is to define the resource widely and based on a certain kind of harvest activity, such as 'hunting', 'fishing', or 'mushrooming', and include all species affected by that sort of activity. This leaves one with the challenge of defining these activities, and that is not always straightforward. Often, however, harvesters themselves have a quite clear idea of what they include under such umbrella terms. 'Fishing', for example, is a term that often includes not only the harvest of fish, but also of crustaceans such as crabs and shrimps, as well as molluscs such as squid.

Some species provide various different products and may therefore be harvested with different purposes. For example, the palm *Jessenia bataua*, called *ungurahua* in Ecuador, is subject to harvest of palm hearts as well as of its fruit, whereas another palm, *Iriartea deltoidea*, also known as *pambil*, is harvested for the palm heart, as well as for wood for construction and for leaves to make shelters when staying overnight in the forest. Similarly, several other palm species are harvested for various products each. The resource can therefore also be defined as a specific type of product from a limited set of species, such as palm hearts. If the resource is defined in such a way, then the harvest of other products from the same species – in this case fruit, wood, or leaves – should nevertheless be addressed as additional human interventions affecting the resource.

Second, there has to be a defined **geographical area** of interest. Sometimes, one may want to define this as 'the Planet Earth', but in order to gain precision in the diagnosis, it is often in practice useful to focus on a smaller, clearly defined area. It may be an area with precise limits, such as a municipality or a timber concession, but in other cases it can be more feasible to define the area of interest in other terms, such as, 'the hunting grounds used by the people living in the community of Loboyacu' or 'the full range of migrations of the salmon spawning in the Tana River'. Defining the geographical area of interest does not mean that the surroundings should be ignored, but rather that they should be dealt with as part of the overall setting of the case.

Third, it is often necessary to define the **main actors**, or the harvesters and any other actors in the focus of the analysis. The same wild species resources, even in the same geographical area, can be harvested by a variety of different actors for various motives and through distinct harvest systems. Ideally, the harvest situation can be broadly defined, including all harvesters and the whole chain of resource utilisation. This can be the case for an activity such as hunting, irrespective of who is the hunter and for which motives. Sometimes, it can also be interesting or necessary to set the focus on some specific type of harvest, such as subsistence fishing or commercial berry-picking. This brings up the question of who the relevant stakeholders are, what kinds of pressures they pose, and how they relate to each other. In this respect, too, if the focus is set on some subset of actors (e.g. subsistence fishers), this does not mean ignoring other harvesters and related actors; rather, they are dealt with as forming part of the social setting of the case (e.g. commercial fishers as competitors to subsistence fishers).

The outcome of the case definition, so far, is a title that succeeds in communicating the essential components and objectives of the addressed case. An example of such a title could be: 'Diagnosis of the commercial harvest of leaves of the *irapay* palm (*Lepidocaryum tenue*) in the Nanay River catchment, Peruvian Amazonia'.

In addition, the case definition should include a brief description of why it is important to study the chosen case. That is, describe any problems related to wild species harvest that may be perceived by the harvesters themselves or by other actors, including the analyst her or himself. One may tentatively mention what may be the most important drivers of change already at this stage of the procedure because their recognition helps to set the focus on topics that are likely to be critical for the case in question. However, this should not lead to biased treatment of the case but rather help the identification and challenging of pre-existing ways to scrutinise the case, as well as to 'calibrate' the brain and enter the formal procedure. Box 20.1 presents a condensed example of a case definition scheme inspired by one of our example sites (Chapter 7). A clear case definition is necessary in Step 2 for the DWiSH Procedure to be efficient.

Collecting Information about the Case

Once the case is defined, the next step is to systematically collect information about it from any available sources. Although source criticism is needed, no source of information should be disqualified at this stage. To the contrary, information should be systematically searched, emphasising the diversity of sources

Box 20.1 Case Definition for 'Diagnosis of the Commercial Harvest of Leaves of the *Irapay* Palm (*Lepidocaryum tenue*) in the Nanay River catchment, Peruvian Amazonia'.

Resource

The target of harvest is the palm species *Lepidocaryum tenue*, locally called *irapay*. In more detail, what is harvested are the leaves of the palm, and the harvest is, in principle, nonlethal to the plant.

Geographical area

The geographical area of interest is the Nanay River catchment area in the region of Loreto in Peruvian Amazonia. The selection of the geographical area is justified by at least two specific reasons. First, the Nanay catchment is the main source of *irapay* leaves for the city of Iquitos, found directly in the mouth of the Nanay River. Iquitos is also the main market for the leaves. The above facts lead to the second justification of the geographic definition of the study area, more technical in nature: it is relatively easy to monitor the movement of *irapay* shipments from the Nanay catchment area arriving in the city through the two ports located near the mouth of the Nanay River.

Main actors

The analysis is targeted to commercial harvest, and it should cover the whole chain from the forest to the end users, including direct harvesters living in the communities of the Nanay Basin and possible outsider collectors, as well as people involved in transporting the leaves, commercialising them, and using them as roofing material in the city of Iquitos.

and their types and carefully attaching to them metadata (information about their sources and other important characteristics with implications for their further use). Because of the breadth of the variety of possible sources of information, the value of a clear case definition becomes visible here because it helps to set limits to the information to be collected.

Scientific articles often provide high-quality information but will rarely cover all the diagnostic perspectives of the DWiSH Procedure. Many kinds of formal registers and statistics maintained by authorities or other actors can also offer valuable information. Maps, satellite images, and geographic information system layers in digital form may reveal spatial realities and phenomena; such information is today plentiful and available even free of charge from public sources. There is often relevant grey literature from different sources too, as well as theses made by students and other miscellaneous sources that can include invaluable pieces of information. Although many other sources may not have the same objectivity compared to a scientific article or a technical report, they may reveal something that academic researchers, professional experts, administrative officials, or authorities might omit.

The researcher's own observations and oral communications by, for example, local harvesters, government officials, merchants, or other people well acquainted with the on-the-ground situation also form a valuable source of information. Without the need to use any systematic sampling at this stage, methods of respondent-driven sampling (e.g. Heckathorn, 1997) can help to speed up the identification of interesting key informants with varying roles in the harvest system through chains of referral, either by personal communications or using media or other second-hand sources. Sometimes, fragments of information such as above can be found through the exploration of newspaper articles or the like. The different kinds of mass media are thus well worth examining; articles in newspapers or magazines may be useful as a piece of news, or a letter to the editor in a local newspaper may reveal an important opinion, possible conflicts of interest, cultural practices, or other issues that might otherwise be overlooked.

Although formal in-depth interviews can be put off until a later stage, the experiences and perceptions of the harvesters themselves are particularly important. If one stays or has the opportunity to visit the chosen area of interest during this step of the procedure, it can be very useful to interview harvesters or any other group of people whose actions or attitudes are relevant to the case. Elders often remember such events from the past whose legacies and significances to the present circumstances are important to understand. Younger generations should not, however, be omitted either. Even their possible ignorance about some particular topics in some of the thematic perspectives can be indicative of something that is worth considering. Because people's opinions are diverse, it is good always to actively search for opposing views among the different stakeholders.

Exhaustive and detailed information about a specific case is often unavailable, increasing the necessity to extrapolate in space and time. For example, a study of the ecology of the Brazil nut tree (*Bertholletia excelsa*) made somewhere in Bolivia is quite likely to have some relevance also to a study site in Peru. Likewise, information from a nearby community that still does not have a road connection can be useful for understanding past resource use patterns in a community that recently got a road connection. In a similar manner research findings on the ecology of, e.g. the black-headed spider monkey (*Ateles fusciceps*) can probably to a large extent be applicable also to some closely related species, in this case e.g. the whitefronted spider monkey (*Ateles belzebuth*), if studies on the latter species are not available. Extrapolation must, however, always be done with caution, carefully scrutinising whether it is justified or not, and remembering that this approach always implies a certain level of uncertainty. Extrapolation should also be explicitly indicated using references and keeping account of metadata.

Because the internet is becoming increasingly widespread, reaching even many remote areas, interesting possibilities for information sharing and communication as well as for the involvement of different actors can also be found in the social media and in different kinds of crowd-sourcing methods (Fielding, 2013; Kelly et al., 2012; Stewart et al., 2012). The output from this step of the DWiSH Procedure should be a well-organised collection of information with systematic metadata indicating sources and their context. Particular emphasis should be placed on this phase of the procedure because it often involves managing very distinct types of data and information needing different handling – some of which may not be all that familiar to the person following the procedure. In practice, this collection of information can be anything from documents in print format to a network of shared information resources. The format is not critical but the organisation is: the better the systematisation and metadata attached to the materials collected are, the more useful and accessible the gathered information becomes in the following steps of the procedure and the easier is the further addition of new materials.

Explorative Diagnostics of Seven Thematic Perspectives

This step of the DWiSH Procedure is made up of an analytical assessment of the accumulated information, using as a structural frame the seven thematic perspectives presented in Chapters 13–19 of Part 3. The workload of this phase can be extensive and the details of its content will inevitably vary from case to case. Closer examination of the collected information is also likely to lead to new sources and more material that should be dealt with in the same way as in the initial information collection phase. Although it is likely that the materials that are used in the assessment cover the seven perspectives in an uneven manner, it is important in this phase that the source information is used in such a way that a balanced view of the case is generated, emphasising both the unique characteristics and the relatedness of all the different thematic perspectives. The exact methods used in the assessments depend on the context of the particular case in question, but we point below to some tools that can be useful.

Explorative diagnostics aim to build a holistic understanding, but in order to facilitate the process we provide a list of examples of issues or questions that can be useful to address in each of the perspectives. Note, however, that the lists in this section should not be seen as cookbook recipes or questionnaires. The idea of the DWiSH Procedure is not to produce simple or definite answers to a list of predefined questions. Instead, the topics here should be considered as coarse checklists of things that one should not forget to consider. For any particular case, some items on the lists may be more relevant than others. Whereas some may deserve to be explored in depth; for others, just a brief mention may be enough.

The tangible outcome of this phase of the DWiSH Procedure may vary. It may be a report, a policy briefing document, a slideshow, a thesis chapter, a speech, a book, a documentary film, or anything else found suitable. What is important is that it has a logical structure and is based on a systematic work that covers all of the seven perspectives presented in this book. A written report becomes necessary when the depth of the diagnostic work increases. Emphasis should always be put on proper and sufficient descriptions of the sources and uses of data and analytical methods, just like in any other research effort. Because of the highly variable sources of data and information useful for the diagnostics procedure, accurate and detailed metadata is also of primary importance as described in the above step. Well-documented and described sources make their revisiting meaningful. The analytical work should also be appropriately performed and described, helping to critically assess any conclusions made.

A short and precise reporting style helps the users of a DWiSH assessment to efficiently grasp the main diagnostic results. A good aid for reaching this goal is to use tables, maps, figures and other tools that help to concentrate information and systematise its presentation. Sometimes it is useful to employ **conceptual maps** to identify the most important components of the system under analysis as well as the interrelatedness of these components (e.g. Figure 15.3; see also Ostrom, 2007; Prell et al., 2007). The use of **function graphs** as a way to visualise conceptual models over the interrelations between different variables can also be very useful (such as those presented in Chapters 13, 14 and 18). Moreover, many existing frameworks and checklists of principles, criteria, and indicators developed for different kinds of sustainability purposes can contribute with more details and orientation for the exploration depending on the case (e.g. Allison and Horemans, 2006; Berkes et al., 2003; Blom and van Bueren, 1997; Charles, 2008; Lindenmayer et al., 2006).

In the following sections, we further outline some topics that are likely to be important in the explorative diagnostics of a case of wild species harvest using the seven thematic perspectives presented in Chapters 13–19.

Resource Dynamics

This thematic perspective, presented in Chapter 13, addresses the ecology of the resource system and the harvester's transactions with nature – in other words, this perspective includes an assessment of the dynamics of the resource and the effects of harvester interventions (Figures 13.1 and 20.2). The exploration of this thematic perspective should thus aim at the identification of the critical components of the harvest system for the sustainability of harvest from an ecological viewpoint. Important issues to address are, for example, the following:

- Basic biological characteristics of the harvested species, including reproduction and growth, and how these depend on population density and structure
- Direct effects of harvest and management at the levels of individual organisms, populations, ecological communities, and ecosystems
- Relevant interactions between the harvested species and other species and with the abiotic environment; any known or potential indirect ecological effects mediated through such interactions
- Current rate of harvest compared to the potential of the ecosystem for long-term sustainable provision of the resource; harvest rate in comparison to the current rate of resource regeneration; the effects of harvest on the provision from nature
- Trends of change in the abundance and structure of the populations subject to harvest; abundance of the harvested species in comparison with the potential abundance if not subject to harvest

Depending on the case, the findings of this assessment can be presented in many forms. In addition to descriptive textual presentations, it can also be very useful to summarise the above in a series of graphs. In Chapter 13, several types of function graphs are presented that may be useful. This, however, requires careful thinking and assessment of the ecological characteristics of the particular species in questions and accordingly adaptation of the graphs to fit the case in question. Formulating mathematical models can also be helpful.

Costs and Benefits

This thematic perspective, presented in Chapter 14, deals with the harvester's immediate transactions with society (Figures 14.1 and 20.2), involving issues related to the direct incentives to the harvester in terms of costs and benefits. Some key issues to address include the following:

- Benefits obtained by harvesting the resource; subsistence needs and commercialisation of harvested goods; any other benefits that may function as motivating factors, such as pleasure or social appreciation when giving the product away
- Main costs involved in harvest; fixed costs, search costs, handling costs, or transport costs, as well as any other costs that perhaps may not fit into this scheme, but that nevertheless may be present and important; any potential costs due to detection in case of rule breaking
- Characteristics of harvesters as economic actors; how these affect their economic decisions; profit maximisation and other economic objectives; rationality and information as basis of harvesters' decisions as assumed in many economic models
- Market conditions that potential harvesters face; for the harvested goods as well as for labour, and how these may affect harvest behaviour; availability of capital, credit and debt; income irregularity and seasonality
- Discount rates in comparison with rates of resource growth
- Elasticity of consumer demand of the harvested goods; changes in consumer demand related to income level changes, availability of substitute products; dependence of supply of the harvested goods on price, differing between short-term supply from an individual harvester and long-term aggregated supply

Again, it can be useful to summarise some of the findings in the form of graphs. For example, at least for commercial harvest for sale on the market, supply-demand graphs, such as those in Figures 14.8 and 14.13, can be sketched. Again, the use of mathematical models can be helpful in order to express one's thoughts in a stringent way and gain insights into the consequences of alternative assumptions. All cases are unique, so care is needed again to assess the utility of different tools. The main point here, however, is to get an overview of how the incentive structure facing the harvester shapes the costs and benefits associated with harvest and, thus, the harvester's decision about such things as how much to harvest as well as where and how to do it. Absolute figures of costs and benefits can be useful if they allow making comparisons that aid in this, but usually such figures are not readily available and they are not necessarily indispensable.

Management

This thematic perspective, presented in Chapter 15, describes the managerial aspect of a case – or how the harvester's immediate acts and decisions can potentially improve the overall balance of the transactions with nature and society (Figures 15.1 and 20.2). In the assessment of this thematic perspective, any purposeful manipulations of the resource system in order to increase the benefits of harvest should be identified, including those that target the harvested species, those addressing other species, and those directed to abiotic components of the ecosystem. Some potentially central issues include the following:

- Basic description of the harvesting activity: who are the harvesters; which parts of the targeted species they harvest; how (using which methods and technology) they harvest; where they harvest; when they harvest; how much they harvest
- Existence of any refraining from harvest, thus forsaking immediate own benefits in order to benefit others or ensure own benefits in the future
- Management actions taken in order to optimise population structure; sex ratio, age structure; abundance of preferred types of individuals
- Management actions that involve direct increment of the harvested species; introductions, planting, sowing
- Management actions that target other species than those harvested; removal of competitors or predators, promotion of species that support the harvested species populations
- Management actions involving the manipulation of abiotic ecosystem components; drainage, regulation of water levels
- Management actions that aim at reduced undesired effects of harvest on biodiversity and the ecosystem; the negative consequences of management

In this thematic perspective, the explorative diagnostics should lead to the identification of existing and/or needed management activities in the studied harvest system. Some management practices can be aimed at improving the benefits of harvest and others at decreasing the negative effects of the harvest interventions. Potential undesired impacts of management should also be identified.

Governance

This thematic perspective, presented in Chapter 16, explores the ways in which the harvester's transactions with nature and society are governed and how the incentive structure of governance is embedded in larger societal contexts (Figures 16.1 and 20.2). The focus should thus be in the details of the overall incentive structure, which through different kinds of mechanisms influences the harvesters' decisions, including those that weigh the costs and benefits of harvest. On a more general level, this thematic perspective also includes considerations of the underlying societal context, including general policies and ideologies. Some issues to treat here include the following:

- Property rights to the resource, in terms of access, withdrawal, management, exclusion, and alienation; different levels of property rights, such as the national government granting collective property rights to a community which, in turn, has its own internal system of property rights
- Type of property rights in comparison with the characteristics of the resource in terms of subtractability and excludability
- For common-pool resources: the degree of fulfilment of those attributes of the harvesters and the resource that are known to be conducive to self-organising for common-pool resource management and governance
- The existence of any rules regulating, for example, who is allowed to harvest the resource; timing, quantity, location and technology used in harvest; who is obliged to contribute to labour to managing the resource
- How harvest and obligation to activities are monitored and enforced, how conflicts are resolved; how the rules about harvest and management of the resource are changed over time
- Any other policies, formal laws, or informal norms and codes of conduct related to harvest and management of the resource; fit of the formal and the informal regulations
- Policy instruments that indirectly affect harvest by somehow influencing the costs and benefits associated with harvest, such as any fees, subsidies, trade restrictions, or alike; institutions regulating land use, zoning
- Fit of policies and policy instruments; the reception of policy instruments by the harvesters in terms of compliance and noncompliance; legitimacy, transparency, accountability of harvest institutions and associated actors
- The general underlying societal context in which the harvest takes place, including general macroeconomic policies, dominant cultural context and ideologies; the role of gender

In addition to descriptive textual presentation of results in this thematic perspective, there are many possible ways to enhance the presentation of the diagnostic results. For example, systemic diagrams such as conceptual maps, organisational and flow charts, pathway diagrams, or Venn diagrams such as those in Figure 13.1, and its sister figures in the other chapters of the Part 3, can be helpful for visualising the institutional and social settings of the harvest situation. Different kinds of stakeholder analysis and mapping methods can also be useful to identify the actors and their relationships (e.g. Prell et al., 2009; Reed et al., 2009).

Knowledge

This thematic perspective, presented in Chapter 17, addresses the kinds and sufficiency of information and knowledge managed by different stakeholders, as well as communication flows that are relevant in the harvest system. The focus should be set on the different actors with a stake in the harvest system (e.g. harvesters, merchants, consumers, administrators, researchers, and policy makers), addressing a broad range of issues from operative knowledge to cultural aspects of knowledge and information. Issues to explore also include traditional knowledge, social capital within the community level or among groups of extractors, and the adoption of new practices by those involved in the concrete harvest. It is important to keep in mind that here the focus is on the evaluation of the targeted case from the above perspective and not on the overall quality of information collected in the early phases of the DWiSH Procedure. The list below outlines some specific topics that are likely to hold importance in most cases:

- Types of operative knowledge (as described in Chapter 17) needed and managed by different actors in the harvest system; flows of information among different actors; information and knowledge asymmetries and their implications on the behaviour of different actors; different kinds of knowledge systems; gendered knowledge
- Absolute sufficiency, availability and intelligibility of scientific and other research-based data, information and knowledge from stakeholder perspective; awareness about sustainability issues related to the harvested resource among different stakeholders
- Communication about resource management issues among different harvesters and other stakeholders; communication among harvesters about management issues related to a common resource base; cognitive proximity or distance among different actors; barriers to efficient communication
- Diffusion of ideas and spread of good practices; good experiences that could be followed; awareness of harvesters about successful practices elsewhere; barriers to the spread of ideas based on good experiences; knowhow about resource use and practices
- Uncertainty; major uncertainties involved, considering whether they are model uncertainties, parameter uncertainties, or inherent uncertainties
- Role of scientific knowledge and associated information in decision-making by harvesters and other stakeholders relevant to the case; spatial planning

Because this part of the explorative diagnostics focuses on what knowledge and communication-related resources are generated held and exchanged among different actors involved in the studied harvest case, it can be illustrating to sketch a diagram showing the most important actors, indicating their needs of information and knowledge as well as communication and its barriers. One should also critically evaluate the quality requirements, sufficiency, and accuracy of information in relation to the actual needs of information, technology, and know-how.

Spatiality

All of the above thematic perspectives often touch on issues related to spatial processes and variability, topics of the thematic perspective presented in

Chapter 18. Here, spatiality should be dealt with in more detail, addressing the forms of interplay among different spatial processes and variability in nature as well as in society. Issues to discuss here include the following:

- Basic geographical setting of the area of interest; its natural and human features and their spatial characteristics
- Accessibility in terms of cost-distance; variability in space
- Spatial patterns of harvest; correlations with specific natural or societal features of a spatial character
- Resource abundance; distribution of the harvested species, or a particular part thereof subject to harvest; abundance variations and patterns
- Habitat quality; spatial variability in terms of habitat quality caused by natural environmental heterogeneity and by human interventions
- Mobility of the harvested species; dispersal, migrations, home ranging
- Source-sink dynamics; caused by natural spatial variability and/or by human activities
- Fit or misfit between the spatial scales and extents of natural and social systems

It can be worthwhile to explore map material such as, for example, data showing the limits of property rights areas or administrative jurisdictions, various types of thematic maps about many different physical and human characteristics in the region, as well as remotely sensed imageries, including satellite data that often reveals large-scale environmental variability, related to natural factors as well as land use and other human activities in a way no other information source does. Aerial photographs are useful as well because individual big trees, houses, trails, and other landscape structuring features can often be distinguished in those imageries so that they are understandable and eye-opening for anyone. In the absence of such materials, hand-drawn map sketches can serve the same purposes too. It can also be very useful to draw function graphs representing how different features, such as harvesting labour, harvest rates, and resource abundance, vary with distance.

Legacies

This thematic perspective, presented in Chapter 19, concentrates on the temporal dimensions of wild species harvest, being essentially a cross-section of all of the other thematic perspectives. The harvest system in general, and the harvester's transactions with nature and society in particular, change over time, forming legacies from the past. Similarly, all present actions and events have after effects that shape the future. In order to explore the legacies from the past, each of the previous thematic perspectives should be revisited, in order to analyse deeper the respective historical developments that have led to the current situation and how current trends can be projected into the future. Thus, some of the main topics to explore are the following:

- Changes in the extent and quality of habitat through time; changes in resource abundance and harvest rates
- Changes in costs of harvest, including opportunity cost of time; changes in market price; changes in the availability of substitutes
- Changes in management practices
- Institutional changes affecting harvest, as well as other societal changes affecting the incentive structure facing harvesters
- Changes in knowledge about the resource system, among harvesters as well as society in general; changes affecting communication and diffusion of ideas among harvesters
- Changes in the spatial organisation of the geographical area in focus; changes in accessibility; changes in spatial pattern of harvest

Functional illustrations are helpful, such as drawing a **timeline** and using it to indicate relevant natural and human processes or occurrences. Timelines can be linear, but when the temporal scales are long, it is preferable to use a logarithmic scale. Because getting information from the past can often be difficult and the quality of the data varies greatly, particular care with source documentation and metadata should be applied here so that it is clear how the conclusions have been drawn and what uncertainties there are. In particular, it is seldom possible to give a very detailed treatment of the distant past because of the scarcity of data. However, the closer one gets to the present, the more data tends to be available, and the more relevant it also becomes to include an increasing degree of detail. This is not a bias because, as shown by the narrowing tip of the time cone (Figure 19.3), those events that are closest in time are also responsible for many of the most immediate details around us.

Preliminary Diagnostic Conclusions

The findings under the different perspectives are intricately related to each other. Thus, the explorative diagnostics of the seven different thematic perspectives should be followed by a more synthetic analysis that aims to draw the somewhat separate findings back together. This can be done verbally, but a well-documented analytic comparison is preferred because it can always be revisited and interpretations changed if needed. One useful approach can be to employ cross-tabulations that overlay one or more perspectives (e.g. in a way similar to that presented in Table 20.1). Illustrations can also be used whenever appropriate, such as using two-dimensional space–time diagrams (as in Figure 18.16) to distinguish what is momentary and local from those occurrences that are parts of a bigger whole.

Sometimes scoring may be applied, such as to define the relevance or importance of each perspective in relation to the actual challenges in the case that is examined

TABLE 20.1 A Simple Cross-Tabulation of the Thematic Perspectives under Examination in the Exploratory Diagnostics of the DWiSH Procedure

Thematic perspective against which to examine

Thematic perspective under examination	*Resource dynamics*	*Costs and benefits*	*Management*	*Governance*	*Knowledge*	*Spatiality*	*Legacies*
Resource dynamics	*...behind the provision from nature*	Harvest levels and pressure	Outcomes in populations and environment	Methods, quantities and timing of subtraction	Appropriateness to relevant ecological constraints	State of populations in different areas	Human impact in state of resource
Costs and benefits	Resource availability and quality, income irregularity	*...weighted by the harvesters*	Resource availability and quality	Sanctions, fees, subsidies, credit	Price awareness, knowledge-asymmetry	Cost-distance patterns	Cultural settings for markets
Management	Manageable components of the resource base	Types and costs of management	*...of resource systems*	Incentives and rules for management	Knowhow including adaptive mechanisms	Distribution and accessibility of habitats	Options available for management
Governance	Resource specificity of institutions	Property rights and benefit sharing	Organization and self-organization	*...shaping incentive structures*	Information needs of institutions	Spatial fit of institutions and resource system, zoning	Institutional stability and change
Knowledge	Modeling and understanding of the resource	Know-how to decrease costs and increase benefits	Means supporting adaptive management	Research, education, media	*...for action and interaction*	Barriers to communication, spatial planning	Accumulated knowledge and harvest culture
Spatiality	Resource distribution and mobility	Distribution of harvest pressure	Resource distribution	Land tenure and ownership, usufruct	Mapping and monitoring	*...in nature and society*	Spatial variability of human impact
Legacies	Delayed effects in resource	Trends in harvest, consumption and markets	Anthropogenic influences in the resource	Power-relations, codes of conduct, informal norms	Harvest culture and tradition	Spatial patterns in the state of environment	*...from the past and for the future*

The rows represent the thematic perspectives under analysis and the columns the thematic perspectives against which these are analysed. Examples of relevant phenomena under such simple cross-analysis are presented in the cells.

(e.g. Johnston et al., 2013; Clark and Kozar, 2011). Such scoring is always to some degree subjective and therefore also problematic – and in many ways, case specific. However, as part of a learning process, such techniques may help direct attention to those thematic perspectives that need the most additional attention.

The explorative diagnostics phase of the procedure should help to identify a series of issues of particular interest, such as issues considered to be critical for the sustainability of harvest or simply being deficiently known. These findings should also be such that the interrelatedness of the different thematic perspectives is taken into account, integrating dimensions from the different perspectives into the findings as seen necessary and appropriate. The explorative diagnostics thus help reveal and identify such characteristics of the studied socioecological system of wild species harvest that are critical for any attempts to improve its sustainability (overall ecological balance of the harvester's transactions with nature) or to improve its societal performance (overall benefits from the harvester's transactions with society), or ideally both.

This procedure also allows one to discuss alternative future scenarios. Such scenarios, as they are related to potential goals of intervention, are useful for directing follow-up research and for assessing the outcomes of any eventual interventions with the harvest system. This can take place either as part of an adaptive cycle of management or governance, or independently. Given the complexities inherent in most socioecological systems, scenarios built on the basis of an analysis and information accumulated in the DWiSH Procedure do not have much predictive potential. However, their value is found elsewhere: they identify possible futures in gross qualitative terms, particularly regarding the sustainability of the harvest system.

For example, one scenario can project specific current trends a few decades into the future. Such a scenario may be based simply on current trends of harvest rates, or, alternatively, estimating future harvest rates based on trends of market demand, human demography, development of transportation infrastructure, or other features of society. In addition to this, one may make a worst-case and a best-case scenario, according to one's own values about what is desirable and what is not. Some people may put more emphasis on the conservation of biodiversity and ecosystem services, whereas others may consider the socioeconomic benefits of wild species harvest to be the most important issue. The risk of falling into the trap of wishful thinking is apparent in the development of such scenarios and they must be based on realism. The point is to highlight which changes in management or governance, or in underlying society as a whole, can make the difference between a worsening or an improvement of the situation. Below, we outline some basic components of these future scenarios:

- Foreseeable risks that harvest practices will change for the worse; foreseeable changes in harvest practices that would improve the situation; potential outcomes of both
- Foreseeable risks that management practices will change for the worse; foreseeable changes in management practices that would improve the situation; potential outcomes of both

- Foreseeable risks that the incentive structure will change for the worse; foreseeable changes in the incentive structure that would improve the situation; potential outcomes of both
- Foreseeable risks that the underlying societal structure will change for the worse; foreseeable changes in underlying societal factors that would improve the situation; potential outcomes of both

Based on the explorative diagnostics, including the identification of critical issues for sustainability, information gaps and uncertainties, and the generation of simple future scenarios, there should be enough insight and holistic understanding about the studied case for sketching a plan for follow-up and continuous learning.

Focussed Research-based Diagnostics

Finally, one may want to deepen the analytical work and make it increasingly research based. Such research contributions are essentially case specific; thus, it is beyond the scope of the present book to explore them in detail. Considering that scientific research usually comes up with three new questions after solving one, research can never reach a fully satisfactory level. We may aim to reach rather high levels in the information triangle (Figure 17.2), but the top may be inaccessible. There are three basic ways how the DWiSH Procedure can be helpful in this reality. First, when explorative diagnostics are performed in a balanced way so that each of the seven aspects are dealt with at comparable levels of detail, the overall picture that the diagnostic process produces is likely to be realistic (yet many details may remain open, and some are probably wrong). Second, if closely followed, the systematic basis of the DWiSH Procedure helps to compare different cases of wild species harvest. Thus, a case compared to another elsewhere in space or in another time and under different ecological and societal circumstances may reveal interesting differences and similarities. This allows benchmarking (i.e. comparing the case to other similar ones elsewhere) and thus facilitates learning from the experiences accumulated in other locations and through sharing of good practices. To facilitate such benchmarking, comparative tools are useful. These may involve, for example, tables and associated graphical illustrations, or some sort of statistical or semistatistical approaches that help to compare different cases or cluster them into more or less similar groups.

Third, and finally, the DWiSH Procedure is never-ending and it is always possible to revisit earlier diagnostic stages. Because the data sources, methods, and even goals are changeable in the DWiSH Procedure whenever needed, such continued application of the procedure resembles to a degree adaptive management and governance schemes, and it can potentially be integrated with these approaches.

Chapter 21

Whose Interest? Whose Action?

HARVESTERS IN THE SPOTLIGHT

The harvester's position is unique in wild species harvest because he or she is the only actor that in effect stands at the interface of the natural and human realms, thus forming a physical link between the two. In many cases, the actual decisions about harvest are entirely made by such individuals, in spite of them being influenced by a wide variety of direct and indirect incentives provided by the surrounding society. This makes the definition of the harvester ambiguous (Chapter 2).

The hunters in Sarayaku (Chapter 4), for example, can be seen as individual harvesters who themselves decide what, where, when, how, and how much to hunt. When their consumption is for household subsistence, beyond that level the harvester's transactions take place only with the natural world that provides game in exchange for the harvester's work. In other cases, however, the person who performs the act of extraction may not be the one who in effect makes the **decision about harvest**. In the case of the fishermen of *Mi Marido* (Chapter 5), for example, the owner of the boat stayed in the city whereas the men on board were paid workers whose momentary decision-making was essentially conditional on what the boat owner had decided. Or, consider a man hired by a timber company to use a chainsaw to fell a tree. Should he or the timber company be considered as the harvester? In this case, the concept of a harvester making sovereign decisions subject to incentives provided by the surrounding society is in one way an abstraction. Company owners or officials may take decisions at one level, such as about which concession areas to bid on, while lower-ranking technical personnel may decide upon criteria to use when selecting which trees to fell. Little choice is left for the chainsaw operator, because sometimes even the direction to which the tree is to be felled is decided by someone else.

Harvesters can also be examined in relation to their **motivation**. A fair assumption is that they often seek to optimise the outcome of extraction aiming at getting maximal benefit using minimal effort. In this consideration, any broader-level aspects about the rest of the world, or its rights or wrongs, may not matter much (or at all). Many harvesters live in conditions under which there is not much choice among alternative decisions – for pure economic reasons, for working as employed labour, for not being free individuals, or for some other reason. They may also consider themselves as such small actors that their interventions do not have significant effects on anything beyond the immediate

Diagnosing Wild Species Harvest. http://dx.doi.org/10.1016/B978-0-12-397204-0.00021-8

place of harvest. They may also confront the provision from nature alone and in rather isolated conditions, which makes it tempting to take some liberty at the very moment of acting, such as the moment of shooting a protected animal or taking home a fish that is smaller than the official regulations would allow.

The harvester, as we consider him or her in this book, can also be seen to represent a group of people whose decisions lead to concrete resource harvest. Such an organised group can be, for instance, a cooperative, a company, or any other type of association of decision-makers that ‘presses the button’. Within such a framework, there may be an internal culture of harvest that differs from the ways in which similar things are made in other places. As a consequence, the concept of the harvester as well as the ways in which she or he behaves are at the same time both deliberately vague and easy to understand.

In many different contexts, all around the world, both men and women take part in wild species harvest. However, they often have different roles. Women typically gather, for example, wild vegetables and berries, whereas hunting and felling of timber are mostly male domains. This **gender-based differentiation** of the kind of wild species resources harvested is common in a wide variety of cultures all over the world, but it is by no means absolute. That women hunt or that men gather edible plants is nothing exceptional, but it is clearly less common than the opposite. Some researchers have even suggested that evolutionary processes selecting for women’s capabilities associated with gathering and for men’s capabilities associated with hunting could explain to some degree not only the existing physical differences between the sexes but also some cognitive differences that have been detected between the genders (e.g. Silverman et al., 2007).

In Amazonia specifically, wild nature tends to be the domain of men, whereas women are in charge of the home and, in many cases, also of the gardens. Women do, for example, gather wild mushrooms growing in gardens or secondary forests, and sometimes men and women fish together. The collection of wild fruit or ants can be family outings for men, women, and children all together. But nevertheless, today in this particular part of the world, wild species harvest is mostly carried out by men. This is the reason behind the apparent bias in the ‘stories from the forest floor’, where most of the main characters were men. Women should, in any case, not be disregarded when the management and governance of wild species are discussed, not even in cases where they are not directly involved in the harvest. They are then often in charge of processing the harvested goods, and therefore they both hold important knowledge related to the harvested species and resources and may have an interest in their sustainable use.

An anecdote from the community of Sarayaku described in Chapter 4 may illustrate this. One of the authors of this book was about to start fieldwork in the community and held a number of workshops in order to talk with the people about their perceptions of the state of wild species and other local natural resources (Sirén, 2004; Chapters 4 and 7). Some men around the age of 50, who had no previous experience of research projects, had perceived that the purpose of this particular project was to abolish hunting altogether. Therefore, they vehemently opposed the whole project and claimed that there was plenty of game in the forest and no reason to worry. Finally, however, the wife of one of these men stood up and exclaimed that her husband was 'talking rubbish'. 'Before, he used to go out to hunt and returned home at noon with bunches of birds,' she said, and continued: 'Nowadays, when he goes out to hunt he does not come back until after dark, and nevertheless returns empty-handed. This is because the game is getting depleted!' General laughter and no more opposition to the research project was heard.

INTERACTING WITH NATURE AND SOCIETY

Having the harvester's position clarified, all of the other actors, as important as their roles might be, are only indirectly connected to the ecological interface of the harvest system. Thus, seen from the harvester's perspective, all of the other human beings forming part of any harvest-based commodity chain depend on his or her work and decisions. Each **harvester's decision**, then, is based on an often invisible and even intuitive process of evaluation of different social and ecological dimensions of this whole that simultaneously allows harvest to take place and limits the harvester's freedom of choice. This evaluation always involves a considerable degree of uncertainty. Although sometimes there is not much choice between alternative options, most of the times there is, and these decisions have an effect.

The provision from nature is based on a **network** of different interrelated ecological processes that together provide the conditions for flows of harvestable biological entities to appear. Likewise, when the harvest supplies wild species goods to complex markets after the harvested matter is added into society, its flows form a network along chains of trade, processing, and consumption that shape the harvesters' transactions with society. It is easy to realise that the further one goes from the act of harvest, the less clear all of these networks start to look and the less direct **control** the actors have over the conditions that regulate harvest in both nature and society. The ways in which wild species harvest ties together the natural world and the embedded human societies thus often form a manifold network that, if thought as an imaginary flow chart, may be seen as assuming fractal-like forms (Merkx, 2007; Hamilton et al., 2007; Figure 21.1).

These ever branching forms help to emphasise how the complexity of nature is inevitably far beyond the comprehension of any individual. Although he or she may understand relatively well the dynamics of the resource that is under use, including its abundances and qualities as well as the conditions affecting them, the entire ecological system behind them is but an abstraction. Each harvester interacts with nature in such ways that decision-making inevitably involves some degree of assumptions, beliefs or guesstimates. The harvesters also meet provision from nature individually, as schematically illustrated in Figure 21.2A. However, when there are many harvesters that work together, it may also be seen that it is a group of people rather than an individual that interacts with nature

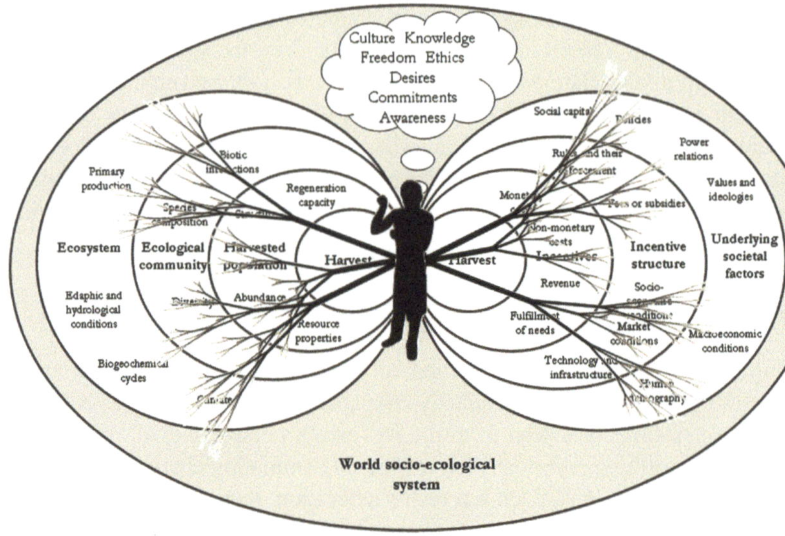

FIGURE 21.1 A harvester forms a 'node' that connects society with the provision of resources from nature. From the harvester's perspective, looking towards any direction, the chains of interaction both within the natural domain and society soon start to diverge and get fuzzier. Also, the concept of a harvester is to a degree an abstraction; harvest decisions are sometimes made by an individual, but often they are the result of several decisions made by different people. The harvester's transactions with nature and with society can be seen as a continuous negotiation process between the harvester and the ecosystem on one hand and society on the other. For every intervention and flow of matter, there is an effect, an influence, and a feedback – both on the side of the ecological system and in society. This reciprocity is an inherent characteristic and a driving force of a socio-ecological system.

(Figure 21.2B). Although these figures are simple as schematical theorisations, they help in the successful diagnosis of a case by illustrating where and how the important transactions between nature and the harvester take place and how they may little by little fade out to finally become almost unperceivable.

The harvester's transactions with society are equally multifaceted and vary from one situation to another. Sometimes, it is only the harvester and his or her stomach that are involved, as is the case, for instance, in picking a fruit and getting rewarded only through the satisfaction of eating it. However, often there are two or more parties involved. When wild species goods are traded and processed, harvest decisions involve multiple actors, either directly or indirectly (Figure 21.2C and D). Moreover, the range of choices available to harvesters making decisions is affected by a wide variety of societal arrangements involving actors from different places and at different levels of governance. Furthermore, the actions and decisions of people affect the options of harvesters; they are in direct or indirect interaction with neighbours, urban consumers, politicians, and policymakers, to name a few. This is how we all are, to some degree, accountable when the decisions made by harvesters are evaluated.

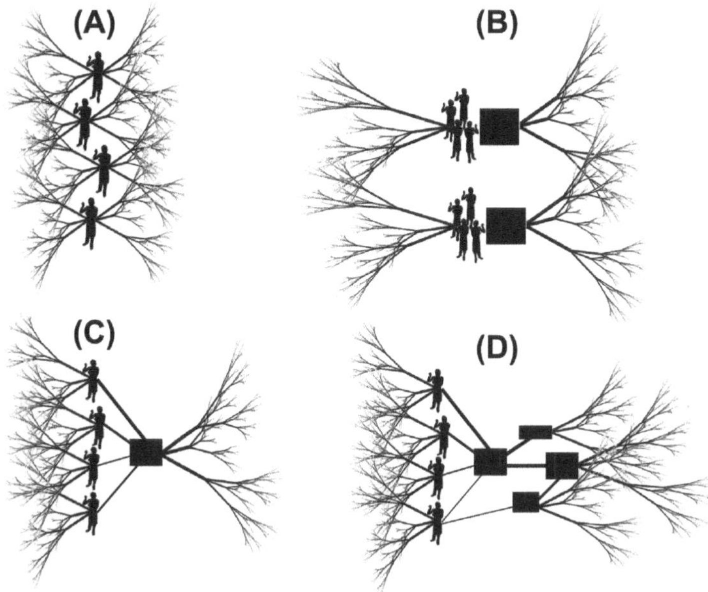

FIGURE 21.2 (A) Harvesters form nodes that distribute extracted matter from nature into society. Each harvester faces a more or less unique setting of natural and societal factors when the harvested resources are consumed locally without much further processing or chains of trade. (B) A group of harvesters working collectively can also form the interface between nature and society. (C) Sometimes, intermediary buyers control the harvested goods purchased from various harvesters and take care of further processing and local trade through a centralised node. (D) More complex commodity chains increase the network of interlinked actors. In this case, for the harvester as well as the end consumer of the wild species goods, it may be very difficult to see the other end and to comprehend how the whole system works.

When the harvest-based commodity chain is looked at from the other end, from the perspective of the **end consumer** of wild species goods, the reality often looks very different. The biological matter may have come a long way to arrive at that point. In addition, the more complex the chain, the more things have influenced the form in which, and the way how, the wild species goods finally end up with the consumer. As a consequence of the long commodity chains involving trade, processing, marketing, and other processes, harvest can thus have an important cumulative overall contribution to society. However, looking down the commodity chain all the way to the harvester and beyond – to the ecological system – can be difficult. An overseas consumer of products certified as fair trade, indigenous, or organic may associate them with very different values compared to the original harvester of the biological matter or to the different intermediary actors comprising the commodity chain.

It is not easy to get an overview of all of the different stakeholders and their interests. There are households, rural and indigenous communities, small and large companies, nongovernmental organisations, different governing bodies, policymakers, political and economic elites, individual consumers at different levels, beneficiaries of ecosystem services, the scientific community, and many others who have unique positions, and within such groups there is also diversity. Because of their divergent interests, values, power, and means of making decisions, the same policies do not represent all actors equally and rules or incentives may not have the same efficiency in relation to all of the different actors.

What all these stakeholders have in common, however, is that all actors can potentially benefit from the provision from nature. This provision can be harnessed and made useful through the work of humans, be that work manual, mechanised, or intellectual. As a consequence, some forms of wild species harvest give rise to large industries such as forestry and fisheries, with their associated **commodity chains** and technologies. Some of these forms of wild species harvest have even constituted the cornerstone for the construction of nations, such as forestry in Finland or fisheries in Iceland. Even if all of the material needs of humanity one day become available from other sources, the existence of wild species harvest as an activity that ties together the natural and human systems can be a value in itself. Harvesting wild species of plants and animals is, after all, something very essential to being human, something that connects us not only to nature, but also to our own deepest roots.

STAKEHOLDER NETWORKS, COMMITMENT, AND TRUST

Because there is a great diversity of stakeholders, their interests also vary from those that have relevance on the forest floor, the high seas, lakes, or rivers to those that move consumers and, finally, those actors found at the levels of policy formulation and governance. These interests shape the ways in which harvesters can work. Although some of them are mutually reinforcing, many others are antagonistic or competing. There may be various kinds of networks and feedback. On the grassroots level, harvesters try to improve the overall performance of their extractivism in the presence of other actors that may strive for antagonistic goals through such activities as harvest of different wild species, agriculture, or infrastructure development. Moreover, the goals that different actors pursue may affect how harvesters can improve their way of life.

Pressures fall upon people making harvest decisions. One may simultaneously have to seek answers for a variety of questions, such as whether to maximise extraction in the short term or the long term, how to allocate benefits generated by harvest and in the commodity chain, and whether to promote harvest as a low-impact cash generator or as a motor of regional or national economic growth, to list a few possible dilemmas. All of the involved actors cannot be reasonably expected to make their interests converge with the interests of others. Not even the long-term sustainability of extractive activities is necessarily a common goal shared by everyone. The inevitable existence of a degree of antagonism should not be seen as an excuse for one not to act, however.

Identification of links between different actors that have a stake in wild species harvest should show where the reconciliation of goals takes place most efficiently.

Different actors along the commodity chains have different degrees of freedom of choice, however. Some people depend on wild species harvest for subsistence from one day to another and have little other choice than to continue with what they are doing. These harvesters can therefore hardly be expected to bear alone the responsibility of taking alternative action with long-term sustainability in mind if this implies a loss of benefits in the short term. On the other hand, although actors involved in the trade of the harvested goods often reap larger benefits compared to the harvesters, their roles also vary greatly, spanning from those intermediary merchants that depend to a great extent on the harvest of a handful of extractors of a single wild species resource to those traders that sell wild species goods just as an almost insignificant item among a myriad of other products.

Finally, among the end consumers, those who heavily depend on a few available wild species goods are in a completely different position compared to those consumers that buy selected wild species products just for the pleasure of it and do not depend on them in any significant way. This last type of end consumer probably has the largest **freedom of choice** because they can increase their consumption or refrain from buying the products without suffering any major negative consequences. All of this points to the reality that the actions of those individual people with the most freedom of choice can often have little direct effect at the level of the everyday life of the harvesters – and thus points to the need for joint actions of many actors involved and at different levels.

Stakeholder consultations are today, at least in theory, promoted as an essential part of any institutional reform targeting natural resource use and governance. The stakeholders form a social network in which the actors are linked to one another through socially meaningful, trust-based relationships that can be either direct or indirect and whose strengths vary (Crona and Parker, 2012; Crona and Hubacek, 2010; Prell et al., 2009). For this reason, the DWiSH Procedure involves the identification of any relevant actors, but this should be done with sufficient care. It is important to avoid prematurely ranking different stakeholders based on one's preconceived ideas about what their roles and interests are.

The simultaneous commitment of people to sustainability goals and to compliance with institutional constraints is a question of legitimacy and trust. Commitment to common goals is impossible if there are no goals that, to a sufficient degree, are shared by all of the relevant stakeholders involved. Mutual interest and trust can nevertheless be generated among actors that share some interests related to wild species harvest. Incentive structures should promote the implementation of policies that have been formulated to reflect common values, but in practice such values are not always easy to identify and thus it is often difficult to devise institutions that would make people simultaneously adhere to the rules that are in place and commit to the goals defined through public policies.

TABLE 21.1 Tradeoffs between Livelihood and Biodiversity Conservation Goals in Wild Species Harvest

		Livelihoods	
		Win	**Lose**
Biodiversity conservation	**Win**	Management actions increase socioeconomic benefits from harvest and enhance biodiversity conservation	Management actions decrease socioeconomic benefits from harvest but enhance biodiversity conservation
	Lose	Management actions increase socioeconomic benefits from harvest but cause losses of biodiversity	Management actions decrease socioeconomic benefits from harvest and cause losses of biodiversity

Source: Modified from Shackleton et al. (2011b).

It is often possible to identify the potential for win–win solutions to problems related to wild species harvest. Only in rare cases where wild species are either in a completely pristine state or managed extremely well may it be impossible to increase the socioeconomic benefits from harvest without harming the biodiversity at least a little, or vice versa. Making win–win solutions a reality is, however, much more difficult to achieve. Especially if harvesters discount future benefits at a high rate, conservation and production goals may not only be divergent but so antagonistic to each other that win–win solutions in practice are impossible. The tradeoff between livelihood goals and conservation goals are at the centre of concerns regarding sustainability (Table 21.1).

WORRIES AND DISPUTES

Scientists studying environmental problems often feel that their role in society is to provide recommendations, as simple as possible, on how to handle complex and multifaceted phenomena. They also believe that, by providing facts and evidence-based alternative options to decision-makers, they contribute to solving problems in human–environment relationships. Often, this is how it works. Yet, providing well-founded recommendations in an efficient way is not always a straightforward process.

For instance, when researchers provide information regarding recommended levels of hunting in tropical forests, one may ask who the addressee of such recommendations is. In many cases, hunters work in near-open-access situations, and there is no individual or group of people that actually has the power to determine or even significantly influence the practices of others. National laws that in theory regulate hunting may have little or no influence on the praxis of harvest on the forest floor, and less so in remote areas where the state's presence is not only weak but sometimes almost nonexistent. In such situations, any recommendations from researchers to be implemented by authorities are likely to be just words in the air and letters on paper.

There is also a risk that strong **preconceptions** or long commodity chains blur objective judgment. A belief in the Brazil nut (*Bertholletia excelsa*) harvest's positive contribution to the conservation of tropical forests may lead to purchasing certified nut packs as a sign of consumer concern. Nothing is wrong there, but it may be difficult for the consumer to see those vicious poverty traps in which many Brazil nut collector families live, despite a fair trade premium that may form part of the consumer price. The commodity chain is long; in addition, even if the premium was genuinely addressed to the harvesters, it may have a very small direct contribution to their overall income levels. Although the Brazil nut collectors earn money, their work in the forest is hard, sometimes dangerous, and always involves uncertainties. There is a risk of nuts deteriorating during transportation, buyers not paying as agreed or as much as the harvester expected, or market prices fluctuating unpredictably because of reasons far beyond the control of local harvesters and buyers.

The expressed interest to support and promote low-impact harvest is entangled with implicit (and involuntary) market-driven consequences in harvester livelihoods. Without disregarding in any way the fair trade or other paid premiums, better livelihoods should be and also are constantly pursued through improving the overall working conditions of the harvesters and the harvest system as a whole. Additional contributions, such as company–community partnerships involving capacity-building and transfer of know-how, can improve the harvester economies (Ros-Tonen and Kusters, 2011), for example, by establishing guaranteed contract prices that can decrease uncertainty and income irregularity during the year (Morsello et al., 2012) or by empowering women in harvest economies (Shackleton et al., 2011b); however, these effects can be difficult to perceive by the end consumers. Sometimes, price hikes in booming harvest economies can also mean increasing hardship for marginal populations that depend on the booming wild species products, as seen in the case of açaí palm (*Euterpe oleracea*) (Chapter 11).

There are also other kinds of contradictions that may seem irreconcilable. If the external observer thinks that roads in Amazonia are evil because they usually lead to deforestation, it may be unpleasant for him or her to dedicate thought to the ways in which improved logistics and consequent trading conditions could help Brazil nut extractors to achieve better livelihoods. Alternatively, if one's interest is in commercial logging and in the associated economic opportunities, it will be hard to tolerate those policies that dedicate large forest areas to nontimber forest product concessions with strict limits to timber extraction. Naturally, there are also people whose standpoint is that letting some areas be used intensively will ultimately open the possibility to leave other forests protected elsewhere.

Whose interest is valued most and prevails? Whose action would be needed if some specific change is set as a goal? These are not just questions of academic interest but also form part of much wider debates shaping opinions and perceptions and, through action, the world in concrete ways too. Much has happened since the assassination of Chico Mendes in 1988 (Chapters 3 and 8). Besides being a wild species harvester, Mendes was an outspoken environmentalist, an active labour union leader, and a human rights defender whose work contributed to the development of cooperative systems to support the livelihoods of Amazonian extractivists. In the 1980s, his work became widely known reference for local struggles over traditional **extractivist lifestyles**, coinciding in time with increasing international concern about accelerating deforestation in Amazonia. Many of those people whose interests were reinforced by broad-scale development efforts, such as road construction and land conversion for agricultural production and cattle ranching, disliked the kinds of criticism and activism represented by Mendes. However, some of the deepest Amazonian wilderness areas became, along with these confrontations, frontiers that accommodated both traditional extractivism and expansionist, market-driven economies.

The murder of Mendes marked a turning point in Amazonian development policies and politics, but probably in ways different from those the people behind the assassination had expected. The Chico Mendes Extractive Reserve was created in 1990 to support the livelihoods of rural dwellers that made their living collecting wild rubber and Brazil nuts in the forests in the region where Mendes had worked, and, meanwhile, the establishment of **extractive reserves** was becoming a popular conservation and development tool also elsewhere in Amazonia (e.g. Dourojeanni, 2012). All of these developments were based on the hope that providing clearly defined conditions for extractive resource use, including better guaranteed rights to land, could help overcome fundamental social and environmental confrontations (Fearnside, 1989). Many wild species resources have been important trade commodities throughout the history of the world and, it was hoped, pure market mechanisms could thus reveal the efficiency of this approach.

However, due to the relatively high human density in some extractive reserves, the livelihoods of the harvesters have, to a certain degree, shifted to agriculture and livestock husbandry. This has occurred partly in order to produce meat for subsistence consumption because the wild game has largely been hunted out from the reserves, and partly for commercial purposes, given that agriculture and livestock production has turned out to be the most profitable use of readily accessible land (Peres, 2011). Moreover, smallholders engaged in extractive activities in rural Amazonia have also often failed in market terms and frequently depend on continued external support (Pokorny et al., 2012). When the harvest, processing, and exporting costs are high and market prices are low or plagued by unpredictable fluctuations, problems of a socioeconomic nature cannot be escaped even if the harvest itself was made sustainable.

Local dwellers are well aware of their realities but often have limited knowledge of how the harvest situation connects to the larger world. Similarly, people living far from the everyday life of the forest-dwelling population may promote conservation without a clear notion that it can be at the cost of reduced local development options, finally limiting the growth possibilities of sustainable economies based on renewable natural resources.

FROM AWARENESS TO ACTION

Diagnostic assessment of any particular case of wild species harvest should help to improve the understanding and awareness of the ways in which the actual harvest and the surrounding realities interact. Evidence-based examination facilitates the evaluation of different alternatives and supports the stakeholders' understanding of how multifaceted the picture finally is. This should help one comprehend the ecological effects of harvest in relation to the benefits obtained, and vice versa, as well as facilitate the evaluation of the perceived potential of wild species resources as a source of economic incentives for the protection of natural habitats. A broad comprehension also opens windows to see if and how overharvest of wild species contributes to impoverishment of human populations.

However, knowledge and improved understanding are not sufficient on their own. The provision of recommendations in the form of ready-to-use recipes for action is of little use unless they have a clear addressee and are provided in such a manner that they actually reach that addressee. Furthermore, unless the addressee shares the view on the goals of the recommendations with those providing them, they will not heed the recommended action. There is an ongoing (and sometimes heated) debate between those conservationists who are convinced that national governments are those who have the best will and abilities to protect nature (e.g. Terborgh, 1999) and others who instead put their trust in indigenous people and forest and rural dwellers in general (e.g. Schwartzmann, 2000). In reality, there is no reason to take it for granted that either governments or local populations are fully committed to conservation, or that they have the know-how to formulate efficient management strategies and the capacity to implement them. However, neither is there reason to believe that the situation is always the contrary. Different actors have different interests and possess different resources, and the issue is about whether forces can be joined in order to strive for such interests that are shared.

The best that researchers can provide may not be recommendations as such but rather capacity-building that helps to generate knowledge about the potential for improvements among the actors involved and in society at large (Chapter 17). The more the different stakeholders have access to reliable sources of information and the more they share relevant knowledge, the easier it will be to make decisions that are understood and accepted by all parties involved. For this reason, in the application of the DWiSH Procedure in development-related contexts, efforts

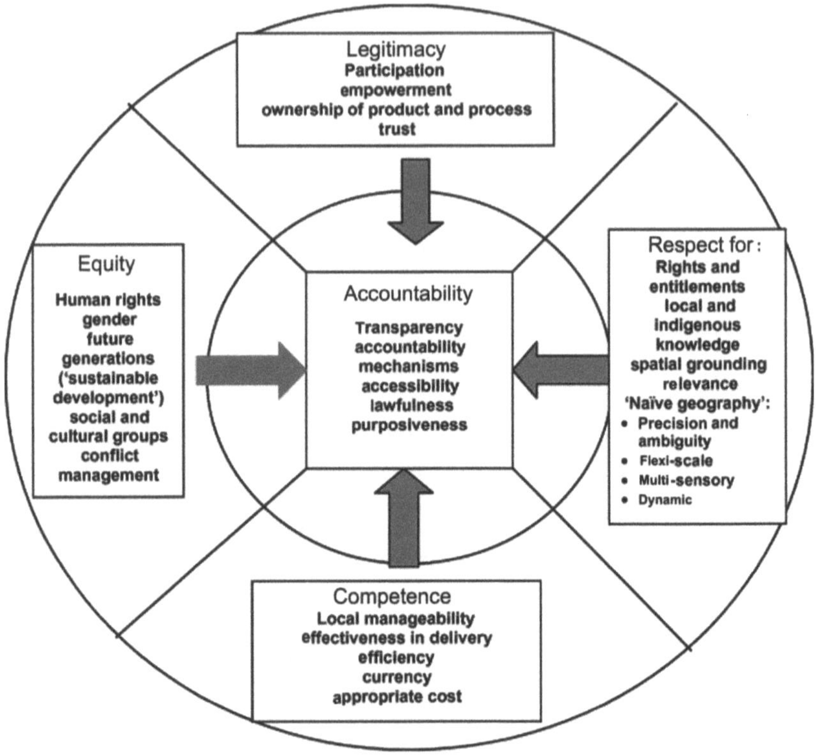

FIGURE 21.3 Components of good governance in participatory spatial planning. *From McCall and Dunn (2012).*

should also be made to avoid accumulating valuable information into the hands of a few people only. If data resources are instead transparent and easy to access, information is shared and facilitates evidence-based argumentation. Stakeholder involvement can also be facilitated by the use of the same conceptualisations and terminologies and thereby creating better possibilities to identify similarities or dissimilarities in the expressed opinions. Consultations and public involvement in decision-making furthermore supports local empowerment (Reed, 2008), and it also contributes to **accountability** or good functional relationships between the governing and the governed (Chapter 16) in terms of legitimacy, respect, equity, and competence (McCall and Dunn, 2012; Figure 21.3).

Chapter 22

Savage, Ravaged, or Managed?

FIGHTING THE FATES OF OVERHARVEST AND OBLIVION

Trial and error is the prime way through which civilisations have generated their first-hand knowledge related to wild species harvest. The modern version of this simple approach involves flexibility in evidence-based decision-making. This can be fuelled by structured and iterative processes in which new inputs from real-world experiences and scientific understanding can be incorporated at any time, and where feedback from both the ecological and social systems is addressed with care. These principles are implemented in natural resource management according to the principles of **adaptive management** (Westgate et al., 2013), which implies the recognition of one's limited knowledge about the system being managed, and therefore includes **learning**, in itself, as one of the management objectives. Adaptive management involves critical thinking before interventions are made and it also facilitates comparisons by considering alternative management measures and then monitoring and evaluating their outcomes, after which the accumulated knowledge is used to plan how to continue management. All these characteristics make it suitable to be combined with **participatory planning**, and, of course, with the **DWiSH** Procedure, thus providing local resource users with possibilities to learn directly by personal experience with resource management and governance.

This all sounds very good but, unfortunately, adaptive management is more famous than it is common. This situation prevails despite adaptive management being often considered as the best available approach for the management of complex socioeconomic systems in the presence of considerable uncertainty (Westgate et al., 2013). Ideally applied adaptive management is rare in practice because of a number of problems that make its application difficult; as a result, many processes that have been labeled adaptive management have, in reality, not fulfilled all the requirements of the approach.

A review of adaptive management literature by Westgate et al. (2013) lists six steps that are widely agreed to form the basis of adaptive management: (1) the goals of management interventions must be defined with the involved actors; (2) multiple alternative management options must be identified, possibly including no action at all; (3) quantitative conceptual models have to be developed so that the outcomes of the alternative interventions can be statistically evaluated; (4) alternative management interventions have to be implemented; (5) the results of the interventions must be monitored in order to use the models; and (6) the

Diagnosing Wild Species Harvest. http://dx.doi.org/10.1016/B978-0-12-397204-0.00022-X

management interventions must be adjusted reflecting the conclusions derived from the analysis of the outcomes (Figure 22.1).

Because the process of adaptive management simultaneously intends to improve the management interventions and the knowledge about their outcomes, there is a risk that these goals are antagonistic. For example, collecting all of the seeds of the Brazil nut trees (*Bertholletia excelsa*) as one alternative and leaving all uncollected in another would both have serious effects. In the first case, the long-term regeneration of the population would be interrupted, whereas in the second case no extraction could be carried out at all. Both of these alternative interventions are thus antagonistic to sustainable harvest. In the adaptive management literature, dilemmas like this are called a **dual-control problem** (Walters and Hilborn, 1978).

Adaptive management should be distinguished from approaches based on trial and error only, as well as from reactive management, because both superficially resemble adaptive management. In the first, management alternatives are developed further based on their outcomes without explicit conceptual models developed *a priori* to monitor their outcomes. In the second, management actions, in turn, are adjusted as reactions to the system as information about new problems appear. Another major challenge in achieving genuine adaptive management processes is that they must be based on systematic monitoring that extends long enough in time in order to detect relevant outcomes of the different management alternatives in the studied system. Furthermore, the assessment of adaptive management processes should also take into account spatial scales and geographical variability in order to make the alternative management interventions comparable (Westgate et al., 2013).

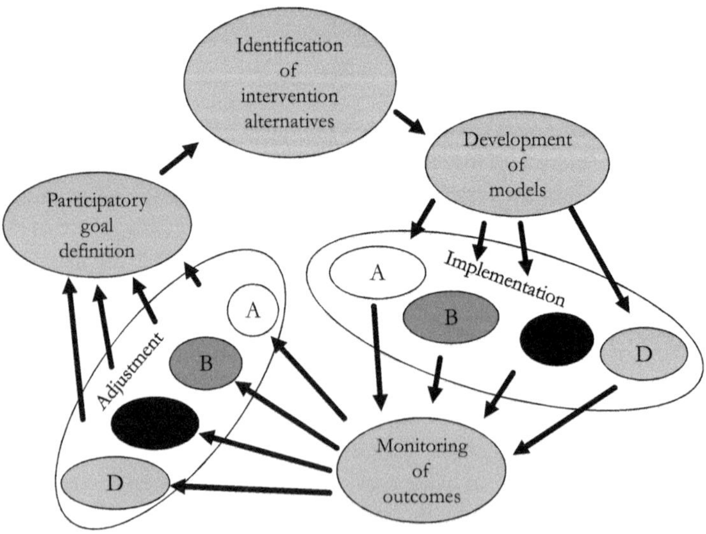

FIGURE 22.1 A simple schematic representation of the adaptive management cycle. See text for details.

Adaptive management in the field of wild species harvest would considerably benefit from the **participation** of harvesters and other stakeholders that manage information and knowledge. Such participation requires that all stakeholders have access to any relevant information as far as possible. This requirement should be considered as an inherent part of the process to facilitate good communication (Chapter 17). The systematic collection of operative knowledge by the involved actors would help the monitoring of alternative management interventions in the long term (e.g. for forestry; see Salo and Toivonen, 2009). There must also be a degree of understanding about the area's ecological characteristics and about the needed management efforts in practically any setting where the populations of wild species are used. Small actors often possess direct information on the resource stocks, regeneration potential, and other such variables, but this knowledge is not reflected in larger-scale planning (Salo, 2009).

The needs of **research-based facts** are endless because scientifically reliable information should cover a range of topics, from the harvester's household level to a multitude of complex systems within nature and society. That said, it is good to remember that the use of scientific knowledge is only one step further. For example, Brunner (2005) highlighted the need to move from scientific management, in which science aspires to take its place above politics, to **adaptive governance**, which integrates distinct forms of knowledge into different decision-making processes and is also based on continuous learning (Crona and Parker, 2012).

THE BIOGEOPOLITICS OF WILD SPECIES HARVEST

In the aftermath of the 1992 Earth Summit in Rio de Janeiro, many politicians and researchers alike highlighted the riches hidden in the largely unstudied biodiversity found in many tropical countries (Chapters 2 and 12; Box 2.1). According to the popular political discourse, biodiversity was a potential source of wealth for those countries that were lucky enough to have the promising species within their borders – and wise enough to conserve and maintain their living environments. This was an argument in favour of not only conservation but also of stricter measures of control regarding the use of different organisms and the biological materials they may provide, and particular interest was focused on those countries and areas that were labelled as 'megadiverse' (Mittermeier et al., 1997).

Counterarguments to the mainstream biodiversity discourse also emerged, often centring around the intuitively attractive question about why diversity should be conserved in poor tropical countries if it happens at the cost of efficient resource use, particularly because it seems that many of the rich countries have already destroyed much of their own diversity in the very processes of making more efficient use of land and ecosystems. A related question is, then, whether it is not also true that it is easier to manage simple systems in order to maximise profits and growth. Furthermore, can it be so that the extraction of raw materials based on wild species is often merely a nationwide poverty trap? In this way, through questions about the allocation of power and wealth in relation to the geographical distribution of biological resources, biodiversity has become an explicit topic in global **geopolitics**.

These concerns add to a long tradition of how tropical nature has been seen from the outside, commonly oscillating between feelings of a paradise and a green hell (e.g. Nygren, 2006). The value of biodiversity itself may have been overlooked, but many particularly important wild species resources have been subject to harvest and even struggle through the centuries. But, although tropical nature has always caused curiosity and wonder in the minds of people from less biologically diverse parts of the world, the imagery of a paradise blessed with untold riches has had its dark sides too. Since colonial times and well into the twentieth century, the tropics have often been described as places of hardship and pestilence, where nature has dominated humans and not the other way around (Arnold, 2002). This dualism has remained present, yet in subtler forms, in many subsequent debates on the role of biological diversity in development in general and about how nations choose their paths in particular (Swanson, 1999).

Biodiversity has thus entered global politics and has been adopted into discourses promoting such things as sustainable development, economic growth, and international trade. Whenever biodiversity resources are highly valued, such as offshore fisheries along the important fishing coasts of basically all continents from polar areas to the tropics, these resources have potential to become entangled with power struggles and even conflict. A well-known example is from the horn of Africa, where many local Somali fishermen – who later became the famous Somali pirates operating in the Gulf of Aden – claimed to function as an extraofficial force of coast guards against foreign trawlers illegally fishing in the area (Hansen, 2009). Although the root causes and reasons for the persistence of piracy in the area in the early twenty-first century have been far more complex than that (e.g. Pham, 2010), it is clear that the struggle over fisheries (Guha, 2012) has played an important role in the developments that have involved not only the local dwellers of the coast and foreign commercial fishers but also the European Union, the United States, and great maritime cargo transport and oil corporations, among other actors.

Tropical forests have sometimes formed scenes of armed conflict and war (e.g. the situation in the Putumayo River described in Chapter 9). Although violence often threatens biodiversity, peace that opens doors to more intensive exploitation of natural resources can also contribute to further degradation (McNeely, 2003). However, it is not the resources in themselves (not even lucrative and easy-to-transport ones, such as valuable timber) that create conflict. The picture is far more complex and involves historical, political, institutional, and cultural factors that are case-specific (e.g. Snyder, 2006). Similarly, it is difficult to establish any globally significant direct causal links between biodiversity and socioeconomic development. Considering that high biological diversity in general contains numerous potentially useful species and that there are indigenous peoples with millennia-long persistency in these regions, **megadiverse** tropical areas are blessed with unique possibilities. But is it really so – and even if it was, is this reflected in how the nations have developed? Is the great abundance of species, economically important wild species in particular, a blessing rather than a curse? And, what would a biogeopolitical map of the world look like, and what would be the criteria used?

To our knowledge, such maps have not been produced and published, but as a thought experiment they can be illustrating. First, let us see how biodiversity is distributed on the surface of Earth. If a global map depicting the number of species found per unit area existed, it would certainly show some specific concentrations of species richness in some places of the planet. Again, such maps for the whole diversity of species are still beyond the capacities of science today, but many have been published to represent selected groups of organisms, such as plants (e.g. Kreft and Jetz, 2007). Many of the concentration areas of biodiversity shown by such maps are found in tropical rainforests, but also other centres of diversity emerge, such as some subtropical mountain ranges and coastal areas. Clearly, biodiversity is not evenly distributed on this planet.

If the distribution of biodiversity is globally aggregated, so are human populations. Thematic atlases usually contain a map showing the density of human population globally, clearly confirming that our species is not distributed with equal density over the globe (e.g. CIESIN et al., 2000). Likewise, the intensity of economic activity is certainly not similar in different parts of the planet. Where there are more people, generally (but not always) there is an associated higher gross domestic product density per unit area as a surrogate for economic activity (Gallup et al., 1999). When the densely populated areas with high economic activity are compared to the areas where biological diversity is concentrated, we can observe that if there is a link between the two, it is not straightforward.

There are some regions that harbour high levels of biological diversity as well as very densely populated hubs of human activity, such as some parts of China. Then there are other such concentrations of human population and economic activity that are surrounded by much less biologically diverse environments; among these places could be many locations in Europe and North America. At the other end of the human density spectrum are places that are very sparsely inhabited and characterised by low economic activity per area unit, such as large parts of the Saharan desert, northern tundra, or the Amazonian rainforest. When we compare the biological diversity found in these regions, which are among the largest wilderness areas on Earth, we can easily notice that the Sahara and tundra are home to fewer species than Amazonia.

Even these simple comparisons of **biopolitical maps** of diversity reveal the fact that biodiversity *per se* cannot be easily portrayed as a blessing or as a curse for humans inhabiting places that vary in relation to the diversity of other species. High levels of biodiversity neither guarantee that there are high human population density and economic prosperity, nor do they necessarily imply low population levels with people condemned to poverty. In addition, relatively low biodiversity does not seem to indicate either poverty or affluence. Thus, defining the role of biological diversity in the development of nations clearly escapes any simple generalisations. The biogeopolitics of biodiversity are apparently affected by more factors than the mere biodiversity resources themselves.

The distributions of the remaining wilderness areas on Earth indicate that there are only three areas that are both large and with high biodiversity: Papua New Guinea, the Congo Basin, and Amazonia. Just a few decades ago, the islands of Borneo and Sumatra would probably also have qualified, but they have since then been largely converted into agricultural lands and oil palm plantations; what little forest remains is criss-crossed by roads and heavily logged. Another characteristic shared by the three large wilderness areas above is that they mostly belong to countries that have experienced and often still experience major governance problems related to deficient law enforcement and compliance (e.g. Reboredo, 2013), insufficient political participation of marginal populations (e.g. Colfer, 2011), high levels of corruption (e.g. Wright et al., 2007), and even violent conflict (e.g. Mcneely, 2003). Even within countries, there can be considerable variability in governance, and remote and sparsely populated regions may have lower levels of governance than the urban centres where most of the population lives. This combination of being a large wilderness area with high biodiversity and low level of governance can lead to very particular challenges when it comes to management of wild species as natural resources in Amazonia.

Looking back, two decades after the 1992 Rio de Janeiro Earth Summit, the Convention on Biological Diversity (CBD) has become one of the most widely ratified multilateral environmental agreements of all times. This is not a surprise considering the fact that its central principles offer something for everyone: 'the conservation of biological diversity, the sustainable use of its components and the fair and equitable sharing of the benefits arising out of the utilisation of genetic resources' are objectives that are difficult to oppose. CBD lifted the hopes of wild species use, being a path for development and growth. As access rights to biological resources, transfer of relevant technologies, and appropriate funding were also listed as objectives, it is clear that an agreement of this magnitude raised high hopes throughout the globe, not least in many tropical countries with diverse ecosystems and rich in biological resources.

CBD has been outstandingly successful in making biological diversity politically addressed as a common concern of the humankind (Chapter 2). The convention's premises are incorporated into mainstream politics and policies on all continents, not only in environmental administration but also in other sectors of society. The convention has lifted many important issues on the political agenda, making such topics as indigenous peoples and traditional knowledge, protected areas, gender issues, benefit-sharing, and rights to biodiversity-based intellectual property part of development discussions worldwide. Its principles are constantly implemented through new programmes, initiatives, and protocols, adding more components to its already overarching nature. However, there has also been room for severe criticism.

Precisely because CBD has something for everyone, it is also often vague in its wordings and leaves implementation to the mercy of varying political interests. The use of genetic resources and the rights to such resources and associated intellectual property have often become contentious questions, too (Box 2.1). Although the unauthorised appropriation of genetic material and traditional knowledge – biopiracy – can be real problems, overemphasising such risks can also diminish interest in scientific research. Finally, the most overarching issue of criticism may be the one that points to the success of the convention in the field of conservation. There is an increasing body of empirical evidence about the state of biological diversity worldwide, showing that there is no reason to celebrate yet (e.g. Butchart et al., 2010). Whether this can be attributed to failures of CBD is another debate, however. It can be argued that, without the convention, things would probably be worse.

WILD OR NOT-SO-WILD NATURE

A basic requisite for any economy based on wild species harvest is that there are ecosystems for the harvested species to thrive. However, the wilderness beyond human influence is a utopia. All places on Earth, maybe with the exception of some curiosities such as ancient closed caves and Antarctic lakes under continental ice, have been affected by humans to some degree. In most places, human activities have changed the natural conditions through direct or indirect mechanisms. This implies that many species are actually harvested from human-modified environments all over the globe. These species are wild, however, in the sense that they keep reproducing in nature, be it that they suffer or benefit from existing human activities. Wild species can actually occur and are harvested anywhere, even in heavily human-modified landscapes as evidenced by, for example, lines of fishers along many rivers flowing through completely built-up urban areas or people picking berries in intensively managed production forests. Wild species are also invading cities at an increasing pace (Niemelä, 1999). Even some edible plants could be easily collected in cities, yet there may be doubts about their quality for human consumption due to pollution. Even the most urbanised populations of many species can, nevertheless, be considered wild if their populations survive and distribute through natural mechanisms. They are just adapted to live in a new type of environment.

Habitat modification and land conversion suppress some species, but they also offer a variety of new opportunities for many species. This does not occur equally for different parts of the planet, however. For example, many regions such as Western Europe, the coastal United States, Asia, and other very densely inhabited regions have seen such historical levels of conversion that there is not much more to convert; thus the pace of further changes has remained slow and is likely to stay so in the future. In contrast to this, many tropical areas are currently experiencing advancing processes of land conversion in which little or moderately modified ecosystems are harnessed to intensive human uses. This diminishes the survival potential of many kinds of species that provide the basis for extractive resource use practices. However, some of these species and their uses may still persist. In addition, some species that thrive in secondary habitats created and maintained by human intervention may become increasingly abundant due to human actions and have important use value. Species characterising secondary growth vegetation are actually widely used in many parts of the world.

In the light of what is said above, the distinction we have made in many illustrations of this book, with the harvester in the middle and with natural factors to the left and societal issues to the right, becomes somewhat problematic. Human actions create and maintain conditions in which some wild species opportunistically thrive on the one hand; on the other, there might be increasingly intensively managed populations of harvested species, blurring the line between what is wild and what is human-made (Figure 22.2). A species that is considered wild may in fact be such that it has been introduced into the area, either purposefully or by accident. Or, it can be otherwise considered as not being natural because its vigour, production potential, and population structure may depend to a large degree on management actions. These considerations are

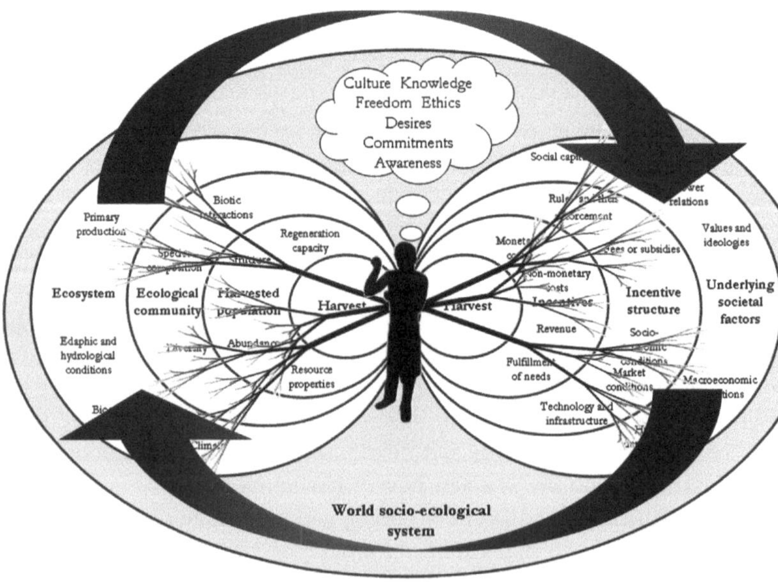

FIGURE 22.2 Harvest systems are embedded in the complex world socioecological system. Natural changes in ecosystems can influence human society, including the incentives of wild species harvest. On the other hand, any such human interventions that modify the provision from nature can impact wild species harvest. What is "wild" in each specific situation is also an a result of a wide variety of factors surrounding the harvest itself.

also linked to questions about the economic value of biodiversity resources as forming an incentive against deforestation.

After more than two decades of active debate on the feasibility of the 'use it or lose it' style of productive conservation (Chapter 1), it is clear the initial high hopes of the late 1980s and the 1990s have been tuned down; this is particularly so in the case of the nontimber forest products that initially presented the 'silver bullet' for productive conservation (Sunderland et al., 2011). This suggests that if extractivism is a third way that combines development and conservation, it is so only under certain conditions and truly successful cases seem to be few (Figure 22.3). In many situations, the significance of wild species harvest tends to decrease as a result of overharvest or because other competing economic activities appear to be more attractive, or for both of these reasons together.

For sure, there is a value in maintaining some areas of nature in a state of wildness, or **savage**, because that is the best (if not the only) way to ensure that there is a possibility for all forms of life, that make up the biodiversity we have inherited, to persist into the future. The harsh reality is, however, that too much nature is instead getting **ravaged**, either through being outrightly converted to agricultural land or pastures, by other intensive human activities, or by being

FIGURE 22.3 Where does the path lead this Brazil nut collector in the community of Sonene in Peruvian Amazonia – and where does the path lead wild species harvesters in general? Is wild species harvest a thing of the past that will progressively be replaced by cultivated or industrially produced raw materials and commodities? Is wild species harvest a threat to biodiversity and ecosystem services, or can it help to conserve them *Photo: Natalia Aravena.*

subject to uncoordinated harvest of wild species. What ultimately is left behind after too intense harvest are those species that are in such inaccessible places, or of so low value to humans, that the costs of harvesting them exceeds the potential benefits to gain – still, but not necessarily forever. Strict protection, including the prohibition of any kind of wild species harvest that may have significant ecological impacts is, however, feasible only up to a certain limit. This is because often such strict protection can be costly when considering the direct costs of surveillance and enforcement, as well as the forgone socioeconomic benefits of harvest.

An alternative way is the one that involves nature being **managed** for wild species harvest, in order to optimise the harvesters' transactions with nature, including the harvested species and the ecosystem at large. This has the double purpose of nature conservation and the provision of socio-economic benefits. These two purposes are compatible with each other up to a certain point only; beyond that point, a tradeoff must be made between the two. Wild species harvest will never be a panacea – not for solving problems of biodiversity loss or those of poverty among people living in remote communities in the Amazonian forest or elsewhere. But, in one form or another, harvest is going to continue.

It will not be possible to save all nature as savage. However, if we do not want it all to become ravaged, let us make sure that as much as possible of it at least becomes well managed – for biodiversity conservation and for the socioeconomic and cultural benefits that the harvest of wild species can provide.

Photo Credits

In addition to our own photographs, a number of others have provided their photos for this book, and we are very grateful for their contributions. All photograph authors, including the authors of the book, are indicated below:

Ammann, Karl (www.karlammann.com): 14.2, 14.4B, 14.9

Aravena, Natalia: 3.10, 8.4A, 8.5B 8.6A, 8.7A, 8.7B, 8.7C, 9.2, 9.4B, 9.4C, 9.6A, 9.6B, 11.1B, 11.5A, 15.6A, 15.6B, 22.3 (cover photo)

Arfman, Tommy: 2.1

BIODAMAZ–IIAP: 12.5

Castello, Leandro: 17.12

Gualinga, Eriberto: 4.4, 4.6, 4.8, 4.9, 5.8, 5.10, 7.1, 7.6, 7.7, 7.9, 16.8

Gualinga, Franklin: 4.1

Hardenburg, Walter: 3.9

Isomäki, Aarno: 12.7

Johansson, Olof: 13.18, 15.2, 15.7

Kalliola, Risto: 18.2

Lilley, Thomas: 2.4

Machoa, José: 15.5

Mikkolainen, Petra: 7.11A

Mäki, Sanna: 12.3A, 12.3B, 12.3C

Norman, Hans-Gunnar (www.vojman.se): 1.1C

Nybrand, Tomas (www.nybrand.se): 15.4

Östlund, Lars: 19.4

Pyhäla, Aili: 16.7

Pärssinen, Martti: 3.4

Salo, Matti: 1.1 A, 3.2, 3.3C, 3.3D, 3.3E, 3.3F, 3.3G, 3.3.H, 3.8, 3.11A, 3.11B, 3.14A, 3.14B, 3.15A, 3.15B, 3.16A, 3.16B, 3.16C, 5.1A, 5.1B, 5.3A, 5.3B, 5.4A, 5.4B, 5.5A, 5.5B, 5.5C, 5.6, 5.7A, 5.7B, 6.2, 6.4A, 6.4B, 6.5A, 6.5B, 6.6, 6.7A, 6.7B, 7.11B, 7.11C, 7.12, 8.2A, 8.2B, 8.3A, 8.3B, 8.4B, 8.5A, 8.6B, 8.8A, 8.8B, 9.1A, 9.1B, 9.4A, 9.5A, 9.5B, 9.7, 9.8A, 9.8B, 11.1A, 11.5B, 12.1A, 12.1B, 12.1C, 12.1D, 13.2, 13.5A, 13.5B, 15.9, 16.4A, 16.4B, 16.5, 16.6, 16.9A, 16.9B, 17.3, 17.4, 17.5, 17.6A, 17.6B, 17.6C, 17.11

Sillanpää, Matti: 20.3

Sirén, Anders: 1.1D, 3.3A, 3.3B, 3.13, 4.3. 4.5, 4.7, 7.2, 7.4, 7.5, 7.8, 10.2, 10.3, 10.4, 10.5, 10.6, 13.8, 13.9, 13.21, 14.3, 14.4A, 14.4C, 14.10, 15.8, 17.10A, 17.10B, 18.13

Sääksjärvi, Ilari E: 3.11C, 12.4A, 12.4B

Tuomisto, Hanna: 1.1B

von Wright, Benedict: 16.2A, 16.2B

References

Addabbo, T., Facchinetti, G., Mastroleo, G., Lang, T., 2009. Evaluating Firms' Gender Equity by Fuzzy Logic. Proceedings IFSA-EUSFLAT, Lisbon.

Agrawal, A., Chhatre, A., Hardin, R., 2008. Changing governance of the world's forests. Science 320 (5882), 1460–1462.

Alavi, M., Leidner, D.E., 2001. Review: knowledge management and knowledge management systems: conceptual foundations and research issues. MIS Quarterly 25, 107–136.

Alexiades, M.N., Lacaze, D., 1996. FENAMADs program in traditional medicine: an integrated approach to health care in the Peruvian Amazon. Medicinal Resources of the Tropical Forest, 341–365.

Alho, C.J.R., 1985. Conservation and management strategies for commonly exploited Amazonian turtles. Biological Conservation 32, 291–298.

Aliaga-Rossel, E., 2002. Distribution and abundance of the river dolphin (Inia geoffrensis) in the Tijamuchi River, Beni, Bolivia. Aquatic Mammals 28 (3), 312–323.

Allendorf, F.W., England, P.R., Luikart, G., Ritchie, P.A., Ryman, N., 2008. Genetic effects of harvest on wild animal populations. Trends in Ecology & Evolution 23 (6), 327–337.

Allendorf, F.W., Hard, J.J., 2009. Human-induced evolution caused by unnatural selection through harvest of wild animals. Proceedings of the National Academy of Sciences 106 (Suppl. 1), 9987–9994.

Allison, E.H., Horemans, B., 2006. Putting the principles of the sustainable livelihoods approach into fisheries development policy and practice. Marine Policy 30 (6), 757–766.

Alvard, M.S., 1993. Testing the "ecologically noble savage" hypothesis: Interspecific prey choice by Piro hunters of Amazonian Peru. Human Ecology 21 (4), 355–387.

Álvarez Alonso, J., 1997. Estado actual de la fauna silvestre en la propesta Reserva Comunal Del Pacacuro. Manejo de fauna silvestre en la Amazonía, Instituto de Ecología, La Paz, Bolivia, pp. 93–104.

Álvarez Alonso, J., 2012. Conservación productiva y gestión comunal de la biodiversidad en la Amazonía peruana. In: Brañas, M.M. (Ed.), Articulando la Amazonía: Una mirada al mundo rural amazónico, Programa de Cooperación Hispano Peruano & Agencia Española de Cooperación Internacional para el Desarrollo, Lima, pp. 7–38.

Álvarez Alonso, J., Araujo Tuesta, A., Rojas Grández, F., 2007. Sistematización de las experiencias de manejo de recursos en las comunidades locales de la RNAM. IIAP–BIODAMAZ, Iquitos.

Amacher, G.S., Merry, F.D., Bowman, M.S., 2009. Smallholder timber sale decisions on the Amazon frontier. Ecological Economics 68 (6), 1787–1796.

Ananda, J., Herath, G., 2009. A critical review of multi-criteria decision making methods with special reference to forest management and planning. Ecological Economics 68, 2535–2548.

Anderson, E.P., Montoya, M., Soto, A., Flores, H., McClain, M., 2009. Challenges and opportunities for co-management of a migratory fish (Prochilodus nigricans) in the Peruvian Amazon. American Fisheries Society Symposium, vol. 69.

Angelsen, A., 1999. Agricultural expansion and deforestation: modelling the impact of population, market forces and property rights. Journal of Development Economics 58 (1), 185–218.

Appelstrand, M., 2007. Miljömålet i skogsbruket - styrning och frivillighet (The Environmental Goal of Swedish Forestry - regulation and voluntariness). Lund Studies in Sociology of Law 26.

Arnason, R., 2012. Property rights in fisheries: how much can individual transferable quotas accomplish? Review of Environmental Economics and Policy 6 (2), 217–236.

Arnold, M., Powell, B., Shanley, P., Sunderland, T.C.H., 2011. Editorial: forests, biodiversity and food security. International Forestry Review 13 (3), 259–264.

Arnold, D., 2002. "Illusory riches": representations of the tropical world, 1840–1950. Singapore Journal of Tropical Geography 21 (1), 6–18.

Asner, G.P., Powell, G.V., Mascaro, J., Knapp, D.E., Clark, J.K., Jacobson, J., Hughes, R.F., 2010. High-resolution forest carbon stocks and emissions in the Amazon. Proceedings of the National Academy of Sciences 107 (38), 16738–16742.

Asner, G.P., Knapp, D.E., Broadbent, E.N., Oliveira, P.J., Keller, M., Silva, J.N., 2005. Selective logging in the Brazilian Amazon. Science 310 (5747), 480–482.

Asner, G.P., Keller, M., Lentini, M., Merry, F., Souza Jr., C., 2009. Selective Logging and its Relation to Deforestation. Geophysical Monograph Series 186, 25–42.

Asner, G., Keller, M., Pereira Jr., R., Zweede, J.C., 2002. Remote sensing of selective logging in Amazonia: assessing limitations based on detailed field observations, Landsat ETM+, and textural analysis. Remote Sensing of Environment 80, 483–496.

Aswani, S., Christie, P., Muthiga, N.A., Mahon, R., Primavera, J.H., Cramer, L.A., Hacker, S., 2012. The way forward with ecosystem-based management in tropical contexts: reconciling with existing management systems. Marine Policy 36 (1), 1–10.

Avital, E., Jablonka, E., 2000. Animal Traditions: Behavioural Inheritance in Evolution. Cambridge University Press.

Ayala, F.J., Gilpin, M.E., Ehrenfeld, J.G., 1973. Competition between species: theoretical models and experimental tests. Theoretical Population Biology 4 (3), 331–356.

Baland, J.M., Platteau, J.P., 1996. Halting degradation of natural resources: is there a role for rural communities? FAO.

Baluarte, J., Vasquez, M., 2000. El intercambio de productos forestales diferentes de la madera en el ámbito de Iquitos-Peru. Parte I. 23. Bosques Amazonicos, Perú.

Barham, B.L., Coomes, O.T., 1994. Reinterpreting the Amazon rubber boom: investment, the state, and Dutch disease. Latin American Research Review 29 (2), 73–109.

Barham, B.L., Coomes, O.T., 1996. Prosperity's Promise: The Amazon Rubber Boom and Distorted Economic Development. Westview Press, Inc.

Barlow, J., Gardner, T.A., Lees, A.C., Parry, L., Peres, C.A., 2012. How pristine are tropical forests? An ecological perspective on the pre-Columbian human footprint in Amazonia and implications for contemporary conservation. Biological Conservation 151 (1), 45–49.

Barnosky, A.D., Koch, P.L., Feranec, R.S., Wing, S.L., Shabel, A.B., 2004. Assessing the causes of late Pleistocene extinctions on the continents. Science 306 (5693), 70–75.

Barnosky, A.D., Lindsey, E.L., 2010. Timing of quaternary megafaunal extinction in South America in relation to human arrival and climate change. Quaternary International 217 (1), 10–29.

Barnosky, A.D., Hadly, E.A., Bascompte, J., Berlow, E.L., Brown, J.H., Fortelius, M., Smith, A.B., 2012. Approaching a state shift in Earth's biosphere. Nature 486 (7401), 52–58.

Barros, A.C., Uhl, C., 1995. Logging along the Amazon River and estuary: patterns, problems and potential. Forest Ecology and Management 77 (1), 87–105.

Basler, E., 1977. Der forstwirtschafliche Begriff der "Nachhaltigkeit" als Orientierungshilfe in der Zivilisatorische Entwicklung. Schweitzerische Zeitschrift für Forstwesen 128, 480–489.

Beck, H., 2006. A review of peccary-palm interactions and their ecological ramifications across the Neotropics. Journal of Mammalogy 87 (3), 519–530.

Beckerman, S., 1983. Carpe diem: an optimal foraging approach to Bari fishing and hunting. Adaptive Responses of Native Amazonians, Academic Press, New York, pp. 269–299.

Bekerman, Z., Kopelowitz, E. (Eds.), 2008. Cultural Education-Cultural Sustainability: Minority, Diaspora, Indigenous and Ethno-Religious Groups in Multicultural Societies, Lawrence Erlbaum.

Berkes, F., 2009. Indigenous ways of knowing and the study of environmental change. Journal of the Royal Society of New Zealand 39, 151–156.

Berkes, F., Colding, J., Folke, C. (Eds.), 2003. Navigating Social-Ecological Systems: Building Resilience for Complexity and Change, Cambridge University Press.

Berryman, A.A., 1992. The origins and evolution of predator-prey theory. Ecology 1530–1535.

Betts, R.A., Malhi, Y., Roberts, J.T., 2008. The future of the Amazon: new perspectives from climate, ecosystem and social sciences. Philosophical Transactions of the Royal Society B: Biological Sciences 363 (1498), 1729–1735.

Bharucha, Z., Pretty, J., 2010. The roles and values of wild foods in agricultural systems. Philosophical Transactions of the Royal Society B: Biological Sciences 365 (1554), 2913–2926.

Birkeland, C., Dayton, P.K., 2005. The importance in fishery management of leaving the big ones. Trends in Ecology & Evolution 20 (7), 356–358.

Bjørndal, T., Herrero, I., Newman, A., Romero, C., Weintraub, A., 2012. Operations research in the natural resource industry. International Transactions in Operational Research 19 (1–2), 39–62.

Blüthgen, N., Klein, A.M., 2011. Functional complementarity and specialisation: the role of biodiversity in plant–pollinator interactions. Basic and Applied Ecology 12 (4), 282–291.

Bodmer, R.E., Robinson, J.G., 2004. Evaluating the Sustainability of Hunting in the Neotropics. People in Nature: Wildlife Conservation in South and Central America. Columbia University Press, New York 299–323.

Bonach, K., Miranda-Vilela, M.P., Alves, M.C., Verdade, L.M., 2003. Effect of translocation on egg viability of the giant Amazon river turtle, Podocnemis expansa. Zoology 69, 82–90.

Bonacic, C., Feber, R.E., Macdonald, D.W., 2006. Capture of the vicuña (Vicugna vicugna) for sustainable use: animal welfare implications. Biological Conservation 129 (4), 543–550.

Bonenfant, C., Gaillard, J.M., Coulson, T., Festa-Bianchet, M., Loison, A., Garel, M., Duncan, P., 2009. Empirical evidence of density-dependence in populations of large herbivores. Advances in Ecological Research 41, 313–357.

Boscolo, M., Dijk, K.V., Savenije, H., 2010. Financing sustainable small-scale forestry: lessons from developing national forest financing strategies in Latin America. Forests 1 (4), 230–249.

Botlhole, T., Asafu-Adjaye, J.O.H.N., Carmignani, F., 2012. Natural resource abundance, institutions and tax revenue mobilisation in Sub-Sahara Africa. South African Journal of Economics 80 (2), 135–156.

Brashares, J.S., Arcese, P., Sam, M.K., Coppolillo, P.B., Sinclair, A.R., Balmford, A., 2004. Bushmeat hunting, wildlife declines, and fish supply in West Africa. Science 306 (5699), 1180–1183.

Brondízio, E.S., 2008. The Amazonian Caboclo and the Açaí Palm: Forest Farmers in the Global Market. vol. 16. New York Botanical Garden Pr Dept.

Brondízio, E.S., 2011. Forest resources, family networks and the municipal disconnect: examining recurrent underdevelopment in the Amazon estuary. The Amazon Várzea, Springer, Netherlands, pp. 207–229.

Brondizio, E.S., Siqueira, A.D., 1997. From extractivists to forest farmers: changing concepts of caboclo agroforestry in the Amazon estuary. Research in Economic Anthropology 18, 233–279.

Brondízio, E.S., Safar, C.A., Siqueira, A.D., 2002. The urban market of açaí fruit (Euterpe oleracea Mart.) and rural land use change: ethnographic insights into the role of price and land tenure constraining agricultural choices in the Amazon estuary. Urban Ecosystems 6 (1–2), 67–97.

Brown, L., 1988. And today we're going to talk about biodiversity... that's right, biodiversity. In: Wilson, E.O. (Ed.), Biodiversity, National Academy Press, Washingon D.C, pp. 446–449.

Brunner, R.D., 2005. Adaptive Governance: Integrating Science, Policy, and Decision Making. Columbia University Press, New York.

Brunnschweiler, C.N., 2008. Cursing the blessings? Natural resource abundance, institutions, and economic growth. World Development 36 (3), 399–419.

Buck, M., Hamilton, C., 2011. The Nagoya Protocol on access to genetic resources and the fair and equitable sharing of benefits arising from their utilisation to the convention on biological diversity. Review of European Community & International Environmental Law 20 (1), 47–61.

Buckland, S.T., Anderson, D.R., Burnham, K.P., Laake, J.L., 1993. Distance Sampling: Estimating Abundance of Biological Populations. Chapman & Hall.

Burke, B.E., 2001. Hardin revisited: a critical look at perception and the logic of the commons. Human Ecology 29 (4), 449–476.

Burkhard, B., Petrosillo, I., Costanza, R., 2010. Ecosystem services–bridging ecology, economy and social sciences. Ecological Complexity 7 (3), 257.

Butchart, S.H., Walpole, M., Collen, B., van Strien, A., Scharlemann, J.P., Almond, R.E., Watson, R., 2010. Global biodiversity: indicators of recent declines. Science 328 (5982), 1164–1168.

Camacho Nassar, C., 2010. Entre el etnocidio y la extinción. Pueblos indígenas aislados, en contacto inicial e intermitente en las tierras bajas de Bolivia.

Cañas, C.M., Pine, W.E., 2011. Documentation of the temporal and spatial patterns of pimelodidae catfish spawning and larvae dispersion in the madre de Dios River (Peru): insights for conservation in the Andean–Amazon headwaters. River Research and Applications 27 (5), 602–611.

Caputo, F.P., Canestrelli, D., Boitani, L., 2005. Conserving the terecay (Podocnemis unifilis, Testudines: Pelomedusidae) through a community-based sustainable harvest of its eggs. Biological Conservation 126 (1), 84–92.

Cardinale, M., Doerner, H., Abella, A., Andersen, J.L., Casey, J., Döring, R., Stransky, C., 2013. Rebuilding EU fish stocks and fisheries, a process under way? Marine Policy 39, 43–52.

Carpenter, S.R., Mooney, H.A., Agard, J., Capistrano, D., DeFries, R.S., Diaz, S., Whyte, A., 2009. Science for managing ecosystem services: beyond the Millennium Ecosystem Assessment. Proceedings of the National Academy of Sciences 106 (5), 1305–1312.

Carrillo, E., Wong, G., Cuarón, A.D., 2000. Monitoring mammal populations in Costa Rican protected areas under different hunting restrictions. Conservation Biology 14 (6), 1580–1591.

Castañeda, F., 2000. Criteria and indicators for sustainable forest management: international processes, current status and the way ahead. Unasylva-FAO, 34–40.

Castello, L., Viana, J.P., Watkins, G., Piñedo-Vasquez, M., Luzadis, V.A., 2009. Lessons from integrating fishers of Arapaima in small-scale fisheries management at the Mamirauá Reserve, Amazon. Environmental Management 43, 197–209.

Castello, L., McGrath, D.G., Hess, L.L., Coe, M.T., Lefebvre, P.A., Petry, P., Arantes, C.C., 2013. The vulnerability of Amazon freshwater ecosystems. Conservation Letters.

Caughley, G., 1982. Vegetation complexity and the dynamics of modelled grazing systems. Oecologia 54 (3), 309–312.

Charles, A.T., 2008. Sustainable Fishery Systems. vol. 5. Wiley-Blackwell.

CIESIN, IFPRI, WRI, 2000. Gridded Population of the World (GPW), Version 2. Center for International Earth Science Network (CIESIN), Columbia University.

Chu, C., 2009. Thirty years later: the global growth of ITQs and their influence on stock status in marine fisheries. Fish and Fisheries 10 (2), 217–230.

Clark, C., 1990. Mathematical Bioeconomics: The Optimal Management of Renewable Resources. John Wiley & Sons.

Clark, M.R., Kozar, J.S., 2011. Comparing sustainable forest management certifications standards: a meta-analysis. Ecology and Society 16 (1), 3.

Clarke, J., Cavendish, W., Coote, C., 1996. Rural Households and Miombo Woodlands: Use, Value and Management. CIFOR.

Clay, B.B., 1997. Brazil nuts. The use of a keystone species for conservation and development. In: Freese, C.H. (Ed.), Harvesting Wild Species. Implications for Biodiversity Conservation, John Hopkins Press, Baltimore, pp. 246–282.

Cleary, D., 2001. Towards an environmental history of the Amazon: from prehistory to the nineteenth century. Latin American Research Review, 65–96.

Clement, C.R., de Cristo-Araújo, M., Coppens D'Eeckenbrugge, G., Alves Pereira, A., Picanço-Rodrigues, D., 2010. Origin and domestication of native Amazonian crops. Diversity 2 (1), 72–106.

Cocks, M., López, C., Dold, T., 2011. Cultural importance of non-timber forest products: opportunities they pose for bio-cultural diversity in dynamic societies. Non-Timber Forest Products in the Global Context, Springer, Berlin Heidelberg, pp. 107–128.

Colfer, C.J.P., 2011. Marginalised forest peoples' perceptions of the legitimacy of governance: an exploration. World Development 39 (12), 2147–2164.

Commons, J.R., 1968. Legal Foundations of Capitalism. University of Wisconsin Press, Madison.

Coomes, O.T., 1995. A century of rain forest use in Western Amazonia. Forest and Conservation History 39, 108–120.

Coomes, O.T., Barham, B.L., 1997. Rain forest extraction and conservation in Amazonia. Geographical Journal 180–188.

Cornelius, J.P., Navarro, C.M., Wightman, K.E., Ward, S.E., 2005. Is mahogany dysgenically selected? Environmental Conservation 32 (2), 129–139.

Cowlishaw, G., Mendelson, S., Rowcliffe, J., 2005. Evidence for post–depletion sustainability in a mature bushmeat market. Journal of Applied Ecology 42 (3), 460–468.

Crona, B., Hubacek, K., 2010. The right connections: how do social networks lubricate the machinery of natural resource governance? Ecology and Society 15 (4), 18.

Crona, B.I., Parker, J.N., 2012. Learning in support of governance: theories, methods, and a framework to assess how bridging organisations contribute to adaptive resource governance. Ecology and Society 17 (1), 32.

Cumming, G.S., Cumming, D.H.M., Redman, C.L., 2006. Scale mismatches in social-ecological systems: causes, consequences, and solutions. Ecology and Society 11 (1), 14.

Dalvi, S., Sreenivasan, R., Price, T., 2013. Exploitation in Northeast India. Science 339 (6117), 270.

Davidson, E.A., Araujo, A.C., Artaxo, P., Balch, J.K., Brown, I.F., Bustamante, M.M.C., Coel, M.T., DeFries, R.S., Keller, M., Longo, M., Munger, J.W., Schroeder, W., Soares-Filho, B.S., Souza Jr., C.M., Wofsy, S.C., 2012. The Amazon basin in transition. Nature 481, 321–328.

Da Silveira, R., Thorbjarnarson, J.B., 1999. Conservation implications of commercial hunting of black and spectacled caiman in the Mamirauá sustainable development reserve, Brazil. Biological Conservation 88 (1), 103–109.

Datta, D., Guha, P., Chattopadhyay, R.N., 2010. Application of criteria and indicators in community based sustainable mangrove management in the Sunderbans, India. Ocean & Coastal Management 53 (8), 468–477.

de Alencar Guzman, D., 2009. Mixed Indians, caboclos and curibocas: historical analysis of a process of miscegenation; Rio Negro (Brazil), 18th and 19th centuries. Amazon Peasant Societies in a Changing Environment, Springer, Netherlands, pp. 55–68.

de Oñate Calvín, R., 2012. Caracterización y aprovechamiento de los varillales amazónicos: contribución a la economía de las comunidades locales. Thesis (Doctoral), ETSI de Montes, Universidad Politécnica de Madrid.

Dean, A.M., 1983. A simple model of mutualism. American Naturalist, 409–417.

De Carvajal, G., 1955. Relación del nuevo descubrimiento del famoso río Grande de las Amazonas. In: Hernández Millares, J. (Ed.), Fondo de cultura económica, México.

de Fátima Rossetti, D., de Toledo, P.M., Moraes-Santos, H.M., 2004. Reconstructing habitats in central Amazonia using megafauna, sedimentology, radiocarbon, and isotope analyses. Quaternary Research 61 (3), 289–300.

Dirzo, R., Miranda, A., 1990. Contemporary neotropical defaunation and forest structure, function, and diversity—a sequel to John Terborgh*. Conservation Biology 4 (4), 444–447.

De Groot, R.S., Wilson, M.A., Boumans, R.M.J., 2002. A typology for the classification, description and valuation of ecosystem functions, goods, and services. Ecological Economics 14, 393–408.

De Jesús, M.J., Kohler, C.C., 2004. The commercial fishery of the Peruvian Amazon. Fisheries 29 (4), 10–16.

Denevan, W.M., 1998. Comments on Prehistoric Agriculture in Amazonia. Culture & Agriculture 20, 54–59.

Denevan, W.M., 2011. The "pristine myth" revisited. Geographical Review 101 (4), 576–591.

de Jong, W., Ruiz, S.A., 2012. Strangers among trees: territorialisation and forest policies in the northern Bolivian Amazon. Forest Policy and Economics 16, 65–70.

Desmarchelier, C., Gurni, A., Ciccia, G., Giulietti, A.M., 1996. Ritual and medicinal plants of the Ese'ejas of the Amazonian rainforest (Madre de Dios, Perú). Journal of Ethnopharmacology 52 (1), 45–51.

Dincauze, Dena F., 2000. Environmental Archaeology: Principles and Practice. Cambridge University Press.

Dourojeanni, M., 2012. Amazonía probable y deseable. Ensayo sobre el presente y futuro de la Amazonía. Univ. Inca Garcilaso de la Vega, Lima.

Dourojeanni, M.J., 2009. Crónica forestal del Perú. UNMSM–UNALM, Lima.

Driscoll, C.T., Lambert, K.F., Chapin III, F.S., Nowak, D.J., Spies, T.A., Swanson, F.J., Hart, C.M., 2012. Science and society: the role of long-term studies in environmental stewardship. BioScience 62 (4), 354–366.

Duarte, N., Montúfar, R., 2012. Effects of harvest on wax palm (Ceroxylon echinulatum Galeano) growth, and implications for sustainable management in Ecuador. Tropical Conservation Science 5 (3), 340–351.

Duchelle, A., Almeyda, A., Hoyos, N., Marsik, M., Broadbent, E., Kainer, K.A., 2010. Conservation in an Amazonian Tri-National Frontier: Patterns and Drivers of Land Cover Change in Community-Managed Forests. Proceedings of the conference taking stock of smallholder and community forestry: where do we go from here .

Elías, S.I., 2008. Evaluación de impacto ambiental: Castaña en la Reserva Nacional de Tambopata. (No. 38319) Inter-American Development Bank.

Elujoba, A.A., Odeleye, O.M., Ogunyemi, C.M., 2006. Traditional medicine development for medical and dental primary health care delivery system in Africa. African Journal of Traditional, Complementary and Alternative Medicines (AJTCAM) 2 (1), 46–61.

Endress, B.A., Gorchov, D.L., Berry, E.J., 2006. Sustainability of a non-timber forest product: effects of alternative leaf harvest practices over 6 years on yield and demography of the palm Chamaedorea radicalis. Forest Ecology and Management 234 (1), 181–191.

Escobedo, F.J., Kroeger, T., Wagner, J.E., 2011. Urban forests and pollution mitigation: analysing ecosystem services and disservices. Environmental Pollution 159 (8), 2078–2087.

Evans, K., Murphy, L., de Jong, W., 2013. Global versus local narratives of REDD: a case study from Peru's Amazon. Environmental Science & Policy.

Fa, J.E., Peres, C.A., 2001. Game vertebrate extraction in African and Neotropical forests: an Intercontinental Comparison. In: Reynolds, J.D., Mace, G.M., Redford, K.H., Robinson, J.G. (Eds.), Conservation of Exploited Species. Cambridge University Press, Cambridge, pp. 203–241.

Fagerholm, N., Käyhko, N., Ndumbaro, F., Khamis, M., 2012. Community stakeholders' knowledge in landscape assessments—mapping indicators for landscape services. Ecological Indicators 18, 421–433.

FAO, 2010a. The State of World Fisheries and Aquaculture.

FAO, 2010b. Global Forest Resources Assessment.

FAO, 2012. The State of World Fisheries and Aquaculture.

Fielding, N.G., 2013. Qualitative research in a glocalising world. Global Dimensions of Qualitative Inquiry 47.

Fine, P.V., Kembel, S.W., 2011. Phylogenetic community structure and phylogenetic turnover across space and edaphic gradients in western Amazonian tree communities. Ecography 34 (4), 552–565.

Fearnside, P.M., 1989. Extractive reserves in Brazilian Amazonia. BioScience 39, 387–393.

Flothmann, S., von Kistowski, K., Dolan, E., Lee, E., Meere, F., Album, G., 2010. Closing loopholes: getting illegal fishing under control. Science 328 (5983), 1235–1236.

Folke, C., 2006. Resilience: the emergence of a perspective for social–ecological systems analyses. Global Environmental Change 16 (3), 253–267.

Gallup, J.L., Sachs, J.D., Mellinger, A.D., 1999. Geography and economic development. International Regional Science Review 22 (2), 179–232.

Gamarra, P., 2007. Amazonía Norte de Bolivia. Economía Gomera 1870–1940. Bases económicas de un poder regional. La Casa Suárez, Colegio nacional de Historiadores de Bolivia, La Paz.

Garcia, A., Tello, S., Vargas, G., Duponchelle, F., 2009. Patterns of commercial fish landings in the Loreto region (Peruvian Amazon) between 1984 and 2006. Fish Physiology and Biochemistry 35 (1), 53–67.

Gibson, J.R., 2011. Feeding the Russian fur trade: provisionment of the Okhotsk seaboard and the Kamchatka Peninsula, 1639–1856. University of Wisconsin Pres.

Gilmore, R.M., 1950. Fauna and ethnozoology of South America. In: Steward, J.H. (Ed.), Handbook of South American Indians, vol. 6. Bureau of American Ethnology, Bulletin I43, Washington, D.C., pp. 345–464.

Giudice, R., Soares-Filho, B.S., Merry, F., Rodrigues, H.O., Bowman, M., 2012. Timber concessions in Madre de Dios: are they a good deal? Ecological Economics 77, 158–165.

Glander, K.E., 1992. Dispersal patterns in Costa Rican mantled howling monkeys. International Journal of Primatology 13 (4), 415–436.

Glaser, B., Birk, J.J., 2012. State of the scientific knowledge on properties and genesis of anthropogenic dark earths in Central Amazonia (terra preta de Índio). Geochimica et Cosmochimica Acta 82, 39–51.

Gooding, F., 2005. Of Dodos and Dutchmen: reflections on the nature of history. Critical Quarterly 47 (4), 32–47.

Gómez, L.Á., Torres, S.R., 2009. Evaluación económica de la extracción de castaña (Bertholletia excelsa H: B: K.) en el departamento de Madre de Dios. Instituto de Investigaciones de la Amazonía Peruana.

González, J.A., 2003. Harvesting, local trade, and conservation of parrots in the Northeastern Peruvian Amazon. Biological Conservation 114 (3), 437–446.

Grogan, J., Blundell, A.G., Landis, R.M., Youatt, A., Gullison, R.E., Martinez, M., Rice, R.E., 2010. Over–harvesting driven by consumer demand leads to population decline: big–leaf mahogany in South America. Conservation Letters 3 (1), 12–20.

Guha, B., 2012. Pirates and fishermen: is less patrolling always bad? Journal of Economic Behavior & Organisation 81 (1), 29–38.

Guimarães, P.R., Galetti, M., Jordano, P., 2008. Seed dispersal anachronisms: rethinking the fruits extinct megafauna ate. PLoS One 3 (3), e1745.

Gutierrez-Velez, V.H., MacDicken, K., 2008. Quantifying the direct social and governmental costs of illegal logging in the Bolivian, Brazilian, and Peruvian Amazon. Forest Policy and Economics 10 (4), 248–256.

Hamadeh, S.K., Bistanji, G.N., Darwish, M.R., Abi Said, M., Abi Ghanem, D., 2001. Economic sustainability of small ruminants production in semi-arid areas of Lebanon. Small Ruminant Research 40 (1), 41–49.

Halme, K.J., Bodmer, R.E., 2007. Correspondence between scientific and traditional ecological knowledge: rain forest classification by the non-indigenous ribereños in Peruvian Amazonia. Biodiversity and Conservation 16, 1785–1801.

Hamilton, M.J., Milne, B.T., Walker, R.S., Burger, O., Brown, J.H., 2007. The complex structure of hunter–gatherer social networks. Proceedings of the Royal Society B: Biological Sciences 274 (1622), 2195–2203.

Hansen, L.J., Hoffman, J.R., 2011. Regulating Harvest in a Changing World. In: Climate Savvy. Island Press/Center for Resource Economics, pp. 171–179.

Hansen, S.J., 2009. Piracy in the Greater Gulf of Aden. Norwegian Institute for Urban and Regional Research, London.

Hardin, G., 1968. The tragedy of the commons. Science 162, 1243–1248.

Harley, S.J., Myers, R.A., Dunn, A., 2001. Is catch-per-unit-effort proportional to abundance? Canadian Journal of Fisheries and Aquatic Sciences 58 (9), 1760–1772.

Hartig, G.L., 1804. Anweisung zur Taxation und Beschreibung der Forste:... Theoretischer Theil: Nebst einem illuminirten Forstkarten-Schema und mehreren Tabellen, vol. 5. Heyer.

Hautala, J., 2011. Cognitive proximity in international research groups. Journal of Knowledge Management 15 (4), 601–624.

Heckathorn, D.D., 1997. Respondent-driven sampling: a new approach to the study of hidden populations. Social Problems, 174–199.

Heckenberger, M., Neves, E.G., 2009. Amazonian archaeology. Annual Review of Anthropology 38, 251–266.

Hemming, J., 2008. Tree of Rivers: The Story of the Amazon. Thames & Hudson, London.

Herndon, W.L., Gibbon, L., 1854. Exploration of the Valley of the Amazon. R. Armstrong (etc.) Public Printer.

Hersoug, B., Holm, P., 2000. Change without redistribution: an institutional perspective on South Africa's new fisheries policy. Marine Policy 24 (3), 221–231.

Heywood, V.H. (Ed.), 1995. Global Biodiversity Assessment, Cambridge University Press, Cambridge (UK). p. 1140.

Hiedanpää, J., Kotilainen, J., Salo, M., 2011. Unfolding the organised irresponsibility: ecosystem approach and the quest for forest biodiversity in Finland, Peru, and Russia. Forest Policy and Economics 13 (3), 159–165.

Higgins, M.A., Ruokolainen, K., Tuomisto, H., Llerena, N., Cardenas, G., Phillips, O.L., Räsänen, M., 2011. Geological control of floristic composition in Amazonian forests. Journal of Biogeography 38 (11), 2136–2149.

Hill, K., Kaplan, H., Hawkes, K., Hurtado, A.M., 1987. Foraging decisions among Ache hunter-gatherers: new data and implications for optimal foraging models. Ethology and Sociobiology 8 (1), 1–36.

Hill, M.O., 1973. Diversity and evenness: a unifying notation and its consequences. Ecology 54 (2), 427–432.

Hiraoka, M., 1995. Land use changes in the Amazon estuary. Global Environmental Change 5 (4), 323–336.

Ho, W.C., Cheung, C.K., 2011. Social sustainability for mothers in Hong Kong's low-income communities. Habitat International 35 (4), 529–536.

Holdsworth, A.R., Uhl, C., 1997. Fire in Amazonian selectively logged rain forest and the potential for fire reduction. Ecological Applications 7 (2), 713–725.

Holling, C.S., 1973. Resilience and stability of ecological systems. Annual Review of Ecology and Systematics 4, 1–23.

Holm, J.A., Miller, C.J., Cropper, W.P., 2008. Population dynamics of the dioecious Amazonian palm Mauritia flexuosa: simulation analysis of sustainable harvesting. Biotropica 40 (5), 550–558.

Holt, F.L., 2005. The Catch-22 of conservation: indigenous peoples, biologists, and cultural change. Human Ecology 33 (2), 199–215.

Hooper, D.U., Chapin III, F.S., Ewel, J.J., Hector, A., Inchausti, P., Lavorel, S., Wardle, D.A., 2005. Effects of biodiversity on ecosystem functioning: a consensus of current knowledge. Ecological Monographs 75 (1), 3–35.

Hoorn, C., Wesselingh, F.P., Ter Steege, H., Bermudez, M.A., Mora, A., Sevink, J., Antonelli, A., 2010. Amazonia through time: Andean uplift, climate change, landscape evolution, and biodiversity. Science 330 (6006), 927–931.

Howard, P.L., 2003. Women and Plants: Gender Relations in Biodiversity Management and Conservation. Zed books.

Hubbe, A., Hubbe, M., Neves, W.A., 2013. The Brazilian megamastofauna of the Pleistocene/Holocene transition and its relationship with the early human settlement of the continent. Earth-Science Reviews 118 (2013), 1–10.

Huertas Castillo, B., 2004. Indigenous Peoples in Isolation in the Peruvian Amazon: Their Struggle for Survival and Freedom. vol. 100. Iwgia.

INEI, 2009. II Censo de Comunidades Indígenas de la Amazonía Peruana 2007. Instituto Nacional de Estadística e Informática, Lima.

INRENA, 2000. Plan Maestro de la Reserva Nacional Pacaya Samiria. Iquitos, Perú.

Irion, G., Kalliola, R., 2009. Long-term landscape development processes in Amazonia. In: Hoorn, C., Wesselingh, F. (Eds.), Amazonia, Landscape and Species Evolution. A Look into the Past, Wiley-Blackwell, pp. 185–197.

Isaac, V.J., De Almeida, M.C., 2011. Consumo de pescado y fauna acuática en la Amazonía brasileña. FAO COPESCAALC Documento ocasional 13.

IIAP, 2009. Viabilidad económica de la de la pesca artesanal en el departamento de Loreto. Serie: Avances Económicos N° 13. Instituto de Investigaciones de la Amazonía Peruana, Iquitos.

Jackson, J.B., Kirby, M.X., Berger, W.H., Bjorndal, K.A., Botsford, L.W., Bourque, B.J., Warner, R.R., 2001. Historical overfishing and the recent collapse of coastal ecosystems. Science 293 (5530), 629–637.

Jennings, S., 2013. When can "principles" support advice on fisheries and environmental management? ICES Journal of Marine Science: Journal du Conseil.

Jensen, A.L., 1996. Density-dependent matrix yield equation for optimal harvest of age-structured wildlife populations. Ecological Modelling 88 (1–3), 125–132.

Jizhong, Z., Shijun, M., Changming, C., 1991. An index of ecosystem diversity. Ecological Modelling 59 (3), 151–163.

Johannes, R.E., 1998. The case for data-less marine resource management: examples from tropical nearshore finfisheries. Trends in Ecology & Evolution 13 (6), 243–246.

Johnston, R., Cools, J., Liersch, S., Morardet, S., Murgue, C., Mahieu, M., Uyttendaele, G.P., 2013. WETwin: a structured approach to evaluating wetland management options in data-poor contexts. Environmental Science & Policy.

Jonzén, N., Lundberg, P., Gardmark, A., 2001. Harvesting spatially distributed populations. Wildlife Biology 7 (3), 197–203.

Joppa, L.N., Loarie, S.R., Pimm, S.L., 2009. On population growth near protected areas. PLoS One 4 (1), e4279.

Jordano, P., Forget, P.M., Lambert, J.E., Böhning-Gaese, K., Traveset, A., Wright, S.J., 2011. Frugivores and seed dispersal: mechanisms and consequences for biodiversity of a key ecological interaction. Biology Letters 7 (3), 321–323.

Jost, L., 2006. Entropy and diversity. Oikos 113 (2), 363–375.

Justiniano, O.T., 2010. El Caucho Ignorado. Editorial El País, Santa Cruz de la Sierra.

Juvonen, S.K., Salo, J., 2004. Sustainable use of ornamental fish populations in Peruvian Amazonia. Lyonia 7 (2), 53–59.

Jyväsjärvi, J., Immonen, H., Högmander, P., Högmander, H., Hämäläinen, H., Karjalainen, J., 2013. Can lake restoration by fish removal improve the status of profundal macroinvertebrate assemblages? Freshwater Biology.

Kaennel, M., 1998. Biodiversity: a diversity in definition. Forestry Sciences 51, 71–82.

Kaimowitz, D., Sheil, D., 2007. Conserving what and for whom? Why conservation should help meet basic human needs in the tropics. Biotropica 39 (5), 567–574.

Kalliola, R., Puhakka, M., Salo, J., Tuomisto, H., Ruokolainen, K., 1991. The dynamics, distribution and classification of swamp vegetation in Peruvian Amazonia. Annales Botanici Fennici 28, 225–239.

Kalliola, R., Linna, A., Puhakka, M., Salo, J., Räsänen, M., 1993. Mineral nutrients from fluvial sediments in the Peruvian Amazon. Catena 20, 333–349.

Kalliola, R., Toivonen, T., Miyakawa, V., Mavila, M., 2008. Open access to information bridges science and development in Amazonia: lessons of the SIAMAZONIA service. Environmental Research Letters 3 (3), 034004.

Kalliola, R., Flores, P., 2011. Brazil nut harvesting in Peruvian Amazonia from the perspective of ecosystem services. Fennia 189 (2), 1–13.

Kelly, M., Ferranto, S., Lei, S., Ueda, K.I., Huntsinger, L., 2012. Expanding the table: the web as a tool for participatory adaptive management in California forests. Journal of Environmental Management 109, 1–11.

Killeen, T.J., 2007. A perfect storm in the Amazon wilderness. Advances in Applied Biodiversity Science 7, 1–102.

Kirby, J.S., Stattersfield, A.J., Butchart, S.H., Evans, M.I., Grimmett, R.F., Jones, V.R., Newton, I., 2008. Key conservation issues for migratory land-and waterbird species on the world's major flyways. Bird Conservation International 18 (1), S49.

Kirch, Patrick V., 2005. Archaeology and global change: the Holocene record. Annual Review of Environmental Resources 30, 409–440.

Kishor, N., Lescuyer, G., 2012. Controlling illegal logging in domestic and international markets by harnessing multi-level governance opportunities. International Journal of the Commons 6 (2), 255–270.

Korpisaari, A., Sääksjärvi, I.E., Salo, M., in press. Evidencias de contactos entre la sierra, la ceja de selva y la selva baja en el arte de la cultura Tiwanaku. In: Platt, T., Daillant, I., Harris, M., Rivière, G. (Eds.), Actas del Taller Andes, Amazonas y sus Transformaciones.

Koster, J., 2009. Hunting dogs in the lowland Neotropics. Journal of Anthropological Research, 575–610.

Kot, M., Schaffer, W.M., 1984. The effects of seasonality on discrete models of population growth. Theoretical Population Biology 26 (3), 340–360.

Kreft, H., Jetz, W., 2007. Global patterns and determinants of vascular plant diversity. Proceedings of the National Academy of Sciences 104 (14), 5925–5930.

Kumpula, J., 2001. Productivity of the Semi-Domesticated Reindeer (Rangifer t. tarandus L.) Stock and Carrying Capacity of Pastures in Finland During 1960–1990's. Oulun yliopisto.

Laihonen, P., Kalliola, R., Salo, J., 2004. The biodiversity information clearing-house mechanism (CHM) as a global effort. Environmental Science & Policy 7, 99–108.

Laird, S.A., Wynberg, R., McLain, R.J., 2011. Regulating complexity: policies for the governance of non-timber forest products. Non-Timber Forest Products in the Global Context, Springer, Berlin, Heidelberg, pp. 227–253.

Larson, A.M., Barry, D., Dahal, G.R., 2010. New rights for forest-based communities? Understanding processes of forest tenure reform. International Forestry Review 12 (1), 78–96.

Larson, Anne M., Soto, Fernanda, 2008. Decentralisation of natural resource governance regimes. Annual Review of Environment and Resources 33 (1), 213.

Lathrap, D.W., 1973. The antiquity and importance of long–distance trade relationships in the moist tropics of pre–Columbian South America. World Archaeology 5 (2), 170–186.

Latrubesse, E.M., Rancy, A., 2000. Neotectonic influence on tropical rivers of southwestern Amazon during the late Quaternary: the Moa and Ipixuna river basins, Brazil. Quaternary International 72, 67–72.

Lavsund, S., Nygrén, T., Solberg, E.J., 2003. Status of moose populations and challenges to moose management in Fennoscandia. Alces 39, 109–130.

Levis, C., de Souza, P.F., Schietti, J., Emilio, T., da Veiga Pinto, J.L.P., Clement, C.R., Costa, F.R., 2012. Historical human footprint on modern tree species composition in the Purus-Madeira Interfluve, Central Amazonia. PLoS One 7 (11), e48559.

Lewis, J.A., 2008. The power of knowledge: information transfer and açaí intensification in the peri-urban interface of Belém, Brazil. Agroforestry Systems 74, 293–302.

Liermann, M., Hillborn, R., 2001. Depensation: evidence, models and implications. Fish and Fisheries 2, 33–58.

Lim, K., Treitz, P., Wulder, M., St-Onge, B., Flood, M., 2003. LiDAR remote sensing of forest structure. Progress in Physical Geography 27 (1), 88–106.

Lindenmayer, D.B., Franklin, J.F., Fischer, J., 2006. General management principles and a checklist of strategies to guide forest biodiversity conservation. Biological Conservation 131 (3), 433–445.

López Ríos, J., 2010. El mercado de productos pesqueros en la ciudad de Iquitos 2010. Proyecto Mejoramiento del Acceso a los Mercados de Productos Pesqueros y Acuícolas de la Amazonia, CFC/FAO/INFOPESCA CFC/FSCFT/28.

Ludwig, D., Walker, B., Holling, C.S., 1997. Sustainability, stability, and resilience. Conservation Ecology 1 (1), 7. Available from: http://www.consecol.org/vol1/iss1/art7/.

MacLachlan, C., 1972. The Indian Directorate: forced acculturation in Portuguese America (1757–1799). The Americas 28 (4), 357–387.

MacMillan, D.C., Marshall, K., 2006. The Delphi process—an expert-based approach to ecological modelling in data-poor environments. Animal Conservation 9 (1), 11–19.

Magnin, G., 1991. Hunting and persecution of migratory birds in the Mediterranean region. Conserving migratory birds. International Council for Bird Preservation (Techn. Publ. 12), Cambridge, UK, pp. 63–75.

Maslin, M., Malhi, Y., Phillips, O., Cowling, S., 2005. New views on an old forest: assessing the longevity, resilience and future of the Amazon rainforest. Transactions of the Institute of British Geographers 30 (4), 477–499.

Mäki, S., Kalliola, R., 2000. Visualisation and communication in map-making: a case study of mapping a complex rainforest environment in Peruvian Amazonia. Cartographica 37 (2), 27–42.

Mäki, S., Kalliola, R., Vuorinen, K., 2001. Road construction in the Peruvian Amazon: process, causes and consequences. Environmental Conservation 28 (3), 199–214.

Maués, M.M., 2002. Reproductive phenology and pollination of the Brazil nut tree (Bertholletia excelsa Humb. & Bonpl. Lecythidaceae) in Eastern Amazonia. In: Kevan, P., Imperatriz Fonseca, V.L. (Eds.), Pollinating Bees: The Conservation Link between Agriculture and Nature, Ministério do Meio Ambiente, Brasília, pp. 245–254.

Maunder, M.N., Sibert, J.R., Fonteneau, A., Hampton, J., Kleiber, P., Harley, S.J., 2006. Interpreting catch per unit effort data to assess the status of individual stocks and communities. ICES Journal of Marine Science: Journal du Conseil 63 (8), 1373–1385.

Maxim, L., Spangenberg, J.H., O'Connor, M., 2009. An analysis of risks for biodiversity under the DPSIR framework. Ecological Economics 69 (1), 12–23.

Malhi, Y., Roberts, J.T., Betts, R.A., Kileen, T.J., Li, W., Nobre, C.A., 2008. Climate change, deforestation, and the fate of the Amazon. Science 319, 169–172.

Matisoo-Smith, E., Ramirez, J.M., 2010. Human skeletal evidence of Polynesian presence in South America? Metric analyses of six crania from Mocha Island, Chile. Journal of Pacific Archaeology 1 (1), 76–88.

McCay, B.J., Brandt, S., Creed, C.F., 2011. Human dimensions of climate change and fisheries in a coupled system: the Atlantic surfclam case. ICES Journal of Marine Science: Journal du Conseil 68 (6), 1354–1367.

McCall, M.K., Dunn, C.E., 2012. Geo-information tools for participatory spatial planning: fulfilling the criteria for 'good' governance? Geoforum 43, 81–94.

McConkey, K.R., Drake, D.R., 2006. Flying foxes cease to function as seed dispersers long before they become rare. Ecology 87 (2), 271–276.

McDermott, C.L., 2012. Certification and equity: Applying an "equity framework" to compare certification schemes across product sectors and scales. Environmental Science & Policy.

McDonald, J., Veth, P. (Eds.), 2012. A Companion to Rock Art, Wiley-Blackwell.

McMichael, C.H., Piperno, D.R., Bush, M.B., Silman, M.R., Zimmerman, A.R., Raczka, M.F., Lobato, L.C., 2012. Sparse pre-Columbian human habitation in western Amazonia. Science 336 (6087), 1429–1431.

McNeely, J.A., 2003. Conserving forest biodiversity in times of violent conflict. Oryx 37 (02), 142–152.

McNeely, J.A., 2003. Biodiversity, war, and tropical forests. Journal of Sustainable Forestry 16 (3–4), 1–20.

McSweeney, K., Arps, S., 2005. A "demographic turnaround": the rapid growth of indigenous populations in Lowland Latin America. Latin American Research Review 40 (1), 3–29.

MEA, 2005. Ecosystems and Human Well-Being. Millennium Ecosystem Assessment. Island Press, Washington, D.C., USA.

Medina, G., Pokorny, B., Campbell, B.M., 2009. Community forest management for timber extraction in the Amazon frontier. International Forestry Review 11 (3), 408–420.

Mendoza, G.A., Martins, H., 2006. Multi-criteria decision analysis in natural resource management: a critical review of methods and new modelling paradigms. Forest Ecology and Management 230, 1–22.

Merkx, G.W., 2007. Constructal models in social processes. Constructal Theory of Social Dynamics, Springer, US, pp. 35–50.

Miller, S.D., Goulden, M.L., Hutyra, L.R., Keller, M., Saleska, S.R., Wofsy, S.C., de Camargo, P.B., 2011. Reduced impact logging minimally alters tropical rainforest carbon and energy exchange. Proceedings of the National Academy of Sciences 108 (48), 19431–19435.

Milner-Gulland, E.J., Mace, R., 2009. Conservation of Biological Resources. Wiley.
Mittermeier, R.A., Mittermeier, C.G., Gil, P.R., 1997. Megadiversity: Earth's Biologically Wealthiest Nations. Cemex, Mexico City, pp. 282–297.
Moegenburg, S.M., Levey, D.J., 2003. Do frugivores respond to fruit harvest? An experimental study of short-term responses. Ecology 84 (10), 2600–2612.
Moll, D., Moll, E.O., 2004. The Ecology, Exploitation and Conservation of River Turtles. Oxford University Press, USA.
Morcillo, J.G., 1982. Del caucho al oro: El proceso colonizador de Madre de Dios. Revista Española de Antropología Americana 255.
Morra, W., Hearn, G., Buck, A.J., 2009. The market for bushmeat: Colobus Satanas on Bioko Island. Ecological Economics 68 (10), 2619–2626.
Morsello, C., Ruiz-Mallén, I., Diaz, M.D.M., Reyes-García, V., 2012. The effects of processing non-timber forest products and trade partnerships on people's well-being and forest conservation in Amazonian societies. PLoS One 7 (8), e43055.
Moxnes, E., 2000. Not only the tragedy of the commons: misperceptions of feedback and policies for sustainable development. System Dynamics Review 16 (4), 325–348.
Nassauer, J.I., 2004. Monitoring the success of metropolitan wetland restorations: cultural sustainability and ecological function. Wetlands 24 (4), 756–765.
Nepstad, D., Soares-Filho, B.S., Merry, F., Lima, A., Moutinho, P., Carter, J., Stella, O., 2009. The end of deforestation in the Brazilian Amazon. Science 326 (5958), 1350–1351.
Neves, W.A., Hubbe, M., Correal, G., 2007. Human skeletal remains from Sabana de Bogota, Colombia: a case of Paleoamerican morphology late survival in South America? American Journal of Physical Anthropology 133 (4), 1080–1098.
Neves-Graça, K., 2006. Politics of environmentalism and ecological knowledge at the intersection of local and global processes. Journal of Ecological Anthropology 10 (1), 19–32.
Niemelä, J., 1999. Ecology and urban planning. Biodiversity & Conservation 8 (1), 119–131.
Norse, E.A., Brooke, S., Cheung, W.W., Clark, M.R., Ekeland, I., Froese, R., Watson, R., 2012. Sustainability of deep-sea fisheries. Marine Policy 36 (2), 307–320.
North, D.C., 1994. Economic performance through time. The American Economic Review 84 (3), 359–368.
Nunes, F., Soares-Filho, B., Giudice, R., Rodrigues, H., Bowman, M., Silvestrini, R., Mendoza, E., 2012. Economic benefits of forest conservation: assessing the potential rents from Brazil nut concessions in Madre de Dios, Peru, to channel REDD+ investments. Environmental Conservation 39 (2), 132–143.
Nygren, A., 2006. Representations of tropical forests and tropical forest-dwellers in travel accounts of "National Geographic". Environmental Values, 505–525.
De Oliveira, A.A., Mori, S.A., 1999. A central Amazonian terra firme forest. I. High tree species richness on poor soils. Biodiversity & Conservation 8 (9), 1219–1244.
Ogle, B.M., 2001. Wild Vegetables and Micronutrient Nutrition: Studies on the Significance of Wild Vegetables in Women's Diets in Vietnam. Doctoral dissertation, Uppsala University.
Oliver, J.R., 2008. The archaeology of agriculture in ancient Amazonia. The Handbook of South American Archaeology, Springer, New York, pp. 185–216.
Oliveira, P.J.C., Asner, G.P., Knapp, D.E., Almeyda, A., Galvan-Gildemeister, R., Keene, S., et al., 2007. Landuse allocation protects the Peruvian Amazon. Science 317 (5842), 1233–1236.
Östlund, L., Ahlberg, L., Zackrisson, O., Bergman, I., Arno, S., 2009. Bark-peeling, food stress and tree spirits—the use of pine inner bark for food in Scandinavia and North America. Journal of Ethnobiology 29 (1), 94–112.
Ostrom, E., 1990. Governing the Commons: The Evolution of Institutions for Collective Action. Cambridge University Press.

Ostrom, E., 1999. Coping with tragedies of the commons. Annual Review of Political Science 2, 493–535.

Ostrom, E., 2007. A General Framework for Analysing Sustainability of. In Proc. R. Soc. London Ser. B, vol. 274, p. 1931.

Ostrom, E., Hess, C., 2007. Private and Common Property Rights. Indiana University, Bloomington School of Public & Environmental Affairs Research Paper.

Ostrom, E., Cox, M., 2010. Moving beyond panaceas: a multi-tiered diagnostic approach for social-ecological analysis. Environmental Conservation 37 (4), 451–463.

Owen-Smith, N., 1989. Megafaunal extinctions: the conservation message from 11,000 years BP. Conservation Biology 3 (4), 405–412.

Pacheco, P., Ormachea, E., Cronkleton, P., Albornoz, M.A., Paye, L., 2009. Trayectorias y tendencias de la economía forestal extractiva en el norte Amazónico de Bolivia. CIFOR, CEDLA, RRI, La Paz, Bolivia.

Padoch, C., Brondizio, E., Costa, S., Pinedo-Vasquez, M., Sears, R.R., Siqueira, A., 2008. Urban forest and rural cities: multi-sited households, consumption patterns, and forest resources in Amazonia. Ecology and Society 13 (2), 2.

Paiva, P.M., Guedes, M.C., Funi, C., 2011. Brazil nut conservation through shifting cultivation. Forest Ecology and Management 261 (3), 508–514.

Palo, M., Uusivuori, J., Lehto, E., 1999. Forest-based development in Finland—a unique success? World Forests, Society and Environment, Springer, Netherlands, pp. 300–318.

Peres, et al., 19 December 2003. Demographic threats to the sustainability of Brazil nut exploitation. Science 302 (5653), 2112–2114.

Peres, C.A., 1994. Primate responses to phenological changes in an Amazonian terra firme forest. Biotropica, 98–112.

Peres, C.A., Baider, C., 1997. Seed dispersal, spatial distribution and population structure of Brazilnut trees (Bertholletia excelsa) in southeastern Amazonia. Journal of Tropical Ecology 13 (4), 595–616.

Peres, C.A., Dolman, P.M., 2000. Density compensation in neotropical primate communities: evidence from 56 hunted and nonhunted Amazonian forests of varying productivity. Oecologia 122 (2), 175–189.

Peres, C.A., Lake, I.R., 2003. Extent of nontimber resource extraction in tropical forests: accessibility to game vertebrates by hunters in the Amazon basin. Conservation Biology 17 (2), 521–535.

Peres, C.A., 2011. Conservation in sustainable-use tropical forest reserves. Conservation Biology 25, 1124–1129.

Peters, C.M., Gentry, A.H., Mendelsohn, R.O., 1989. Valuation of an Amazonian rainforest. Nature 339 (29), 655–656.

Pfeiffer, J.M., Butz, R.J., 2005. Assessing cultural and ecological variation in ethnobiological research: the importance of gender. Journal of Ethnobiology 25 (2), 240–278.

Pham, J.P., 2010. Putting Somali piracy in context. Journal of Contemporary African Studies 28 (3), 325–341.

Phillips, N., Sakamoto, L., 2012. Global production networks, chronic poverty and 'slave labour' in Brazil. Studies in Comparative International Development 47 (3), 287–315.

Piperno, D.R., 1990. Aboriginal agriculture and land usage in the Amazon Basin, Ecuador. Journal of Archaeological Science 17 (6), 665–677.

Piperno, D.R., Pearsall, D.M., 1998. The Origins of Agriculture in the Lowland Neotropics. Cambridge University Press, New York (NY).

Pinard, M., 1993. Impacts of Stem Harvesting on Populations of Iriartea deltoidea (Palmae) in an Extractive Reserve in Acre. Biotropica, Brazil, pp. 2–14.

Pineda-Catalan, O., Mendez, M., Gleizer, A., García-Dávila, C., Aguirre, A.A., Pinedo-Vasquez, M., Amato, G., 2012. Conservation genetics of harvested river turtles, Podocnemis expansa and Podocnemis unifilis, in the Peruvian Amazon: all roads lead to Iquitos. Mitochondrial DNA 23 (3), 230–238.

Pinedo-Vasquez, M., Hecht, S., Padoch, C., 2012. Amazonia. Traditional Forest-Related Knowledge, Springer, Netherlands, pp. 119–155.

Pokorny, B., Adams, M., 2003. What do criteria and indicators assess? An analysis of five CI sets relevant for forest management in the Brazilian Amazon. International Forestry Review 5 (1), 20–28.

Pokorny, B., Johnson, J., Medina, G., Hoch, L., 2012. Market-based conservation of the Amazonian forests: revisiting win–win expectations. Geoforum 43 (3), 387–401.

Power, M.J., Bush, M.B., Behling, H., Horn, S.P., Mayle, F.E., Urrego, D.H., 2010. Paleofire Activity in Tropical America During the Last 21 ka: A Regional Synthesis Based on Sedimentary Charcoal.

Pred, A., 1977. The choreography of existence: comments on Hägerstrand's time-geography and its usefulness. Economic Geography, 207–221.

Prell, C., Hubacek, K., Reed, M., 2009. Stakeholder analysis and social network analysis in natural resource management. Society and Natural Resources 22 (6), 501–518.

Prell, C., Hubacek, K., Reed, M., Quinn, C., Jin, N., Holden, J., Sendzimir, J., 2007. If you have a hammer everything looks like a nail: traditional versus participatory model building. Interdisciplinary Science Reviews 32 (3), 263.

Prescott, G.W., Williams, D.R., Balmford, A., Green, R.E., Manica, A., 2012. Quantitative global analysis of the role of climate and people in explaining late Quaternary megafaunal extinctions. Proceedings of the National Academy of Sciences 109 (12), 4527–4531.

Pyhälä, A., Brown, K., Adger, W.N., 2006. Implications of livelihood dependence on non-timber products in Peruvian Amazonia. Ecosystems 9 (8), 1328–1341.

Pärssinen, M., Schaan, D., Ranzi, A., 2009. Pre-Columbian geometric earthworks in the upper Purús: a complex society in western Amazonia. Antiquity 83 (322), 1084.

Quammen, D., 2012. The Song of the Dodo: Island Biogeography in an Age of Extinctions. Vintage Digital.

Quevedo-Gil, A., 2006. Plan de acción nacional para los loros amenazados de Colombia: una iniciativa para garantizar la conservación de nuestros loros. Conservación Colombiana 1, 58–66.

Ramcilovic-Suominen, S., Epstein, G., 2012. Towards an analytical framework for forest law compliance. International Forestry Review 14 (3), 326–336.

Ramula, S., Rees, M., Buckley, Y.M., 2009. Integral projection models perform better for small demographic data sets than matrix population models: a case study of two perennial herbs. Journal of Applied Ecology 46 (5), 1048–1053.

Räsänen, M.E., Salo, J.S., Kalliola, R.J., 1987. Fluvial perturbance in the western Amazon basin: regulation by long-term sub-Andean tectonics. Science 238, 1398–1401.

Räsänen, M.E., Salo, J.S., Jungner, H., Romero Pittman, L., 1990. Evolution of the Western Amazon lowland relief: impact of Andean foreland dynamics. Terra Nova 2, 320–332.

Raygada Guerra, V., Rojas Grández, F., Álvarez Alonso, J., 2007. Plan de manejo adaptativo de irapay para pequeños extractores de la RNAM. IIAP–BIODAMAZ, Iquitos.

Reboredo, F., 2013. Socio-economic, environmental, and governance impacts of illegal logging. Environment Systems and Decisions, 1–10.

Recharte Uscamaita, M., Bodmer, R., 2010. Recovery of the endangered giant otter Pteronura brasiliensis on the Yavari-Mirin and Yavari Rivers: a success story for cites. Oryx 44 (01), 83–88.

Redford, K.H., 1992. The empty forest. Bioscience 42, 412–422.

Reed, M.S., 2008. Stakeholder participation for environmental management: a literature review. Biological Conservation 141, 2417–2431.

Reed, M.S., Graves, A., Dandy, N., Posthumus, H., Hubacek, K., Morris, J., Stringer, L.C., 2009. Who's in and why? A typology of stakeholder analysis methods for natural resource management. Journal of Environmental Management 90 (5), 1933–1949.

Reitz, E.J., Shackley, M., 2012. Environmental Archaeology. Springer.

Revkin, A., 2004. The Burning Season: The Murder of Chico Mendes and the Fight for the Amazon Rain Forest. Island Press.

Rice, R.E., Gullison, R.E., Reid, J.W., 1997. Can sustainable management save tropical forests? Scientific American 276 (4), 44–49.

Rist, J., Milner-Gulland, E.J., Cowlishaw, G., Rowcliffe, M., 2010. Hunter reporting of catch per unit effort as a monitoring tool in a bushmeat–harvesting system. Conservation Biology 24 (2), 489–499.

Rocha, K., Moreira, A.R., Reis, E.J., Carvalho, L., 2006. The market value of forest concessions in the Brazilian Amazon: a real option approach. Forest Policy and Economics 8 (2), 149–160.

Roda, J.M., Rahman, R.A., Ahmad, I., Fui, L.H., Mamat, M.P., 2005. Global timber trade pattern: the cards have changed. EAS Strategic Options 2011 (08), 14–15.

Rodríguez, A.F., 2006. Estrategia nacional de zonificación ecológica económica. CONAM, Lima.

Roosevelt, A.C., Lima da Costa, M., Lopes Machado, C., Michab, M., Mercier, N., Valladas, H., Schick, K., 1996. Paleoindian cave dwellers in the Amazon: the peopling of the Americas. Science 272, 373–384.

Rose, K.A., Cowan, J.H., Winemiller, K.O., Myers, R.A., Hilborn, R., 2001. Compensatory density dependence in fish populations: importance, controversy, understanding and prognosis. Fish and Fisheries 2 (4), 293–327.

Ros-Tonen, M.A., Kusters, K., 2011. Pro-poor governance of non-timber forest products: the need for secure tenure, the rule of law, market access and partnerships. Non-Timber Forest Products in the Global Context, Springer, Berlin Heidelberg, pp. 189–207.

Rowcliffe, J.M., Cowlishaw, G., Long, J., 2003. A model of human hunting impacts in multi-prey communities. Journal of Applied Ecology 40 (5), 872–889.

Rueffler, C., Egas, M., Metz, J.A., 2006. Evolutionary predictions should be based on individual-level traits. The American Naturalist 168 (5), E148–E162.

Runk, J.V., 1998. Productivity and sustainability of a vegetable ivory palm (Phytelephas aequatorialis, Arecaceae) under three management regimes in northwestern Ecuador. Economic Botany 52 (2), 168–182.

Ruokolainen, K., Tuomisto, H., Macía, M.J., Higgins, M.A., Yli-Halla, M., 2007. Are floristic and edaphic patterns in Amazonian rain forests congruent for trees, pteridophytes and Melastomataceae? Journal of Tropical Ecology 23, 13–25.

Sachs, J.D., Warner, A.M., 2001. The curse of natural resources. European Economic Review 45 (4), 827–838.

Saether, B.E., Solberg, E.J., Heim, M., 2003. Effects of altering sex ratio structure on the demography of an isolated moose population. The Journal of Wildlife Management 455–466.

Salati, E., Dourojeanni, M., Mundim, A., Sampaio, G., Lovejoy, T., 2012. The Brazilian Amazon. Climate and Conservation, Island Press/Center for Resource Economics, pp. 45–55.

Sahley, C.T., Torres-Vargas, J., Sanchez-Valdivia, J., 2007. Biological sustainability of live shearing of vicuña in Peru. Conservation Biology 21 (1), 98–105.

Salisbury, D.S., Fagan, C., 2013. Coca and conservation: cultivation, eradication, and trafficking in the Amazon borderlands. GeoJournal 78 (1), 41–60.

Salo, J., Kalliola, R., Häkkinen, I., Mäkinen, Y., Niemelä, P., Puhakka, M., Coley, P.D., 1986. River dynamics and the diversity of Amazon lowland forest. Land Use Policy 322 (6076), 254–258.

Salo, M., 2009. Knowledge Meets Landscape: Interacting People, Science, and Natural Resources in Amazonian Rainforests. PhD thesis, University of Turku, Turku, Finland.

Salo, M., Puro, J.P., Knuuti, K., 2013. Control and Intimacy in the Amazonian Reality: Newspaper Rhetoric on Forest Sector Reform in Peru. Land Use Policy 35, 226–236.

Salo, J., Pyhälä, M., 2004. Falleció el Dr. Pekka Soini, eli investigador autodidacta de la Amazonía. Neotropical Primates 12 (2), 90–91.

Salo, M., Pyhälä, A., 2007. Exploring the gap between conservation science and protected area establishment in the Allpahuayo-Mishana National Reserve (Peruvian Amazonia). Environmental Conservation 34 (01), 23–32.

Salo, M., Helle, S., Toivonen, T., 2011. Allocating logging rights in Peruvian Amazonia—does it matter to be local? PLoS One 6 (5), e19704.

Salo, M., Toivonen, T., 2009. Tropical timber rush in Peruvian Amazonia: spatial allocation of forest concessions in an uninventoried frontier. Environmental Management 44 (4), 609–623.

Salonen, M., Toivonen, T., Cohalan, J.M., Coomes, O.T., 2012. Critical distances: comparing measures of spatial accessibility in the riverine landscapes of Peruvian Amazonia. Applied Geography 32 (2), 501–513.

Santos-Granero, F.S., Barclay, F., 2002. La frontera domesticada: historia económica y social de Loreto, 1850–2000. Fondo Editorial PUCP.

Schaan, D., 2010. Long-term human induced impacts on Marajó island landscapes, Amazon estuary. Diversity 2 (2), 182–206.

Schaan, D., Pärssinen, M., Saunaluoma, S., Ranzi, A., Bueno, M., Barbosa, A., 2012. New radiometric dates for precolumbian (2000–700 bp) earthworks in western Amazonia, Brazil. Journal of Field Archaeology 37 (2), 132–142.

Schwartzman, S., Moreira, A., Nepstad, D., 2000. Rethinking tropical forest conservation: perils in parks. Conservation Biology 14 (5), 1351–1357.

Shanley, P., López, C., 2009. Out of the loop: why research rarely reaches policy makers and the public and what can be done. Biotropica 41 (5), 535–544.

Shanley, P., Rodrigues, G.G., 2002. Equitable ecology: collaborative learning for local benefit in Amazonia. Agricultural Systems 73 (1), 83–97.

Shanley, P., Rosa, N., 2004. Eroding knowledge: an ethnobotanical inventory in Eastern Amazonia's logging frontier. Economic Botany 58, 135–160.

Schleuning, M., Fründ, J., Klein, A.M., Abrahamczyk, S., Alarcón, R., Albrecht, M., Blüthgen, N., 2012. Specialisation of mutualistic interaction networks decreases toward tropical latitudes. Current Biology.

Schneider, J.L., 2012. Sold into Extinction: The Global Trade in Endangered Species. Praeger Pub Text.

Schneider, L., Ferrara, C.R., Vogt, R.C., Burger, J., 2011. History of turtle exploitation and management techniques to conserve turtles in the Rio Negro Basin of the Brazilian Amazon. Chelonian Conservation and Biology 10 (1), 149–157.

Schulman, L., Ruokolainen, K., Junikka, L., Sääksjärvi, I.E., Salo, M., Juvonen, S.K., Higgins, M., 2007. Amazonian biodiversity and protected areas: do they meet? Biodiversity and Conservation 16 (11), 3011–3051.

Schüle, W., 1992. Vegetation, megaherbivores, man and climate in the Quaternary and the genesis of closed forests. Tropical Forests in Transition, Birkhäuser, Basel, pp. 45–76.

Sears, R.R., Pinedo–Vasquez, M., 2011. Forest policy reform and the organisation of logging in Peruvian Amazonia. Development and Change 42 (2), 609–631.

Shackleton, S.E., Shackleton, C.M., Shanley, P., 2011a. Non-Timber Forest Products in the Global Context. Springer.

Shackleton, C.M., Shackleton, S.E., Shanley, P., 2011b. Building a holistic picture: an integrative analysis of current and future prospects for non-timber forest products in a changing world. In: Shackleton, S.E., Shackleton, C.M., Shanley, P. (Eds.), Non-Timber Forest Products in the Global Context, Springer, pp. 255–280.

Shackleton, S., Delang, C.O., Angelsen, A., 2011c. From Subsistence to Safety Nets and Cash Income: Exploring the Diverse Values of Non-Timber Forest Products for Livelihoods and Poverty Alleviation. Springer, Berlin Heidelberg, pp. 55–81.

Sheng-Ji, P., 2001. Ethnobotanical approaches of traditional medicine studies: some experiences from Asia. Pharmaceutical Biology 39 (Suppl. 1), 74–79.

Shepard Jr., G.H., Ramirez, H., 2011. "Made in Brazil": human dispersal of the Brazil Nut (Bertholletia excelsa, Lecythidaceae) in Ancient Amazonia1. Economic Botany 65 (1), 44–65.

Shin, Y.J., Rochet, M.J., Jennings, S., Field, J.G., Gislason, H., 2005. Using size-based indicators to evaluate the ecosystem effects of fishing. ICES Journal of Marine Science: Journal du Conseil 62 (3), 384–396.

Sills, E., Saha, S., 2010. Subsidies for rubber: Conserving rainforests while sustaining livelihoods in the Amazon? Journal of Sustainable Forestry 29 (2–4), 152–173.

Silverman, I., Choi, J., Peters, M., 2007. The hunter-gatherer theory of sex differences in spatial abilities: data from 40 countries. Archives of Sexual Behavior 36 (2), 261–268.

Sioli, H., 1984. The Amazon and its main affluents: hydrography, morphology of the river courses, and river types. In: Sioli, H. (Ed.), The Amazon. Limnology and Landscape Ecology of Mighty Tropical River and its Basin, Dr W. Junk, The Hague, pp. 127–165.

Sirén, A., 2004. Changing interactions between humans and nature in Sarayaku, Ecuadorian Amazon. Uppsala: Sveriges lantbruksuniv, Acta Universitatis agriculturae Sueciae. Agraria, 1401–6249; 447. Doctoral dissertation.

Sirén, A., Hambäck, P., Machoa, J., 2004. Including spatial heterogeneity and animal dispersal when evaluating hunting: a model analysis and an empirical assessment in an Amazonian community. Conservation Biology 18 (5), 1315–1329.

Sirén, A., 2012. Festival hunting by the Kichwa people in the ecuadorian Amazon. Journal of Ethnobiology 32 (1), 30–50.

Sirén, A.H., Cardenas, J.C., Hambäck, P., Parvinen, K., 2013a. Distance friction and the cost of Hunting in tropical forest. Land Economics 89 (3), 558–574.

Sirén, A., Machoa, J., 2008. Fish, wildlife, and human nutrition in tropical forests: a fat gap? Interciencia 33 (3), 186.

Sirén, A., Montúfar, R., Gualinga, J., 2013b. In: Valencia, R., Montúfar, R., Navarrete, H., Balslev, H. (Eds.), Palma de wayuri. En: Palmas ecuatorianas: biología y uso sostenible, pp. 135–143.

Sirén, A., Tuomisto, H., Navarrete, H., 2013c. Mapping environmental variation in lowland Amazonian rainforests using remote sensing and floristic data. International Journal of Remote Sensing 34 (5), 1561–1575.

Siry, J.P., Cubbage, F.W., Ahmed, M.R., 2005. Sustainable forest management: global trends and opportunities. Forest Policy and Economics 7 (4), 551–561.

Smith, J., Colan, V., Sabogal, C., Snook, L., 2006. Why policy reforms fail to improve logging practices: the role of governance and norms in Peru. Forest Policy and Economics 8 (4), 458–469.

Smith, B.T., Klicka, J., 2010. The profound influence of the late Pliocene Panamanian uplift on the exchange, diversification, and distribution of New World birds. Ecography 33 (2), 333–342.

Snyder, R., 2006. Does lootable wealth breed disorder? A political economy of extraction framework. Comparative Political Studies 39 (8), 943–968.

Soini, P., 1999. Un manual para el manejo de quelonio acuáticos en la Amazonía peruana (Charapa, Taricaya y Cupiso). Instituto de Investigaciones de la Amazonía Peruana, Iquitos, Perú.

Soll, D., 2012. Resurrecting the story of the passenger pigeon in Pennsylvania. Pennsylvania History 79 (4), 507–519.

Soriano, M., Kainer, K.A., Staudhammer, C.L., Soriano, E., 2012. Implementing multiple forest management in Brazil nut-rich community forests: effects of logging on natural regeneration and forest disturbance. Forest Ecology and Management 268, 92–102.

Stewart, K., 2009. Effects of bark harvest and other human activity on populations of the African cherry (Prunus africana) on Mount Oku, Cameroon. Forest Ecology and Management 258 (7), 1121–1128.

Starrett, D.A., 2003. Property rights, public goods and the environment. Handbook of Environmental Economics 1, 97–125.

Stewart, A., Ambrose-Oji, B., Morris, J., 2012. Social Media and Forestry: A Scoping Report. The Research Agency of the Forestry Commission Forestry Commission, UK.

Stoner, K.E., Vulinec, K., Wright, S.J., Peres, C.A., 2007. Hunting and plant community dynamics in tropical forests: a synthesis and future directions. Biotropica 39 (3), 385–392.

Stork, N.E., 2010. Re-assessing current extinction rates. Biodiversity and Conservation 19 (2), 357–371.

Suárez, E., Morales, M., Cueva, R., Utreras Bucheli, V., Zapata–Ríos, G., Toral, E., Vargas Olalla, J., 2009. Oil industry, wild meat trade and roads: indirect effects of oil extraction activities in a protected area in north-eastern Ecuador. Animal Conservation 12 (4), 364–373.

Sunderland, T.C., Ndoye, O., Harrison-Sanchez, S., 2011. Non-timber forest products and conservation: what prospects? Non-Timber Forest Products in the Global Context, Springer, Berlin Heidelberg, pp. 209–224.

Svedäng, H., Stål, J., Sterner, T., Cardinale, M., 2010. Consequences of subpopulation structure on fisheries management: cod (Gadus morhua) in the Kattegat and Öresund (North Sea). Reviews in Fisheries Science 18 (2), 139–150.

Swain, D.P., Sinclair, A.F., Hanson, J.M., 2007. Evolutionary response to size-selective mortality in an exploited fish population. Proceedings of the Royal Society B 274 (1613), 1015–1022.

Swanson, T.M., 1992. Economics of a biodiversity convention. Ambio 21, 250–257.

Swanson, T., 1999. Conserving global biological diversity by encouraging alternative development paths: can development coexist with diversity? Biodiversity & Conservation 8 (1), 29–44.

Szabó, P., Hédl, R., 2011. Advancing the integration of history and ecology for conservation. Conservation Biology 25, 680–687.

Tavernor, R., 2007. Visual and cultural sustainability: the impact of tall buildings on London. Landscape and Urban Planning 83 (1), 2–12.

Taylor, A.C., 1986. La evolución demográfica de las poblaciones indígenas de la alta Amazonía del siglo xvi al xx. Cultura 8 (24b), 507–518.

Taylor, B.W., Flecker, A.S., Hall Jr., R.O., 2006. Loss of a harvested fish species disrupts carbon flow in a diverse tropical river. Science 313 (5788), 833–836.

Terborgh, J., 1999. Requiem for Nature. Island Press.

Terborgh, J., Nuñez-Iturri, G., Pitman, N.C., Valverde, F.H.C., Alvarez, P., Swamy, V., Paine, C.T., 2008. Tree recruitment in an empty forest. Ecology 89 (6), 1757–1768.

Ticktin, T., 2004. The ecological implications of harvesting non–timber forest products. Journal of Applied Ecology 41 (1), 11–21.

Ticktin, T., Shackleton, C., 2011. Harvesting non-timber forest products sustainably: opportunities and challenges. Non-Timber Forest Products in the Global Context, Springer, Berlin Heidelberg, pp. 149–169.

Tobler, W.R., 1970. A computer movie simulating urban growth in the Detroit. Economic Geography, vol. 46, pp. 234–240.

Tobler, M.W., 2008. The ecology of the lowland tapir in Madre de Dios, Peru: using new technologies to study large rainforest mammals. Doctoral dissertation, Texas A&M University.

Toivonen, T., Mäki, S., Kalliola, R., 2007. The riverscape of Western Amazonia—a quantitative approach to the fluvial biogeography of the region. Journal of Biogeography 34, 1374–1387.

Tuomisto, H., 2010. A consistent terminology for quantifying species diversity? Yes, it does exist. Oecologia 164, 853–860.

Tuomisto, H., Ruokolainen, K., Kalliola, R., Linna, A., Danjoy, W., Rodriguez, Z., 1995. Dissecting amazonian biodiversity. Science 269, 63–66.

Tuomisto, H., Ruokolainen, K., 2005. Environmental heterogeneity and the diversity of pteridophytes and Melastomataceae in western Amazonia. Biologiske Skrifter 55, 37–56.

Turvey, S.T., Risley, C.L., 2006. Modelling the extinction of Steller's sea cow. Biology Letters 2 (1), 94–97.

Ugarte, R., Sharp, G., 2012. Management and production of the brown algae Ascophyllum nodosum in the Canadian maritimes. Journal of Applied Phycology 24 (3), 409–416.

Upneja, A., Shafer, E.L., Seo, W., Yoon, J., 2001. Economic benefits of sport fishing and angler wildlife watching in Pennsylvania. Journal of Travel Research 40 (1), 68–78.

Vadjunec, J.M., Rocheleau, D., 2009. Beyond forest cover: land use and biodiversity in rubber trail forests of the Chico Mendes Extractive Reserve. Ecology and Society 14 (2), 29.

van Bueren, E.M.L., Blom, E.M., 1997. Hierarchical framework for the formulation of sustainable forest management standards. Tropenbos Foundation.

van der Hammen, T., 1974. The Pleistocene changes of vegetation and climate in tropical South America. Journal of Biogeography, 3–26.

van Vliet, N., Nasi, R., Taber, A., 2011. From the forest to the stomach: bushmeat consumption from rural to urban settings in Central Africa. Non-Timber Forest Products in the Global Context, Springer, Berlin Heidelberg, pp. 129–145.

Vasquez, R., Gentry, A.H., 1989. Use and misuse of forest–harvested fruits in the Iquitos Area. Conservation Biology 3 (4), 350–361.

Veríssimo, A., Barreto, P., Tarifa, R., Uhl, C., 1995. Extraction of a high-value natural resource in Amazonia: the case of mahogany. Forest Ecology and Management 72 (1), 39–60.

Veríssimo, A., Júnior, C.S., Stone, S., Uhl, C., 1998. Zoning of timber extraction in the Brazilian Amazon. Conservation Biology 12, 128–136.

Virtanen, J., Ahvonen, A., Honkanen, A., 2001. Regional socio-economic importance of fisheries in Finland. Fisheries Management and Ecology 8 (4–5), 393–403.

Vormisto, J., 2002. Making and marketing chambira hammocks and bags in the village of Brillo Nuevo, northeastern Peru. Economic Botany 56 (1), 27–40.

Vuorisalo, T., Laihonen, P., 2000. Biodiversity conservation in the north: history of habitat and species protection in Finland. Annales Zoologici Fennici 37 (4), 281–297.

Wallace, A.R., 1878. Tropical Nature and Other Essays. Macmillan and Co., London, p. 365.

Walters, C.J., Hilborn, R., 1978. Ecological optimization and adaptive management. Annual review of Ecology and Systematics 9, 157–188.

WCED, 1987. Our Common Future. Oxford University Press, Oxford.

Weinstein, S., Moegenburg, S., 2004. Açaí palm management in the Amazon estuary: course for conservation or passage to plantations? Conservation & Society 2 (2), 315–346.

Weiss, K., Hamann, M., Marsh, H., 2012. Bridging knowledges: understanding and applying indigenous and western scientific knowledge for marine wildlife management. Society and Natural Resources, 1–18.

Welcomme, R.L., 1999. A review of a model for qualitative evaluation of exploitation levels in multi-species fisheries. Fisheries Management and Ecology 6, 1–19.

Westgate, M.J., Likens, G.E., Lindenmayer, D.B., 2013. Adaptive management of biological systems: a review. Biological Conservation 158, 128–139.

Whaley, W.G., 1948. Rubber-the primary sources for American production. Economic Botany 2, 198–216.

Wiersum, K.F., 1995. 200 years of sustainability in forestry: lessons from history. Environmental Management 19 (3), 321–329.

Wiersum, K.F., Elands, B.H., Hoogstra, M.A., 2005. Small-scale forest ownership across Europe: characteristics and future potential. Small-scale Forest Economics, Management and Policy 4 (1), 1–19.

Wilkie, D.S., Godoy, R.A., 1996. Trade, indigenous rain forest economies and biological diversity: model predictions and directions for research. In: Ruiz Perez, M., Arnold, J.E.M. (Eds.), Current Issues in Non-Timber Forest Products Research, Center for International Forestry Research, Bogor, Indonesia, pp. 83–102.

Wilkie, D.S., Godoy, R.A., 2001. Income and price elasticities of bushmeat demand in lowland Amerindian societies. Conservation Biology 15 (3), 761–769.

Wilkie, D.S., Starkey, M., Abernethy, K., Effa, E.N., Telfer, P., Godoy, R., 2005. Role of prices and wealth in consumer demand for bushmeat in Gabon, Central Africa. Conservation Biology 19 (1), 268–274.

Winkler, D., 2009. Caterpillar fungus (Ophiocordyceps sinensis) production and sustainability on the Tibetan Plateau and in the Himalayas. Asian Medicine 5 (2), 291–316.

Wittemyer, G., Elsen, P., Bean, W.T., Burton, A.C.O., Brashares, J.S., 2008. Accelerated human population growth at protected area edges. Science 321 (5885), 123–126.

Wolverton, S., 2010. Biodiversity Letter: the North American Pleistocene overkill hypothesis and the re-wilding debate. Diversity and Distributions 16 (5), 874–876.

Wright, S.J., Hernandéz, A., Condit, R., 2007. The bushmeat harvest alters seedling banks by favoring lianas, large seeds, and seeds dispersed by bats, birds, and wind. Biotropica 39 (3), 363–371.

Wright, S.J., Sanchez-Azofeifa, G.A., Portillo-Quintero, C., Davies, D., 2007. Poverty and corruption compromise tropical forest reserves. Ecological Applications 17 (5), 1259–1266.

Youatt, A., Cmar, T., 2009. The fight for red gold: ending illegal mahogany trade from Peru. Natural Resources & Environment 23 (3), 19–23.

Zebich-Knos, M., 1997. Preserving biodiversity in Costa Rica: the case of the Merck-INBio agreement. The Journal of Environment Development 6 (2), 180–186.

Zimmermann, Erich W., 1951. World Resources and Industries. Harper and Row, New York.

Zu Ermgassen, P.S.E., et al., 2012. Historical ecology with real numbers: past and present extent and biomass of an imperilled estuarine habitat. Proceedings of the Royal Society B 2012 (279), 3393–3400.

Index

Note: Page numbers with "*b*" denote boxes; "*f*" figures; "*t*" tables.

A

D

E

F

I

M

N

S

T

W

Y

Z

FIGURE 3.4 Chapter 3 presents a review of the history of wild species harvest in Amazonia. *Photo: Martti Pärssinen.*

FIGURE 4.6 Chapter 4 addresses hunting and the efforts of the people in an indigenous village to improve the sustainability of their hunting practices. *Photo: Eriberto Gualinga.*

FIGURE 5.3A Chapter 5 deals with commercial and subsistence fishing, and the problems of managing a migratory resource.

FIGURE 6.7A Chapter 6 presents river turtles as an extraordinary example of recovery of a harvested resource.

FIGURE 7.1 Chapter 7 deals with the harvest and management of palm leaves for roof thatch as an example of how locally designed management systems may not always be fully effective. *Photo: Eriberto Gualinga.*

FIGURE 8.7B Chapter 8 is about the Brazil nut, a wild species product that is well established and traded in considerable volumes on the international market. *Photo: Natalia Aravena.*

FIGURE 9.6B Chapter 9 explores the implications of an institutional reform on logging practices in Peru. *Photo: Natalia Aravena.*

FIGURE 10.5 Chapter 10 focuses on the experiences of an attempt to make business out of Amazonian medicinal plants.

FIGURE 11.1A Chapter 11 presents the story of the açaí palm as an intermediate case between the management of a wild and a domesticated resource.

FIGURE 12.1B Chapter 12 addresses the role of science and planning in relation to the use of land and resources.

FSC
www.fsc.org
MIX
Papier | Fördert gute Waldnutzung
FSC® C083411

Kolibri 360 GmbH
Ferdinand-Jühlke-Straße 7
99095 Erfurt, Deutschland
produktsicherheit@kolibri360.de